MORE PRAISE FOR

# The Language Instinct

"Reading Steven Pinker's book is one of the biggest favors I've ever done my brain. It is the sort of writing that any genuine expert on a subject longs to achieve: highly accessible to the general reader yet at the same time seminal for professionals. Laypeople will be gripped by a lucid and witty introduction to the fascinating subject of linguistics. Orthodox social scientists—and their biological fellow travelers—will find a formidable Darwinian challenge to their cherished dogmas. Word-pedants like me (or those who say 'gender' when they mean 'sex') will retreat chastened. Even if you disagree with it, you'll surely be charmed and engaged by this brilliant work."  —Richard Dawkins, author of *The Selfish Gene*

"A brilliant, witty, and altogether satisfying book. . . . Mr. Pinker has that facility, so rare among scientists, of making the most difficult material accessible to the average reader. Most important, he never talks down to the reader. . . . The fundamental unity of humanity is the theme of this exciting book. Arresting . . . amusing and instructive . . . a useful, compelling book."  —*New York Times Book Review*

"Absorbing. He makes a persuasive, entertaining case for his thesis."  —*Time*

"A remarkably engaging book. The book is packed tight with observations, experimental results, insight and forceful arguments based on what we all know of language but never analyze. This reader finds Professor Pinker's genuinely instructive volume funny as well, a delightful member of that rare genre headed by that classic *Life on the Mississippi*."  —*Scientific American*

"His own use of language is a powerful advertisement for this human ability, as he lays his stall out with clarity and candor. . . . Darwin . . . would surely be impressed by the way in which Pinker sheds light on these questions. . . . A superb book, simply at the level of being a good read: it is packed with fascinating facts and information. . . . Pinker debunks with panache, cuts through the confusion of jargon, and tells a mean anecdote. He does for language what David Attenborough does for animals,

explaining difficult scientific concepts so easily that they are indeed absorbed as a transparent stream of words. . . . I will be astonished if a better science book of any kind, let alone one accessible to the general reader, comes along this year. . . . His book is groundbreaking, exhilarating, fun, and almost certainly correct. Do yourself a favor and read it."
—*Sunday Times* (London)

"A brilliant study of language. . . . Language is full of mysteries, which Pinker excavates like a pig after truffles. Professor Pinker . . . was a brave man to write this book, for who would have taken it seriously if it had been clodhoppingly written? As it happens, he writes splendidly."
—*The Times* (London)

"An excellent book full of wit and wisdom and sound judgment . . . better than most college courses on language and the mind—and a great deal more digestible."          —*Boston Globe Book Review*

"A book to inspire."          —*Perspectives of the Orton Dyslexia Society*

"A dazzling new book. . . . This is all immensely fine and trenchant, and Pinker embarks on his argument with brilliant dash and swagger: 'I want to debauch your mind with learning,' he begins. . . . What a wonderful ambition. Not many writers aim this high. [He is] a canny writer and a bit of a wag. . . . *The Language Instinct* vibrates with delicious asides and poignant discoveries. . . . Words can hardly do justice to the superlative range and liveliness of Pinker's investigations."          —*The Independent*

"Steven Pinker is, I think, engagingly wrong in some of his conclusions, but the operative word here is *engagingly*. He reminds us of the pleasures of reading about language, provided people like him are at the wheel."
—William F. Buckley, Jr.

"Splendid. . . . Not bad for a supposedly stuffy scholar from MIT."
—*Los Angeles Times*

"He writes with authority and grace about the sprawling science of linguistics, making even its thornier branches accessible to general readers."
—*USA Today*

"[A] triumph of common sense over some of the nonsense that has dominated psychology and linguistics for much of this century. . . . A book about language had better be well written, and Mr. Pinker's book is superbly so. Rarely can such a rich harvest of new ideas and profound insights have been made so accessible by one of their inventors. . . . He is unfailingly articulate, funny, and clear. The book is to Chomsky as Shakespeare is to Spenser."

—The Economist

"[A] marvelously readable book about language, written by a real expert. Steven Pinker tackles with wit and erudition the kinds of question everyone asks. . . . [He] brings not only an expertise in linguistics and psychology and a wide knowledge of biology, but also an ability to understand the ordinary person's linguistic hang-ups and to shake them loose with gentle ridicule. . . . Whatever its eventual impact on linguistics and psychology, *The Language Instinct* will undoubtedly be greeted as a distinguished contribution to the lay understanding of science. . . . With its wealth of examples, its flawless typesetting, its wide-ranging bibliography and its irresistible good humor, Pinker's book is certain to increase its readers' respect for the amazing natural phenomena that the author and his colleagues have made their life's study."

—Nature

"Somebody finally got it right. Steve Pinker's thoroughly modern, totally engaging book introduces lay readers to the science of language in ways that are irreverent and hilarious while coherent and factually sound. A delicious read."

—Leila Gleitman, University of Pennsylvania;<br>President, Linguistic Society of America

"A mightily ambitious book. . . . With an unusual and attractive blend of patience and wit, Pinker is extremely good at explaining. . . . Pinker is not yet 40; with his voracious intelligence and his gift for prose, we can expect many more installments from the front lines of neuroscience. I await them—nervously."

—Montreal Gazette

"[An] exciting synthesis—an entertaining, totally accessible study that will regale language lovers and challenge professionals in many disciplines. . . . A beautiful hymn to the infinite creative potential of language."

—Publishers Weekly

"Examples are clear and easy to understand; Pinker's humor and insight make this the perfect introduction to the world of cognitive science and language. Highly recommended."

—Booklist

"Run, don't walk, to your local bookstore and buy *The Language Instinct*. In a dazzling, informative, and funny book, Steven Pinker brings you into the wonderful world of language. He spares the reader the mumbo jumbo of linguistics and directs your attention to an unalterable truth. Language is an instinct, and in the discovery, Pinker reveals the secrets of the mind. It is brilliant."

> —Michael S. Gazzaniga, Director, Center for Neuroscience, University of California, Davis; author of *Nature's Mind*

"[E]xtremely important. . . . The power of the book . . . is in the elegant assembly of a coherent argument, based on a foundation of evolutionary biology. . . . *The Language Instinct* is provocative. But there are no cheap points scored nor is there any intemperate denunciation of opposing views. . . . The case is intelligently structured, forcefully argued, and couched in beautiful prose. Readers may reject Pinker's conclusions, but they will greatly enjoy the experience of the journey through his mind."

> —*New Scientist*

"[A] brilliant exposition . . . he expounds ideas with clarity, wit, and polish."
> —*The Observer*

"Pinker is unfailingly stimulating, as well as writing in a genuinely democratic style that combines elegance with unforced touches of the popular."
> —*New Statesman and Society*

"[A]n impressive book. It is vividly written by a man of great learning . . . full of useful information to impart at cocktail parties."
> —*Literary Review*

"[A]n important and fascinating book. . . . Professor Pinker writes very clearly and wittily. He makes us appreciate the marvelous nature of what we ordinarily take for granted—one of the marks of a good popularizer."
> —*Sunday Telegraph*

"A cracking book . . . marvelous . . . wonderful to read."
> —*The Guardian*

"A great book. . . . Its author is in love with language and revels in its uses. . . . While providing an astonishingly thorough course in psycholinguistics, Pinker also manages to be funnier than I would have thought such a substantive, critical discussion could possibly be. . . . Pinker's biology is impeccably up-to-date. Indeed, he displays a much more sophisticated and critical understanding of issues in evolution and adaptation than most biologists. *The Language Instinct* should be a candidate for best book of the 90's. Or at least a Pullet Surprise."

—*Quarterly Review of Biology*

"The most lucid, charming, and wide-ranging popularization of Noam Chomsky's linguistics ever written." —*Toronto Globe and Mail*

"Pinker writes clearly and engagingly about the most difficult matters."

—*American Airlines Way*

"Steven Pinker has made several landmark contributions to cognitive science in the past, and his latest book *The Language Instinct* constitutes yet another one. [It is] written in an exceptionally clear, engaging, and witty style and directed towards a general audience. . . . One might be tempted to think that this book is simply a rerun of Chomsky's greatest hits, but that would be mistaken. What makes Pinker's version of the story extremely original and valuable is that . . . he makes abundant use of a wide range of different sources of evidence, including developmental psycholinguistics, cognitive neuroscience, and evolutionary psychology. . . . A brilliant piece of work. It succeeds in demonstrating that the recent discoveries about the uniquely human ability to acquire and use language are as elegant and exciting as anything in modern science."

—*Mind and Language*

"Pinker eloquently explains the details of Chomsky's revolutionary theory, and then proceeds to bring us up-to-date on the latest advances in linguistics. . . . Luckily, he's also a user-friendly writer able to transform technical stuff into a fun and informative read. . . . But don't be fooled by the entertainment: Pinker is dead serious about language." —*Kinesis*

"Steven Pinker, one of the leading linguistic researchers in the world, skillfully defends a provocative thesis about the biological bases of language and in the process provides authoritative answers to major questions about the nature of language. *The Language Instinct* is a splendid and indispensable book."

—Howard Gardner, Graduate School of Education,
Harvard University; author of *The Mind's New Science*

"[A]n accessible, entertaining, and authoritative introduction to the modern science of language. . . . It is a joy to witness, at last, the promise of linguistics fulfilled." —*London Review of Books*

OTHER BOOKS BY STEVEN PINKER

steven pinker

# The Language Instinct

## HOW THE MIND CREATES LANGUAGE

HARPER**PERENNIAL** MODERN**CLASSICS**

NEW YORK • LONDON • TORONTO • SYDNEY

HARPER**PERENNIAL** ● MODERN**CLASSICS**

This book was originally published in 1994 by William Morrow and Company.

P.S.™ is a trademark of HarperCollins Publishers.

HarperCollins books may be purchased for educational, business, or sales promotional use. For information please write: Special Markets Department, HarperCollins Publishers, 10 East 53rd Street, New York, NY 10022.

First Harper Perennial edition published 1995.
First Perennial Classics edition published 2000.
First Harper Perennial Modern Classics edition published 2007.

*Designed by Stanley S Drate/Folio Graphics Co. Inc.*

Library of Congress Cataloging-in-Publication Data is available upon request.

ISBN 978-0-06-133646-1

09 10 11 12 RRD 12 11 10 9 8 7 6

*for*
*Harry and Roslyn Pinker*
who gave me language

# Contents

# Preface

꧁

*I have never met a person who is not interested in language. I wrote this* book to try to satisfy that curiosity. Language is beginning to submit to that uniquely satisfying kind of understanding that we call science, but the news has been kept a secret.

For the language lover, I hope to show that there is a world of elegance and richness in quotidian speech that far outshines the local curiosities of etymologies, unusual words, and fine points of usage.

For the reader of popular science, I hope to explain what is behind the recent discoveries (or, in many cases, nondiscoveries) reported in the press: universal deep structures, brainy babies, grammar genes, artifically intelligent computers, neural networks, signing chimps, talking Neanderthals, idiot savants, feral children, paradoxical brain damage, identical twins separated at birth, color pictures of the thinking brain, and the search for the mother of all languages. I also hope to answer many natural questions about languages, like why there are so many of them, why they are so hard for adults to learn, and why no one seems to know the plural of *Walkman*.

For students unaware of the science of language and mind, or worse, burdened with memorizing word frequency effects on lexical decision reaction time or the fine points of the Empty Category Principle, I hope to convey the grand intellectual excitement that launched the modern study of language several decades ago.

For my professional colleagues, scattered across so many disciplines and studying so many seemingly unrelated topics, I hope to offer a semblance of an integration of this vast territory. Although I am an opinionated, obsessional researcher who dislikes insipid compromises that fuzz up the issues, many academic controversies remind me of the blind men palpating the elephant. If my personal synthesis seems to embrace both sides of debates like "formalism versus functionalism" or "syntax versus semantics versus pragmatics," perhaps it is because there was never an issue there to begin with.

For the general nonfiction reader, interested in language and human beings in the broadest sense, I hope to offer something different from the airy platitudes—Language Lite—that typify discussions of language (generally by people who have never studied it) in the humanities and sciences alike. For better or worse, I can write in only one way, with a passion for powerful, explanatory ideas, and a torrent of relevant detail. Given this last habit, I am lucky to be explaining a subject whose principles underlie wordplay, poetry, rhetoric, wit, and good writing. I have not hesitated to show off my favorite examples of language in action from pop culture, ordinary children and adults, the more flamboyant academic writers in my field, and some of the finest stylists in English.

This book, then, is intended for everyone who uses language, and that means everyone!

I owe thanks to many people. First, to Leda Cosmides, Nancy Etcoff, Michael Gazzaniga, Laura Ann Petitto, Harry Pinker, Robert Pinker, Roslyn Pinker, Susan Pinker, John Tooby, and especially Ilavenil Subbiah, for commenting on the manuscript and generously offering advice and encouragement.

My home institution, the Massachusetts Institute of Technology, is a special environment for the study of language, and I am grateful to the colleagues, students, and former students who shared their expertise. Noam Chomsky made penetrating criticisms and helpful suggestions, and Ned Block, Paul Bloom, Susan Carey, Ted Gibson,

Morris Halle, and Michael Jordan helped me think through the issues in several chapters. Thanks go also to Hilary Bromberg, Jacob Feldman, John Houde, Samuel Jay Keyser, John J. Kim, Gary Marcus, Neal Perlmutter, David Pesetsky, David Pöppel, Annie Senghas, Karin Stromswold, Michael Tarr, Marianne Teuber, Michael Ullman, Kenneth Wexler, and Karen Wynn for erudite answers to questions ranging from sign language to obscure ball players and guitarists. The Department of Brain and Cognitive Sciences' librarian, Pat Claffey, and computer system manager, Stephen G. Wadlow, those most admirable prototypes of their professions, offered dedicated, expert help at many stages.

Several chapters benefited from the scrutiny of real mavens, and I am grateful for their technical and stylistic comments: Derek Bickerton, David Caplan, Richard Dawkins, Nina Dronkers, Jane Grimshaw, Misia Landau, Beth Levin, Alan Prince, and Sarah G. Thomason. I also thank my colleagues in cyberspace who indulged my impatience by replying, sometimes in minutes, to my electronic queries: Mark Aronoff, Kathleen Baynes, Ursula Bellugi, Dorothy Bishop, Helena Cronin, Lila Gleitman, Myrna Gopnik, Jacques Guy, Henry Kučera, Sigrid Lipka, Jacques Mehler, Elissa Newport, Alex Rudnicky, Jenny Singleton, Virginia Valian, and Heather Van der Lely. A final thank you to Alta Levenson of Bialik High School for her help with the Latin.

I am happy to acknowledge the special care lavished by John Brockman, my agent, Ravi Mirchandani, my editor at Penguin Books, and Maria Guarnaschelli, my editor at William Morrow; Maria's wise and detailed advice vastly improved the final manuscript. Katarina Rice copy-edited my first two books, and I am delighted that she agreed to my request to work with me on this one, especially considering some of the things I say in Chapter 12.

My own research on language has been supported by the National Institutes of Health (grant HD 18381) and the National Science Foundation (grant BNS 91-09766), and by the McDonnell-Pew Center for Cognitive Neuroscience at MIT.

# The Language Instinct

# 1

# An Instinct to Acquire an Art

*As you are reading these words, you are taking part in one of the wonders* of the natural world. For you and I belong to a species with a remarkable ability: we can shape events in each other's brains with exquisite precision. I am not referring to telepathy or mind control or the other obsessions of fringe science; even in the depictions of believers these are blunt instruments compared to an ability that is uncontroversially present in every one of us. That ability is language. Simply by making noises with our mouths, we can reliably cause precise new combinations of ideas to arise in each other's minds. The ability comes so naturally that we are apt to forget what a miracle it is. So let me remind you with some simple demonstrations. Asking you only to surrender your imagination to my words for a few moments, I can cause you to think some very specific thoughts:

> When a male octopus spots a female, his normally grayish body suddenly becomes striped. He swims above the female and begins caressing her with seven of his arms. If she allows this, he will quickly reach toward her and slip his eighth arm into her breathing tube. A series of sperm packets moves slowly through a groove in his arm, finally to slip into the mantle cavity of the female.

Cherries jubilee on a white suit? Wine on an altar cloth? Apply club soda immediately. It works beautifully to remove the stains from fabrics.

When Dixie opens the door to Tad, she is stunned, because she thought he was dead. She slams it in his face and then tries to escape. However, when Tad says, "I love you," she lets him in. Tad comforts her, and they become passionate. When Brian interrupts, Dixie tells a stunned Tad that she and Brian were married earlier that day. With much difficulty, Dixie informs Brian that things are nowhere near finished between her and Tad. Then she spills the news that Jamie is Tad's son. "My what?" says a shocked Tad.

Think about what these words have done. I did not simply remind you of octopuses; in the unlikely event that you ever see one develop stripes, you now know what will happen next. Perhaps the next time you are in a supermarket you will look for club soda, one out of the tens of thousands of items available, and then not touch it until months later when a particular substance and a particular object accidentally come together. You now share with millions of other people the secrets of protagonists in a world that is the product of some stranger's imagination, the daytime drama *All My Children*. True, my demonstrations depended on our ability to read and write, and this makes our communication even more impressive by bridging gaps of time, space, and acquaintanceship. But writing is clearly an optional accessory; the real engine of verbal communication is the spoken language we acquire as children.

In any natural history of the human species, language would stand out as the preeminent trait. To be sure, a solitary human is an impressive problem-solver and engineer. But a race of Robinson Crusoes would not give an extraterrestrial observer all that much to remark on. What is truly arresting about our kind is better captured in the story of the Tower of Babel, in which humanity, speaking a single language, came so close to reaching heaven that God himself felt threatened. A common language connects the members of a community into an information-sharing network with formidable collective powers. Anyone can benefit from the strokes of genius, lucky accidents, and trial-and-error wisdom accumulated by anyone else, present or past. And people can work in teams, their efforts coordinated by negotiated agreements. As a result, *Homo sapiens* is a species, like blue-green algae and earthworms, that has wrought far-reaching changes on the planet. Archeologists have discovered the bones of ten thousand wild horses at the bottom of a cliff in France, the remains of herds stampeded over the clifftop by groups of paleolithic hunters seventeen thousand years ago. These fossils of ancient cooperation and shared ingenuity may shed light on why saber-tooth tigers, mastodons, giant woolly rhinoceroses, and dozens of other large mammals went extinct around the time that modern humans arrived in their habitats. Our ancestors, apparently, killed them off.

Language is so tightly woven into human experience that it is scarcely possible to imagine life without it. Chances are that if you find two or more people together anywhere on earth, they will soon be exchanging words. When there is no one to talk with, people talk to themselves, to their dogs, even to their plants. In our social relations, the race is not to the swift but to the verbal—the spellbinding orator, the silver-tongued seducer, the persuasive child who wins the battle of wills against a brawnier parent. Aphasia, the loss of language following brain injury, is devastating, and in severe cases family members may feel that the whole person is lost forever.

This book is about human language. Unlike most books with "language" in the title, it will not chide you about proper usage, trace the origins of idioms and slang, or divert you with palindromes, anagrams, eponyms, or those precious names for groups of animals like "exaltation of larks." For I will be writing not about the English language or any other language, but about something much more basic: the instinct to learn, speak, and understand language. For the first time in history, there is something to write about it. Some thirty-five years ago a new science was born. Now called "cognitive science," it combines tools from psychology, computer science, linguistics, philosophy, and neurobiology to explain the workings of human intelligence. The science of language, in particular, has seen spectacular advances in the years since. There are many phenomena of language that we are coming to understand nearly as well as we understand how a camera works or what the spleen is for. I hope to communicate these exciting discoveries, some of them as elegant as anything in modern science, but I have another agenda as well.

The recent illumination of linguistic abilities has revolutionary implications for our understanding of language and its role in human affairs, and for our view of humanity itself. Most educated people already have opinions about language. They know that it is man's most important cultural invention, the quintessential example of his capacity to use symbols, and a biologically unprecedented event irrevocably separating him from other animals. They know that language pervades thought, with different languages causing their speakers to construe reality in different ways. They know that children learn to talk from role models and caregivers. They know that grammatical sophistication used to be nurtured in the schools, but sagging educational standards and the debasements of popular culture have led to a frightening decline in the ability of the average person to construct a grammatical sentence. They also know that English is a zany, logic-defying tongue, in which one drives on a parkway and parks in a drive-way, plays at a recital and recites at a play. They know that English spelling takes such wackiness to even greater heights—George Bernard Shaw complained that *fish* could just as sensibly be spelled *ghoti* (*gh* as in *tough*, *o* as in *women*, *ti* as in *nation*)—and that only institutional inertia prevents the adoption of a more rational, spell-it-like-it-sounds system.

In the pages that follow, I will try to convince you that every one of these common opinions is wrong! And they are all wrong for a single reason. Language is not a cultural artifact that we learn the way we learn to tell time or how the federal government works. Instead, it is a distinct piece of the biological makeup of our brains. Language is a complex, specialized skill, which develops in the child spontaneously, without conscious effort or formal instruction, is deployed without awareness of its underlying logic, is qualitatively the same in every individual, and is distinct from more general abilities to process information or behave intelligently. For these reasons some cognitive scientists have described language as a psychological faculty, a mental organ, a neural system, and a computational module. But I prefer the admittedly quaint term "instinct." It conveys the idea that people know how to talk in more or less the sense that spiders know how to spin webs. Web-spinning was not invented by some unsung spider genius and does not depend on having had the right education or on having an aptitude for architecture or the construction trades. Rather, spiders spin spider webs because they have spider brains, which give them the urge to spin and the competence to succeed. Although there are differences between webs and words, I will encourage you to see language in this way, for it helps to make sense of the phenomena we will explore.

Thinking of language as an instinct inverts the popular wisdom, especially as it has been passed down in the canon of the humanities and social sciences. Language is no more a cultural invention than is upright posture. It is not a manifestation of a general capacity to use symbols: a three-year-old, we shall see, is a grammatical genius, but is quite incompetent at the visual arts, religious iconography, traffic signs, and the other staples of the semiotics curriculum. Though language is a magnificent ability unique to *Homo sapiens* among living species, it does not call for sequestering the study of humans from the domain of biology, for a magnificent ability unique to a particular living species is far from unique in the animal kingdom. Some kinds of bats home in on flying insects using Doppler sonar. Some kinds of migratory birds navigate thousands of miles by calibrating the positions of the constellations against the time of day and year. In nature's talent show we are simply a species of primate with our own act, a knack for communicating information about who did what to whom by modulating the sounds we make when we exhale.

Once you begin to look at language not as the ineffable essence of human uniqueness but as a biological adaptation to communicate information, it is no longer as tempting to see language as an insidious shaper of thought, and, we shall see, it is not. Moreover, seeing language as one of nature's engineering marvels—an organ with "that perfection of structure and co-adaptation which justly excites our admiration," in Darwin's words—gives us a new respect for your ordinary Joe and the much-maligned English language (or any language). The complexity of language, from the scientist's point of view, is part of our biological birthright; it is not something that parents teach their children or something that must be elaborated in school—as Oscar Wilde said, "Education is an admirable thing, but it is well to remember from time to time that nothing that is worth knowing can be taught." A preschooler's tacit knowledge of grammar is more sophisticated than the thickest style manual or the most state-of-the-art computer language system, and the same applies to all healthy human beings, even the notorious syntax-fracturing professional athlete and the, you know, like, inarticulate teenage skateboarder. Finally, since language is the product of a well-engineered biological instinct, we shall see that it is not the nutty barrel of monkeys that entertainer-columnists make it out to be. I will try to restore some dignity to the English vernacular, and will even have some nice things to say about its spelling system.

The conception of language as a kind of instinct was first articulated in 1871 by Darwin himself. In *The Descent of Man* he had to contend with language because its confinement to humans seemed to present a challenge to his theory. As in all matters, his observations are uncannily modern:

> As . . . one of the founders of the noble science of philology observes, language is an art, like brewing or baking; but writing would have been a better simile. It certainly is not a true instinct, for every language has to be learned. It differs, however, widely from all ordinary arts, for man has an instinctive tendency to speak, as we see in the babble of our young children; while no child has an instinctive tendency to brew, bake, or write. Moreover, no philologist now supposes that any language has been deliberately invented; it has been slowly and unconsciously developed by many steps.

Darwin concluded that language ability is "an instinctive tendency to acquire an art," a design that is not peculiar to humans but seen in other species such as song-learning birds.

A language instinct may seem jarring to those who think of language as the zenith of the human intellect and who think of instincts as brute impulses that compel furry or feathered zombies to build a dam or up and fly south. But one of Darwin's followers, William James, noted that an instinct possessor need not act as a "fatal automaton." He argued that we have all the instincts that animals do, and many more besides; our flexible intelligence comes from the interplay of many instincts competing. Indeed, the instinctive nature of human thought is just what makes it so hard for us to see that it is an instinct:

> It takes . . . a mind debauched by learning to carry the process of making the natural seem strange, so far as to ask for the *why* of any instinctive human act. To the metaphysician alone can such questions occur as: Why do we smile, when pleased, and not scowl? Why are we unable to talk to a crowd as we talk to a single friend? Why does a particular maiden turn our wits so upside-down? The common man can only say, "*Of course* we smile, *of course* our heart palpitates at the sight of the crowd, *of course* we love the maiden, that beautiful soul clad in that perfect form, so palpably and flagrantly made for all eternity to be loved!"
>
> And so, probably, does each animal feel about the particular things it tends to do in presence of particular objects. . . . To the lion it is the lioness which is made to be loved; to the bear, the she-bear. To the broody hen the notion would probably seem monstrous that there should be a creature in the world to whom a nestful of eggs was not the utterly fascinating and precious and never-to-be-too-much-sat-upon object which it is to her.
>
> Thus we may be sure that, however mysterious some animals' instincts may appear to us, our instincts will appear no less mysterious to them. And we may conclude that, to the

animal which obeys it, every impulse and every step of every instinct shines with its own sufficient light, and seems at the moment the only eternally right and proper thing to do. What voluptuous thrill may not shake a fly, when she at last discovers the one particular leaf, or carrion, or bit of dung, that out of all the world can stimulate her ovipositor to its discharge? Does not the discharge then seem to her the only fitting thing? And need she care or know anything about the future maggot and its food?

I can think of no better statement of my main goal. The workings of language are as far from our awareness as the rationale for egg-laying is from the fly's. Our thoughts come out of our mouths so effortlessly that they often embarrass us, having eluded our mental censors. When we are comprehending sentences, the stream of words is transparent; we see through to the meaning so automatically that we can forget that a movie is in a foreign language and subtitled. We think children pick up their mother tongue by imitating their mothers, but when a child says *Don't giggle me!* or *We holded the baby rabbits,* it cannot be an act of imitation. I want to debauch your mind with learning, to make these natural gifts seem strange, to get you to ask the "why" and "how" of these seemingly homely abilities. Watch an immigrant struggling with a second language or a stroke patient with a first one, or deconstruct a snatch of baby talk, or try to program a computer to understand English, and ordinary speech begins to look different. The effortlessness, the transparency, the automaticity are illusions, masking a system of great richness and beauty.

In this century, the most famous argument that language is like an instinct comes from Noam Chomsky, the linguist who first unmasked the intricacy of the system and perhaps the person most responsible for the modern revolution in language and cognitive science. In the 1950s the social sciences were dominated by behaviorism, the school of thought popularized by John Watson and B. F. Skinner. Mental terms like "know" and "think" were branded as unscientific; "mind" and "innate" were dirty words. Behavior was explained by a few laws of stimulus-response learning that could be studied with rats pressing bars and dogs salivating to tones. But Chomsky called attention to two fundamental facts about language. First, virtually every sentence that a person utters or understands is a brand-new combination of words, appearing for the first time in the history of the universe. Therefore a language cannot be a repertoire of responses; the brain must contain a recipe or program that can build an unlimited set of sentences out of a finite list of words. That program may be called a mental grammar (not to be confused with pedagogical or stylistic "grammars," which are just guides to the etiquette of written prose). The second fundamental fact is that children develop these complex grammars rapidly and without formal instruction and grow up to give consistent interpretations to novel sentence constructions that they have never before encountered. Therefore, he argued, children must innately be equipped with a plan common to the grammars of all languages, a Universal Grammar, that tells them how to distill the syntactic patterns out of the speech of their parents. Chomsky put it as follows:

> It is a curious fact about the intellectual history of the past few centuries that physical and mental development have been approached in quite different ways. No one would take seriously the proposal that the human organism learns through experience to have arms rather than wings, or that the basic structure of particular organs results from accidental experience. Rather, it is taken for granted that the physical structure of the organism is genetically determined, though of course variation along such dimensions as size, rate of development, and so forth will depend in part on external factors. . . .
>
> The development of personality, behavior patterns, and cognitive structures in higher organisms has often been approached in a very different way. It is generally assumed that in these domains, social environment is the dominant factor. The structures of mind that develop over time are taken to be arbitrary and accidental; there is no "human nature" apart from what develops as a specific historical product. . . .

But human cognitive systems, when seriously investigated, prove to be no less marvelous and intricate than the physical structures that develop in the life of the organism. Why, then, should we not study the acquisition of a cognitive structure such as language more or less as we study some complex bodily organ?

At first glance, the proposal may seem absurd, if only because of the great variety of human languages. But a closer consideration dispels these doubts. Even knowing very little of substance about linguistic universals, we can be quite sure that the possible variety of language is sharply limited. . . . The language each person acquires is a rich and complex construction hopelessly underdetermined by the fragmentary evidence available [to the child]. Nevertheless individuals in a speech community have developed essentially the same language. This fact can be explained only on the assumption that these individuals employ highly restrictive principles that guide the construction of grammar.

By performing painstaking technical analyses of the sentences ordinary people accept as part of their mother tongue, Chomsky and other linguists developed theories of the mental grammars underlying people's knowledge of particular languages and of the Universal Grammar underlying the particular grammars. Early on, Chomsky's work encouraged other scientists, among them Eric Lenneberg, George Miller, Roger Brown, Morris Halle, and Alvin Liberman, to open up whole new areas of language study, from child development and speech perception to neurology and genetics. By now, the community of scientists studying the questions he raised numbers in the thousands. Chomsky is currently among the ten most-cited writers in all of the humanities (beating out Hegel and Cicero and trailing only Marx, Lenin, Shakespeare, the Bible, Aristotle, Plato, and Freud) and the only living member of the top ten.

What those citations *say* is another matter. Chomsky gets people exercised. Reactions range from the awe-struck deference ordinarily reserved for gurus of weird religious cults to the withering invective that academics have developed into a high art. In part this is because Chomsky attacks what is still one of the foundations of twentieth-century intellectual life—the "Standard Social Science Model," according to which the human psyche is molded by the surrounding culture. But it is also because no thinker can afford to ignore him. As one of his severest critics, the philosopher Hilary Putnam, acknowledges,

> When one reads Chomsky, one is struck by a sense of great intellectual power; one knows one is encountering an extraordinary mind. And this is as much a matter of the spell of his powerful personality as it is of his obvious intellectual virtues: originality, scorn for the faddish and the superficial; willingness to revive (and the ability to revive) positions (such as the "doctrine of innate ideas") that had seemed passé; concern with topics, such as the structure of the human mind, that are of central and perennial importance.

The story I will tell in this book has, of course, been deeply influenced by Chomsky. But it is not his story exactly, and I will not tell it as he would. Chomsky has puzzled many readers with his skepticism about whether Darwinian natural selection (as opposed to other envolutionary processes) can explain the origins of the language organ that he argues for; I think it is fruitful to consider language as an evolutionary adaptation, like the eye, its major parts designed to carry out important functions. And Chomsky's arguments about the nature of the language faculty are based on technical analyses of word and sentence structure, often couched in abstruse formalisms. His discussions of flesh-and-blood speakers are perfunctory and highly idealized. Though I happen to agree with many of his arguments, I think that a conclusion about the mind is convincing only if many kinds of evidence converge on it. So the story in this book is highly eclectic, ranging from how DNA builds brains to the pontifications of newspaper language columnists. The best place to begin is to ask why anyone should believe that human language is a part of human biology—an instinct—at all.

# 2

❧

# Chatterboxes

*By the 1920s it was thought that no corner of the earth fit for human* habitation had remained unexplored. New Guinea, the world's second largest island, was no exception. The European missionaries, planters, and administrators clung to its coastal lowlands, convinced that no one could live in the treacherous mountain range that ran in a solid line down the middle of the island. But the mountains visible from each coast in fact belonged to two ranges, not one, and between them was a temperate plateau crossed by many fertile valleys. A million Stone Age people lived in those highlands, isolated from the rest of the world for forty thousand years. The veil would not be lifted until gold was discovered in a tributary of one of the main rivers. The ensuing gold rush attracted Michael Leahy, a footloose Australian prospector, who on May 26, 1930, set out to explore the mountains with a fellow prospector and a group of indigenous lowland people hired as carriers. After scaling the heights, Leahy was amazed to see grassy open country on the other side. By nightfall his amazement turned to alarm, because there were points of light in the distance, obvious signs that the valley was populated. After a sleepless night in which Leahy and his party loaded their weapons and assembled a crude bomb, they made their first contact with the highlanders. The astonishment was mutual. Leahy wrote in his diary:

> It was a relief when the [natives] came in sight, the men . . . in front, armed with bows and arrows, the women behind bringing stalks of sugarcane. When he saw the women, Ewunga told me at once that there would be no fight. We waved to them to come on, which they did cautiously, stopping every few yards to look us over. When a few of them finally got up courage to approach, we could see that they were utterly thunderstuck by our appearance. When I took off my hat, those nearest to me backed away in terror. One old chap came forward gingerly with open mouth, and touched me to see if I was real. Then he knelt down, and rubbed his hands over my bare legs, possibly to find if they were painted, and grabbed me around the knees and hugged them, rubbing his bushy head against me. . . . The women and children gradually got up courage to approach also, and presently the camp was swarming with the lot of them, all running about and jabbering at once, pointing to . . . everything that was new to them.

That "jabbering" was language—an unfamiliar language, one of eight hundred different ones that would be discovered among the isolated highlanders right up through the 1960s. Leahy's first contact repeated a scene that must have taken place hundreds of times in human history, whenever one people first encountered another. All of them, as far as we know, already had language. Every Hottentot, every Eskimo, every Yanomamö. No mute tribe has ever been discovered, and there is no record that a region has served as a "cradle" of language from which it spread to previously languageless groups.

As in every other case, the language spoken by Leahy's hosts turned out to be no mere jabber but a medium that could express abstract concepts, invisible entities, and complex trains of reasoning. The highlanders conferred intensively, trying to agree upon the nature of the pallid apparitions. The leading conjecture was that they were reincarnated ancestors or other spirits in human form, perhaps ones that turned back into skeletons at night. They agreed upon an empirical test that would settle the matter. "One of the people hid," recalls the highlander Kirupano Eza'e, "and watched them going to excrete. He came back and said, 'Those men from heaven went to excrete over there.' Once they had left many men went to take a look. When they saw that it smelt bad, they said, 'Their skin might be different, but their shit smells bad like ours.' "

The universality of complex language is a discovery that fills linguists with awe, and is the first reason to suspect that language is not just any cultural invention but the product of a special human instinct. Cultural inventions vary widely in their sophistication from society to society; within a society, the inventions are generally at the same level of sophistication. Some groups count by carving notches on bones and cook on fires ignited by spinning sticks in logs; others use computers and microwave ovens. Language, however, ruins this correlation. There are Stone Age societies, but there is no such thing as a Stone Age language. Earlier in this century the anthropological linguist Edward Sapir wrote, "When it comes to linguistic form, Plato walks with the Macedonian swineherd, Confucius with the headhunting savage of Assam."

To pick an example at random of a sophisticated linguistic form in a nonindustrialized people, the linguist Joan Bresnan recently wrote a technical article comparing a construction in Kivunjo, a Bantu language spoken in several villages on the slopes of Mount Kilimanjaro in Tanzania, with its counterpart construction in English, which she describes as "a West Germanic language spoken in England and its former colonies." The English construction is called the dative* and is found in sentences like *She baked me a brownie* and *He promised her Arpège,* where an indirect object like *me* or *her* is placed after the verb to indicate the beneficiary of an act. The corresponding Kivunjo construction is called the applicative, whose resemblance to the English

*All the technical terms from linguistics, biology, and cognitive science that I use in this book are defined in the Glossary on pages 503–516.

dative, Bresnan notes, "can be likened to that of the game of chess to checkers." The Kivunjo construction fits entirely inside the verb, which has seven prefixes and suffixes, two moods, and fourteen tenses; the verb agrees with its subject, its object, and its benefactive nouns, each of which comes in sixteen genders. (In case you are wondering, these "genders" do not pertain to things like cross-dressers, transsexuals, hermaphrodites, androgynous people, and so on, as one reader of this chapter surmised. To a linguist, the term *gender* retains its original meaning of "kind," as in the related words *generic, genus,* and *genre.* The Bantu "genders" refer to kinds like humans, animals, extended objects, clusters of objects, and body parts. It just happens that in many European languages the genders correspond to the sexes, at least in pronouns. For this reason the linguistic term *gender* has been pressed into service by nonlinguists as a convenient label for sexual dimorphism; the more accurate term *sex* seems now to be reserved as the polite way to refer to copulation.) Among the other clever gadgets I have glimpsed in the grammars of so-called primitive groups, the complex Cherokee pronoun system seems especially handy. It distinguishes among "you and I," "another person and I," "several other people and I," and "you, one or more other persons, and I," which English crudely collapses into the all-purpose pronoun *we.*

Actually, the people whose linguistic abilities are most badly underestimated are right here in our society. Linguists repeatedly run up against the myth that working-class people and the less educated members of the middle class speak a simpler or coarser language. This is a pernicious illusion arising from the effortlessness of conversation. Ordinary speech, like color vision or walking, is a paradigm of engineering excellence—a technology that works so well that the user takes its outcome for granted, unaware of the complicated machinery hidden behind the panels. Behind such "simple" sentences as *Where did he go?* and or *The guy I met killed himself,* used automatically by any English speaker, are dozens of subroutines that arrange the words to express the meaning. Despite decades of effort, no artificially engineered language system comes close to duplicating the person in the street, HAL and C3PO notwithstanding.

But though the language engine is invisible to the human user, the trim packages and color schemes are attended to obsessively. Trifling differences between the dialect of the mainstream and the dialect of other groups, like *isn't any* versus *ain't no*, *those books* versus *them books*, and *dragged him away* versus *drug him away*, are dignified as badges of "proper grammar." But they have no more to do with grammatical sophistication than the fact that people in some regions of the United States refer to a certain insect as a *dragonfly* and people in other regions refer to it as a *darning needle*, or that English speakers call canines *dogs* whereas French speakers call them *chiens*. It is even a bit misleading to call Standard English a "language" and these variations "dialects," as if there were some meaningful difference between them. The best definition comes from the linguist Max Weinreich: a language is a dialect with an army and a navy.

The myth that nonstandard dialects of English are grammatically deficient is widespread. In the 1960s some well-meaning educational psychologists announced that American black children had been so culturally deprived that they lacked true language and were confined instead to a "non-logical mode of expressive behavior." The conclusions were based on the students' shy or sullen reactions to batteries of standardized tests. If the psychologists had listened to spontaneous conversations, they would have rediscovered the commonplace fact that American black culture is everywhere highly verbal; the subculture of street youths in particular is famous in the annals of anthropology for the value placed on linguistic virtuosity. Here is an example, from an interview conducted by the linguist William Labov on a stoop in Harlem. The interviewee is Larry, the roughest member of a teenage gang called the Jets. (Labov observes in his scholarly article that "for most readers of this paper, first contact with Larry would produce some fairly negative reactions on both sides.")

> You know, like some people say if you're good an' shit, your spirit goin' t'heaven . . . 'n' if you bad, your spirit goin' to hell. Well, bullshit! Your spirit goin' to hell anyway, good or bad.
>
> [Why?]

Why? I'll tell you why. 'Cause, you see, doesn' nobody really know that it's a God, y'know, 'cause I mean I have seen black gods, white gods, all color gods, and don't nobody know it's really a God. An' when they be sayin' if you good, you goin' t'heaven, tha's bullshit, 'cause you ain't goin' to no heaven, 'cause it ain't no heaven for you to go to.

[. . . jus' suppose that there is a God, would he be white or black?]

He'd be white, man.

[Why?]

Why? I'll tell you why. 'Cause the average whitey out here got everything, you dig? And the nigger ain't got shit, y'know? Y'understan'? So—um—for—in order for *that* to happen, you know it ain't no black God that's doin' that bullshit.

First contact with Larry's grammar may produce negative reactions as well, but to a linguist it punctiliously conforms to the rules of the dialect called Black English Vernacular (BEV). The most linguistically interesting thing about the dialect is how linguistically uninteresting it is: if Labov did not have to call attention to it to debunk the claim that ghetto children lack true linguistic competence, it would have been filed away as just another language. Where Standard American English (SAE) uses *there* as a meaningless dummy subject for the copula, BEV uses *it* as a meaningless dummy subject for the copula (compare SAE's *There's really a God* with Larry's *It's really a God*). Larry's negative concord (*You ain't goin' to no heaven*) is seen in many languages, such as French (*ne . . . pas*). Like speakers of SAE, Larry inverts subjects and auxiliaries in nondeclarative sentences, but the exact set of the sentence types allowing inversion differs slightly. Larry and other BEV speakers invert subjects and auxiliaries in negative main clauses like *Don't nobody know*; SAE speakers invert them only in questions like *Doesn't anybody know?* and a few other sentence types. BEV allows its speakers the option of deleting copulas (*If you bad*); this is not random laziness but a systematic rule that is virtually identical to the contraction rule in SAE that reduces *He is* to *He's*, *You are* to *You're*, and *I am* to *I'm*. In both dialects, *be* can erode only in certain kinds of sentences. No SAE speaker would try the following contractions:

Yes he is! → Yes he's!
I don't care what you are. → I don't care what you're.
Who is it? → Who's it?

For the same reasons, no BEV speaker would try the following deletions:

Yes he is! → Yes he!
I don't care what you are. → I don't care what you.
Who is it? → Who it?

Note, too, that BEV speakers are not just more prone to eroding words. BEV speakers use the full forms of certain auxiliaries (*I have seen*), whereas SAE speakers usually contract them (*I've seen*). And as we would expect from comparisons between languages, there are areas in which BEV is more precise than standard English. *He be working* means that he generally works, perhaps that he has a regular job; *He working* means only that he is working at the moment that the sentence is uttered. In SAE, *He is working* fails to make that distinction. Moreover, sentences like *In order for that to happen, you know it ain't no black God that's doin' that bullshit* show that Larry's speech uses the full inventory of grammatical paraphernalia that computer scientists struggle unsuccessfully to duplicate (relative clauses, complement structures, clause subordination, and so on), not to mention some fairly sophisticated theological argumentation.

Another project of Labov's involved tabulating the percentage of grammatical sentences in tape recordings of speech in a variety of social classes and social settings. "Grammatical," for these purposes, means "well-formed according to consistent rules in the dialect of the speakers." For example, if a speaker asked the question *Where are you going?*, the respondent would not be penalized for answering *To the store*, even though it is in some sense not a complete sentence. Such ellipses are obviously part of the grammar of conversational English; the alternative, *I am going to the store,* sounds stilted and is almost never used. "Ungrammatical" sentences, by this definition, include randomly broken-off sentence fragments, tongue-tied hemming and hawing, slips of the tongue, and other forms of word salad. The results of Labov's tabulation are enlightening. The great majority of sentences were grammatical, especially in casual speech, with higher percentages of grammatical sentences in working-class speech than in middle-class speech. The highest percentage of ungrammatical sentences was found in the proceedings of learned academic conferences.

The ubiquity of complex language among human beings is a gripping discovery and, for many observers, compelling proof that language is innate. But to tough-minded skeptics like the philosopher Hilary Putnam, it is no proof at all. Not everything that is universal is innate. Just as travelers in previous decades never encountered a tribe without a language, nowadays anthropologists have trouble finding a people beyond the reach of VCR's, Coca-Cola, and Bart Simpson T-shirts. Language was universal before Coca-Cola was, but then, language is more useful than Coca-Cola. It is more like eating with one's hands rather than one's feet, which is also universal, but we need not invoke a special hand-to-mouth instinct to explain why. Language is invaluable for all the activities of daily living in a community of people: preparing food and shelter, loving, arguing, negotiating, teaching. Necessity being the mother of invention, language could have been invented by resourceful people a number of times long ago. (Perhaps, as Lily Tomlin said, man invented language to satisfy his deep need to complain.) Universal grammar would simply reflect the universal exigencies of human experience and the universal limitations on human information processing. All languages have words for "water" and "foot" because all people need to refer to water and feet; no language has a word a million syllables long because no person would have time to say it. Once invented, language would entrench itself within a culture as parents taught their children and children imitated their parents. From cultures that had language, it would spread like

wildfire to other, quieter cultures. At the heart of this process is won-
drously flexible human intelligence, with its general multipurpose
learning strategies.

So the universality of language does not lead to an innate language instinct as night follows day. To convince you that there is a
language instinct, I will have to fill in an argument that leads from the
jabbering of modern peoples to the putative genes for grammar. The
crucial intervening steps come from my own professional specialty,
the study of language development in children. The crux of the argument is that complex language is universal because *children actually
reinvent it,* generation after generation—not because they are taught,
not because they are generally smart, not because it is useful to them,
but because they just can't help it. Let me now take you down this
trail of evidence.

The trail begins with the study of how the particular languages we
find in the world today arose. Here, one would think, linguistics runs
into the problem of any historical science: no one recorded the crucial
events at the time they happened. Although historical linguists can
trace modern complex languages back to earlier ones, this just pushes
the problem back a step; we need to see how people create a complex
language from scratch. Amazingly, we can.

The first cases were wrung from two of the more sorrowful epi-
sodes of world history, the Atlantic slave trade and indentured servi-
tude in the South Pacific. Perhaps mindful of the Tower of Babel,
some of the masters of tobacco, cotton, coffee, and sugar plantations
deliberately mixed slaves and laborers from different language back-
grounds; others preferred specific ethnicities but had to accept mixtures because that was all that was available. When speakers of
different languages have to communicate to carry out practical tasks
but do not have the opportunity to learn one another's languages,
they develop a makeshift jargon called a pidgin. Pidgins are choppy
strings of words borrowed from the language of the colonizers or
plantation owners, highly variable in order and with little in the way
of grammar. Sometimes a pidgin can become a lingua franca and grad-

ually increase in complexity over decades, as in the "Pidgin English" of the modern South Pacific. (Prince Philip was delighted to learn on a visit to New Guinea that he is referred to in that language as *fella belong Mrs. Queen.*)

But the linguist Derek Bickerton has presented evidence that in many cases a pidgin can be transmuted into a full complex language in one fell swoop: all it takes is for a group of children to be exposed to the pidgin at the age when they acquire their mother tongue. That happened, Bickerton has argued, when children were isolated from their parents and were tended collectively by a worker who spoke to them in the pidgin. Not content to reproduce the fragmentary word strings, the children injected grammatical complexity where none existed before, resulting in a brand-new, richly expressive language. The language that results when children make a pidgin their native tongue is called a creole.

Bickerton's main evidence comes from a unique historical circumstance. Though the slave plantations that spawned most creoles are, fortunately, a thing of the remote past, one episode of creolization occurred recently enough for us to study its principal players. Just before the turn of the century there was a boom in Hawaiian sugar plantations, whose demands for labor quickly outstripped the native pool. Workers were brought in from China, Japan, Korea, Portugal, the Philippines, and Puerto Rico, and a pidgin quickly developed. Many of the immigrant laborers who first developed that pidgin were alive when Bickerton interviewed them in the 1970s. Here are some typical examples of their speech:

> Me capé buy, me check make.
> Building—high place—wall pat—time—nowtime—an' den—a
>     new tempecha eri time show you.
> Good, dis one. Kaukau any-kin' dis one. Pilipine islan' no
>     good. No mo money.

From the individual words and the context, it was possible for the listener to infer that the first speaker, a ninety-two-year-old Japanese immigrant talking about his earlier days as a coffee farmer, was trying to say "He bought my coffee; he made me out a check." But the utterance itself could just as easily have meant "I bought coffee; I made him out a check," which would have been appropriate if he had been referring to his current situation as a store owner. The second speaker, another elderly Japanese immigrant, had been introduced to the wonders of civilization in Los Angeles by one of his many children, and was saying that there was an electric sign high up on the wall of the building which displayed the time and temperature. The third speaker, a sixty-nine-year-old Filipino, was saying "It's better here than in the Philippines; here you can get all kinds of food, but over there there isn't any money to buy food with." (One of the kinds of food was "pfrawg," which he caught for himself in the marshes by the method of "kank da head.") In all these cases, the speaker's intentions had to be filled in by the listener. The pidgin did not offer the speakers the ordinary grammatical resources to convey these messages—no consistent word order, no prefixes or suffixes, no tense or other temporal and logical markers, no structure more complex than a simple clause, and no consistent way to indicate who did what to whom.

But the children who had grown up in Hawaii beginning in the 1890s and were exposed to the pidgin ended up speaking quite differently. Here are some sentences from the language they invented, Hawaiian Creole. The first two are from a Japanese papaya grower born in Maui; the next two, from a Japanese/Hawaiian ex-plantation laborer born on the big island; the last, from a Hawaiian motel manager, formerly a farmer, born in Kauai:

> Da firs japani came ran away from japan come.
> "The first Japanese who arrived ran away from Japan to here."

> Some filipino wok o'he-ah dey wen' couple ye-ahs in filipin islan'.
> "Some Filipinos who worked over here went back to the Philippines for a couple of years."

> People no like t'come fo' go wok.
> "People don't want to have him go to work [for them]."

One time when we go home inna night dis ting stay fly up.
"Once when we went home at night this thing was flying about."
One day had pleny of dis mountain fish come down.
"One day there were a lot of these fish from the mountains that came down [the river]."

Do not be misled by what look like crudely placed English verbs, such as *go, stay,* and *came,* or phrases like *one time.* They are not haphazard uses of English words but systematic uses of Hawaiian Creole grammar: the words have been converted by the creole speakers into auxiliaries, prepositions, case markers, and relative pronouns. In fact, this is probably how many of the grammatical prefixes and suffixes in established languages arose. For example, the English past-tense ending *-ed* may have evolved from the verb *do: He hammered* was originally something like *He hammer-did.* Indeed, creoles *are* bona fide languages, with standardized word orders and grammatical markers that were lacking in the pidgin of the immigrants and, aside from the sounds of words, not taken from the language of the colonizers.

Bickerton notes that if the grammar of a creole is largely the product of the minds of children, unadulterated by complex language input from their parents, it should provide a particularly clear window on the innate grammatical machinery of the brain. He argues that creoles from unrelated language mixtures exhibit uncanny resemblances—perhaps even the same basic grammar. This basic grammar also shows up, he suggests, in the errors children make when acquiring more established and embellished languages, like some underlying design bleeding through a veneer of whitewash. When English-speaking children say

> Why he is leaving?
> Nobody don't likes me.
> I'm gonna full Angela's bucket.
> Let Daddy hold it hit it,

they are unwittingly producing sentences that are grammatical in many of the world's creoles.

Bickerton's particular claims are controversial, depending as they do on his reconstruction of events that occurred decades or centuries in the past. But his basic idea has been stunningly corroborated by two recent natural experiments in which creolization by children can be observed in real time. These fascinating discoveries are among many that have come from the study of the sign languages of the deaf. Contrary to popular misconceptions, sign languages are not pantomimes and gestures, inventions of educators, or ciphers of the spoken language of the surrounding community. They are found wherever there is a community of deaf people, and each one is a distinct, full language, using the same kinds of grammatical machinery found worldwide in spoken languages. For example, American Sign Language, used by the deaf community in the United States, does not resemble English, or British Sign Language, but relies on agreement and gender systems in a way that is reminiscent of Navajo and Bantu.

Until recently there were no sign languages at all in Nicaragua, because its deaf people remained isolated from one another. When the Sandinista government took over in 1979 and reformed the educational system, the first schools for the deaf were created. The schools focused on drilling the children in lip reading and speech, and as in every case where that is tried, the results were dismal. But it did not matter. On the playgrounds and schoolbuses the children were inventing their own sign system, pooling the makeshift gestures that they used with their families at home. Before long the system congealed into what is now called the Lenguaje de Signos Nicaragüense (LSN). Today LSN is used, with varying degrees of fluency, by young deaf adults, aged seventeen to twenty-five, who developed it when they were ten or older. Basically, it is a pidgin. Everyone uses it differently, and the signers depend on suggestive, elaborate circumlocutions rather than on a consistent grammar.

But children like Mayela, who joined the school around the age of four, when LSN was already around, and all the pupils younger than her, are quite different. Their signing is more fluid and compact, and the gestures are more stylized and less like a pantomime. In fact, when their signing is examined close up, it is so different from LSN that it is referred to by a different name, Idioma de Signos Nicara-güense (ISN). LSN and ISN are currently being studied by the psycholinguists Judy Kegl, Miriam Hebe Lopez, and Annie Senghas. ISN appears to be a creole, created in one leap when the younger children were exposed to the pidgin signing of the older children—just as Bickerton would have predicted. ISN has spontaneously standardized itself; all the young children sign it in the same way. The children have introduced many grammatical devices that were absent in LSN, and hence they rely far less on circumlocutions. For example, an LSN (pidgin) signer might make the sign for "talk to" and then point from the position of the talker to the position of the hearer. But an ISN (creole) signer modifies the sign itself, sweeping it in one motion from a point representing the talker to a point representing the hearer. This is a common device in sign languages, formally identical to inflecting a verb for agreement in spoken languages. Thanks to such consistent grammar, ISN is very expressive. A child can watch a surrealistic cartoon and describe its plot to another child. The children use it in jokes, poems, narratives, and life histories, and it is coming to serve as the glue that holds the community together. A language has been born before our eyes.

But ISN was the collective product of many children communicating with one another. If we are to attribute the richness of language to the mind of the child, we really want to see a single child adding some increment of grammatical complexity to the input the child has received. Once again the study of the deaf grants our wish.

When deaf infants are raised by signing parents, they learn sign language in the same way that hearing infants learn spoken language. But deaf children who are not born to deaf parents—the majority of deaf children—often have no access to sign language users as they grow up, and indeed are sometimes deliberately kept from them by educators in the "oralist" tradition who want to force them to master lip reading and speech. (Most deaf people deplore these authoritarian measures.) When deaf children become adults, they tend to seek out deaf communities and begin to acquire the sign language that takes proper advantage of the communicative media available to them. But by then it is usually too late; they must then struggle with sign language as a difficult intellectual puzzle, much as a hearing adult does in foreign language classes. Their proficiency is notably below that of deaf people who acquired sign language as infants, just as adult immigrants are often permanently burdened with accents and conspicuous grammatical errors. Indeed, because the deaf are virtually the only neurologically normal people who make it to adulthood without having acquired a language, their difficulties offer particularly good evidence that successful language acquisition must take place during a critical window of opportunity in childhood.

The psycholinguists Jenny Singleton and Elissa Newport have studied a nine-year-old profoundly deaf boy, to whom they gave the pseudonym Simon, and his parents, who are also deaf. Simon's parents did not acquire sign language until the late ages of fifteen and sixteen, and as a result they acquired it badly. In ASL, as in many languages, one can move a phrase to the front of a sentence and mark it with a prefix or suffix (in ASL, raised eyebrows and a lifted chin) to indicate that it is the topic of the sentence. The English sentence *Elvis I really like* is a rough equivalent. But Simon's parents rarely used this construction and mangled it when they did. For example, Simon's father once tried to sign the thought *My friend, he thought my second child was deaf.* It came out as *My friend thought, my second child, he thought he was deaf*—a bit of sign salad that violates not only ASL grammar but, according to Chomsky's theory, the Universal Grammar that governs all naturally acquired human languages (later in this chapter we will see why). Simon's parents had also failed to grasp the verb inflection system of ASL. In ASL, the verb *to blow* is signed by opening a fist held horizontally in front of the mouth (like a puff of air). Any verb in ASL can be modified to indicate that the action is being done continuously: the signer superimposes an arclike motion on the sign and repeats it quickly. A verb can also be modified to indicate that the action is being done to more than one object (for example, several candles): the signer terminates the sign in one location in space, then repeats it but terminates it at another location. These inflections can be combined in either of two orders: *blow* toward the left and then toward the right and repeat, or *blow* toward the left twice and then *blow* toward the right twice. The first order means "to blow out the candles on one cake, then another cake, then the first cake again, then the second cake again"; the second means "to blow out the candles on one cake continuously, and then blow out the candles on another cake continuously." This elegant set of rules was lost on Simon's parents. They used the inflections inconsistently and never combined them onto a verb two at a time, though they would occasionally use the inflections separately, crudely linked with signs like *then*. In many ways Simon's parents were like pidgin speakers.

Astoundingly, though Simon saw no ASL but his parents' defective version, his own signing was far better ASL than theirs. He understood sentences with moved topic phrases without difficulty, and when he had to describe complex videotaped events, he used the ASL verb inflections almost perfectly, even in sentences requiring two of them in particular orders. Simon must somehow have shut out his parents' ungrammatical "noise." He must have latched on to the inflections that his parents used inconsistently, and reinterpreted them as mandatory. And he must have seen the logic that was implicit, though never realized, in his parents' use of two kinds of verb inflection, and reinvented the ASL system of superimposing both of them onto a single verb in a specific order. Simon's superiority to his parents is an example of creolization by a single living child.

Actually, Simon's achievements are remarkable only because he is the first one who showed them to a psycholinguist. There must be thousands of Simons: ninety to ninety-five percent of deaf children are born to hearing parents. Children fortunate enough to be exposed to ASL at all often get it from hearing parents who themselves learned it, incompletely, to communicate with their children. Indeed, as the transition from LSN to ISN shows, sign languages themselves are surely products of creolization. Educators at various points in history have tried to invent sign systems, sometimes based on the surrounding spoken language. But these crude codes are always unlearnable, and when deaf children learn from them at all, they do so by converting them into much richer natural languages.

Extraordinary acts of creation by children do not require the extraordinary circumstances of deafness or plantation Babels. The same kind of linguistic genius is involved every time a child learns his or her mother tongue.

First, let us do away with the folklore that parents teach their children language. No one supposes that parents provide explicit grammar lessons, of course, but many parents (and some child psychologists who should know better) think that mothers provide children with implicit lessons. These lessons take the form of a special speech variety called Motherese (or, as the French call it, Mamanaise): intensive sessions of conversational give-and-take, with repetitive drills and simplified grammar. ("Look at the *doggie!* See the *doggie?* There's a *doggie!*") In contemporary middle-class American culture, parenting is seen as an awesome responsibility, an unforgiving vigil to keep the helpless infant from falling behind in the great race of life. The belief that Motherese is essential to language development is part of the same mentality that sends yuppies to "learning centers" to buy little mittens with bull's-eyes to help their babies find their hands sooner.

One gets some perspective by examining the folk theories about parenting in other cultures. The !Kung San of the Kalahari Desert in southern Africa believe that children must be drilled to sit, stand, and walk. They carefully pile sand around their infants to prop them upright, and sure enough, every one of these infants soon sits up on its own. We find this amusing because we have observed the results of the experiment that the San are unwilling to chance: we don't teach our children to sit, stand, and walk, and they do it anyway, on their own schedule. But other groups enjoy the same condescension toward us. In many communities of the world, parents do not indulge their children in Motherese. In fact, they do not speak to their prelinguistic children at all, except for occasional demands and rebukes. This is not unreasonable. After all, young children plainly can't understand a word you say. So why waste your breath in soliloquies? Any sensible person would surely wait until a child has developed speech and more gratifying two-way conversations become possible. As Aunt Mae, a woman living in the South Carolina Piedmont, explained to the anthropologist Shirley Brice Heath: "Now just how crazy is dat? White folks uh hear dey kids say sump'n, dey say it back to 'em, dey aks 'em 'gain and 'gain 'bout things, like they 'posed to be born knowin'." Needless to say, the children in these communities, over-hearing adults and other children, learn to talk, as we see in Aunt Mae's fully grammatical BEV.

Children deserve most of the credit for the language they acquire. In fact, we can show that they know things they could not have been taught. One of Chomsky's classic illustrations of the logic of language involves the process of moving words around to form questions. Consider how you might turn the declarative sentence *A unicorn is in the garden* into the corresponding question, *Is a unicorn in the garden?* You could scan the declarative sentence, take the auxiliary *is,* and move it to the front of the sentence:

> a unicorn is in the garden. →
> is a unicorn    the garden?

Now take the sentence *A unicorn that is eating a flower is in the garden.* There are two *is*'s. Which gets moved? Obviously, not the first one hit by the scan; that would give you a very odd sentence:

> a unicorn that is eating a flower is in the garden. →
> is a unicorn that    eating a flower is in the garden?

But why can't you move that *is?* Where did the simple procedure go wrong? The answer, Chomsky noted, comes from the basic design of language. Though sentences are strings of words, our mental algo-rithms for grammar do not pick out words by their linear positions, such as "first word," "second word," and so on. Rather, the algo-rithms group words into phrases, and phrases into even bigger phrases, and give each one a mental label, like "subject noun phrase" or "verb phrase." The real rule for forming questions does not look for the first occurrence of the auxiliary word as one goes from left to right in the string; it looks for the auxiliary that comes after the phrase labeled as the subject. This phrase, containing the entire string of words *a unicorn that is eating a flower*, behaves as a single unit. The first *is* sits deeply buried in it, invisible to the question-forming rule. The second *is,* coming immediately after this subject noun phrase, is the one that is moved:

> [a unicorn that is eating a flower] is in the garden. →
> is [a unicorn that is eating a flower]    in the garden?

Chomsky reasoned that if the logic of language is wired into children, then the first time they are confronted with a sentence with two auxiliaries they should be capable of turning it into a question with the proper wording. This should be true even though the wrong rule, the one that scans the sentence as a linear string of words, is simpler and presumably easier to learn. And it should be true even though the sentences that would teach children that the linear rule is wrong and the structure-sensitive rule is right—questions with a second auxiliary embedded inside the subject phrase—are so rare as to be nonexistent in Motherese. Surely not every child learning English has heard Mother say *Is the doggie that is eating the flower in the garden?* For Chomsky, this kind of reasoning, which he calls "the argument from the poverty of the input," is the primary justification for saying that the basic design of language is innate.

Chomsky's claim was tested in an experiment with three-, four-, and five-year-olds at a daycare center by the psycholinguists Stephen Crain and Mineharu Nakayama. One of the experimenters controlled a doll of Jabba the Hutt, of *Star Wars* fame. The other coaxed the child to ask a set of questions, by saying, for example, "Ask Jabba if the boy who is unhappy is watching Mickey Mouse." Jabba would inspect a picture and answer yes or no, but it was really the child who was being tested, not Jabba. The children cheerfully provided the appropriate questions, and, as Chomsky would have predicted, not a single one of them came up with an ungrammatical string like *Is the boy who unhappy is watching Mickey Mouse?,* which the simple linear rule would have produced.

Now, you may object that this does not show that children's brains register the subject of a sentence. Perhaps the children were just going by the meanings of the words. *The man who is running* refers to a single actor playing a distinct role in the picture, and children could have been keeping track of which words are about particular actors, not which words belong to the subject noun phrase. But Crain and Nakayama anticipated the objection. Mixed into their list were commands like "Ask Jabba if it is raining in this picture." The *it* of the sentence, of course, does not refer to anything; it is a dummy element that is there only to satisfy the rules of syntax, which demand a subject. But the English question rule treats it just like any other subject: *Is it raining?* Now, how do children cope with this meaningless placeholder? Perhaps they are as literal-minded as the Duck in *Alice's Adventures in Wonderland:*

> "I proceed [said the Mouse]. 'Edwin and Morcar, the earls of Mercia and Northumbria, declared for him; and even Stigand, the patriotic archbishop of Canterbury, found it advisable—' "
>
> "Found *what?*" said the Duck.
>
> "Found *it,*" the Mouse replied rather crossly: "of course you know what 'it' means."
>
> "I know what 'it' means well enough, when *I* find a thing," said the Duck: "it's generally a frog, or a worm. The question is, what did the archbishop find?"

But children are not ducks. Crain and Nakayama's children replied, *Is it raining in this picture?* Similarly, they had no trouble forming question with other dummy subjects, as in "Ask Jabba if there is a snake in this picture," or with subjects that are not things, as in "Ask Jabba if running is fun" and "Ask Jabba if love is good or bad."

The universal constraints on grammatical rules also show that the basic form of language cannot be explained away as the inevitable outcome of a drive for usefulness. Many languages, widely scattered over the globe, have auxiliaries, and like English, many languages move the auxiliary to the front of the sentence to form questions and other constructions, always in a structure-dependent way. But this is not the only way one could design a question rule. One could just as effectively move the leftmost auxiliary in the string to the front, or flip the first and last words, or utter the entire sentence in mirror-reversed order (a trick that the human mind is capable of; some people learn to talk backwards to amuse themselves and amaze their friends). The particular ways that languages do form questions are arbitrary, species-wide conventions; we don't find them in artificial systems like computer programming languages or the notation of mathematics. The universal plan underlying languages, with auxiliaries and inversion rules, nouns and verbs, subjects and objects, phrases and clauses, case and agreement, and so on, seems to suggest a commonality in the brains of speakers, because many other plans would have been just as useful. It is as if isolated inventors miraculously came up with identical standards for typewriter keyboards or Morse code or traffic signals.

Evidence corroborating the claim that the mind contains blueprints for grammatical rules comes, once again, out of the mouths of babes and sucklings. Take the English agreement suffix -s as in *He walks*. Agreement is an important process in many languages, but in modern English it is superfluous, a remnant of a richer system that flourished in Old English. If it were to disappear entirely, we would not miss it, any more than we miss the similar -est suffix in *Thou sayest*. But psychologically speaking, this frill does not come cheap. Any speaker commited to using it has to keep track of four details in every sentence uttered:

- whether the subject is in the third person or not: *He walks* versus *I walk.*
- whether the subject is singular or plural: *He walks* versus *They walk.*
- whether the action is present tense or not: *He walks* versus *He walked.*
- whether the action is habitual or going on at the moment of speaking (its "aspect"): *He walks to school* versus *He is walking to school.*

And all this work is needed just to use the suffix once one has learned it. To learn it in the first place, a child must (1) notice that verbs end in -*s* in some sentences but appear bare-ended in others, (2) begin a search for the grammatical causes of this variation (as opposed to just accepting it as part of the spice of life), and (3) not rest until those crucial factors—tense, aspect, and the number and person of the subject of the sentence—have been sifted out of the ocean of conceivable but irrelevant factors (like the number of syllables of the final word in the sentence, whether the object of a preposition is natural or man-made, and how warm it is when the sentence is uttered). Why would anyone bother?

But little children do bother. By the age of three and a half or earlier, they use the -*s* agreement suffix in more than ninety percent of the sentences that require it, and virtually never use it in the sentences that forbid it. This mastery is part of their grammar explosion, a period of several months in the third year of life during which children suddenly begin to speak in fluent sentences, respecting most of the fine points of their community's spoken language. For example, a pre-schooler with the pseudonym Sarah, whose parents had only a high school education, can be seen obeying the English agreement rule, useless though it is, in complex sentences like the following:

> When my mother *hangs* clothes, do you let 'em rinse out in rain?
> Donna *teases* all the time and Donna has false teeth.
> I know what a big chicken *looks* like.
> Anybody *knows* how to scribble.
> Hey, this part *goes* where this one is, stupid.
> What *comes* after "C"?
> It *looks* like a donkey face.
> The person *takes* care of the animals in the barn.
> After it *dries* off then you can make the bottom.
> Well, someone *hurts* hisself and everything.
> His tail *sticks* out like this.
> What *happens* if ya press on this hard?
> Do you have a real baby that *says* googoo gaga?

Just as interestingly, Sarah could not have been simply imitating her parents, memorizing verbs with the -*s*'s pre-attached. Sarah sometimes uttered word forms that she could not possibly have heard from her parents:

> When she *be's* in the kindergarten . . .
> He's a boy so he *gots* a scary one. [costume]
> She *do's* what her mother tells her.

She must, then, have created these forms herself, using an unconscious version of the English agreement rule. The very concept of imitation is suspect to begin with (if children are general imitators, why don't they imitate their parents' habit of sitting quietly in airplanes?), but sentences like these show clearly that language acquisition cannot be explained as a kind of imitation.

One step remains to complete the argument that language is a specific instinct, not just the clever solution to a problem thought up by a generally brainy species. If language is an instinct, it should have an identifiable seat in the brain, and perhaps even a special set of genes that help wire it into place. Disrupt these genes or neurons, and language should suffer while the other parts of intelligence carry on; spare them in an otherwise damaged brain, and you should have a retarded individual with intact language, a linguistic idiot savant. If, on the other hand, language is just the exercise of human smarts, we might expect that injuries and impairments would make people stupider across the board, including their language. The only pattern we would expect is that the more brain tissue that is damaged, the duller and less articulate the person should be.

No one has yet located a language organ or a grammar gene, but the search is on. There are several kinds of neurological and genetic impairments that compromise language while sparing cognition and vice versa. One of them has been known for over a century, perhaps for millennia. When there is damage to certain circuits in the lower parts of the frontal lobe of the brain's left hemisphere—say, from a stroke or bullet wound—the person often suffers from a syndrome called Broca's aphasia. One of these victims, who eventually recovered his language ability, recalls the event, which he experienced with complete lucidity:

> When I woke up I had a bit of a headache and thought I must have been sleeping with my right arm under me because it felt all pins-and-needly and numb and I couldn't make it do what I wanted. I got out of bed but I couldn't stand; as a matter of fact I actually fell on the floor because my right leg was too weak to take my weight. I called out to my wife in the next room and no sound came—I couldn't speak. . . . I was astonished, horrified. I couldn't believe that this was happening to me and I began to feel bewildered and frightened and then I suddenly realized that I must have had a stroke. In a way this rationalization made me feel somewhat relieved but not for long because I had always thought that the effects of a stroke were permanent in every case. . . . I found I could speak a little but even to me the words seemed wrong and not what I meant to say.

As this writer noted, most stroke victims are not as lucky. Mr. Ford was a Coast Guard radio operator when he suffered a stroke at the age of thirty-nine. The neuropsychologist Howard Gardner interviewed him three months later. Gardner asked him about his work before he entered the hospital.

> "I'm a sig . . . no . . . man . . . uh, well, . . . again." These words were emitted slowly, and with great effort. The sounds were not clearly articulated; each syllable was uttered harshly, explosively, in a throaty voice. . . .
>
> "Let me help you," I interjected. "You were a signal . . ."
>
> "A sig-nal man . . . right," Ford completed my phrase triumphantly.
>
> "Were you in the Coast Guard?"
>
> "No, er, yes, yes . . . ship . . . Massachu . . . chusetts . . . Coast-guard . . . years." He raised his hands twice, indicating the number "nineteen."

"Oh, you were in the Coast Guard for nineteen years."

"Oh . . . boy . . . right . . . right," he replied.

"Why are you in the hospital, Mr. Ford?"

Ford looked at me a bit strangely, as if to say, Isn't it patently obvious? He pointed to his paralyzed arm and said, "Arm no good," then to his mouth and said, "Speech . . . can't say . . . talk, you see."

"What happened to you to make you lose your speech?"

"Head, fall, Jesus Christ, me no good, str, str . . . oh Jesus . . . stroke."

"I see. Could you tell me, Mr. Ford, what you've been doing in the hospital?"

"Yes, sure. Me go, er, uh, P.T. nine o'cot, speech . . . two times . . . read . . . wr . . . ripe, er, rike, er, write . . . practice . . . get-ting better."

"And have you been going home on weekends?"

"Why, yes . . . Thursday, er, er, er, no, er, Friday . . . Bar-ba-ra . . . wife . . . and, oh, car . . . drive . . . purnpike . . . you know . . . rest and . . . tee-vee."

"Are you able to understand everything on television?"

"Oh, yes, yes . . . well . . . al-most."

Obviously Mr. Ford had to struggle to get speech out, but his problems were not in controlling his vocal muscles. He could blow out a candle and clear his throat, and he was as linguistically hobbled when he wrote as when he spoke. Most of his handicaps centered around grammar itself. He omitted endings like -ed and -s and grammatical function words like or, be, and the, despite their high frequency in the language. When reading aloud, he skipped over the function words, though he successfully read content words like bee and oar that had the same sounds. He named objects and recognized their names extremely well. He understood questions when their gist could be deduced from their content words, such as "Does a stone float on water?" or "Do you use a hammer for cutting?," but not one that requires grammatical analysis, like "The lion was killed by the tiger; which one is dead?"

Despite Mr. Ford's grammatical impairment, he was clearly in command of his other faculties. Gardner notes: "He was alert, attentive, and fully aware of where he was and why he was there. Intellectual functions not closely tied to language, such as knowledge of right and left, ability to draw with the left (unpracticed) hand, to calculate, read maps, set clocks, make constructions, or carry out commands, were all preserved. His Intelligence Quotient in nonverbal areas was in the high average range." Indeed, the dialogue shows that Mr. Ford, like many Broca's aphasics, showed an acute understanding of his handicap.

Injuries in adulthood are not the only ways that the circuitry underlying language can be compromised. A few otherwise healthy children just fail to develop language on schedule. When they do begin to talk, they have difficulty articulating words, and though their articulation improves with age, the victims persist in a vareity of grammatical errors, often into adulthood. When obvious nonlinguistic causes are ruled out—cognitive disorders like retardation, perceptual disorders like deafness, and social disorders like autism—the children are given the accurate but not terribly helpful diagnostic label Specific Language Impairment (SLI).

Language therapists, who are often called upon to treat several members in a family, have long been under the impression that SLI is hereditary. Recent statistical studies show that the impression may be correct. SLI runs in families, and if one member of a set of identical twins has it, the odds are very high that the other will, too. Particularly dramatic evidence comes from one British family, the K's, recently studied by the linguist Myrna Gopnik and several geneticists. The grandmother of the family is language-impaired. She has five adult children. One daughter is linguistically normal, as are this daughter's children. The other four adults, like the grandmother, are impaired. Together these four had twenty-three children; of them, eleven were language-impaired, twelve were normal. The language-impaired children were randomly distributed among the families, the sexes, and the birth orders.

Of course, the mere fact that some behavioral pattern runs in families does not show that it is genetic. Recipes, accents, and lullabies run in families, but they have nothing to do with DNA. In this case, though, a genetic cause is plausible. If the cause were in the environment—poor nutrition, hearing the defective speech of an impaired parent or sibling, watching too much TV, lead contamination from old pipes, whatever—then why would the syndrome capriciously strike some family members while leaving their near age-mates (in one case, a fraternal twin) alone? In fact, the geneticists working with Gopnik noted that the pedigree suggests a trait controlled by a single dominant gene, just like pink flowers on Gregor Mendel's pea plants.

What does this hypothetical gene do? It does not seem to impair overall intelligence; most of the afflicted family members score in the normal range in the nonverbal parts of IQ tests. (Indeed, Gopnik studied one unrelated child with the syndrome who routinely received the best grade in his mainstream math class.) It is their language that is impaired, but they are not like Broca's aphasics; the impression is more of a tourist struggling in a foreign city. They speak somewhat slowly and deliberately, carefully planning what they will say and encouraging their interlocutors to come to their aid by completing sentences for them. They report that ordinary conversation is strenuous mental work and that when possible they avoid situations in which they must speak. Their speech contains frequent grammatical errors, such as misuse of pronouns and of suffixes like the plural and past tense:

> It's a flying finches, they are.
> She remembered when she hurts herself the other day.
> The neighbors phone the ambulance because the man fall
>   off the tree.
> They boys eat four cookies.
> Carol is cry in the church.

In experimental tests they have difficulty with tasks that normal four-year-olds breeze through. A classic example is the *wug*-test, another demonstration that normal children do not learn language by imitating their parents. The testee is shown a line drawing of a birdlike creature and told that it is a *wug*. Then a picture of two of them is shown, and the child is told, "Now there are two of them; there are two _____." Your typical four-year-old will blurt out *wugs*, but the language-impaired adult is stymied. One of the adults Gopnik studied laughed nervously and said, "Oh, dear, well carry on." When pressed, she responded, "Wug . . . wugness, isn't it? No. I see. You want to pair . . . pair it up. OK." For the next animal, *zat*, she said, "Za . . . ka . . . za . . . zackle." For the next, *sas*, she deduced that it must be "sasses." Flushed with success, she proceeded to generalize too literally, converting *zoop* to "zoop-es" and *tob* to "tob-ye-es," revealing that she hadn't really grasped the English rule. Apparently the defective gene in this family somehow affects the development of the rules that normal children use unconsciously. The adults do their best to compensate by consciously reasoning the rules out, with predictably clumsy results.

Broca's aphasia and SLI are cases where language is impaired and the rest of intelligence seems more or less intact. But this does not show that language is separate from intelligence. Perhaps language imposes greater demands on the brain than any other problem the mind has to solve. For the other problems, the brain can limp along at less than its full capacity; for language, all systems have to be one hundred percent. To clinch the case, we need to find the opposite dissociation, linguistic idiot savants—that is, people with good language and bad cognition.

Here is another interview, this one between a fourteen-year-old girl called Denyse and the late psycholinguist Richard Cromer; the interview was transcribed and analyzed by Cromer's colleague Sigrid Lipka.

> I like opening cards. I had a pile of post this morning and not one of them was a Christmas card. A bank statement I got this morning!
>
> [A bank statement? I hope it was good news.]
>
> No it wasn't good news.
>
> [Sounds like mine.]

I hate . . . , My mum works over at the, over on the ward and
she said "not another bank statement." I said "it's the second
one in two days." And she said "Do you want me to go to the
bank for you at lunchtime?" and I went "No, I'll go this time
and explain it myself." I tell you what, my bank are awful.
They've lost my bank book, you see, and I can't find it anywhere. I belong to the TSB Bank and I'm thinking of changing
my bank 'cause they're so awful.

They keep, they keep losing . . . [someone comes in to bring
some tea] Oh, isn't that nice.

[Uhm. Very good.]

They've got the habit of doing that. They lose, they've lost my
bank book twice, in a month, and I think I'll scream. My mum
went yesterday to the bank for me. She said "They've lost your
bank book again." I went "Can I scream?" and I went, she
went "Yes, go on." So I hollered. But it is annoying when they
do things like that. TSB, Trustees aren't . . . uh the best ones
to be with actually. They're hopeless.

I have seen Denyse on videotape, and she comes across as a
loquacious, sophisticated conversationalist—all the more so, to American ears, because of her refined British accent. (*My bank are awful,*
by the way, is grammatical in British, though not American, English.)
It comes as a surprise to learn that the events she relates so earnestly
are figments of her imagination. Denyse has no bank account, so she
could not have received any statement in the mail, nor could her bank
have lost her bankbook. Though she would talk about a joint bank
account she shared with her boyfriend, she had no boyfriend, and
obviously had only the most tenuous grasp of the concept "joint bank
account" because she complained about the boyfriend taking money
out of her side of the account. In other conversations Denyse would
engage her listeners with lively tales about the wedding of her sister,
her holiday in Scotland with a boy named Danny, and a happy airport
reunion with a long-estranged father. But Denyse's sister is unmar-

ried, Denyse has never been to Scotland, she does not know anyone named Danny, and her father has never been away for any length of time. In fact, Denyse is severely retarded. She never learned to read or write and cannot handle money or any of the other demands of everyday functioning.

Denyse was born with spina bifida ("split spine") a malformation of the vertebrae that leaves the spinal cord unprotected. Spina bifida often results in hydrocephalus, an increase in pressure in the cerebrospinal fluid filling the ventricles (large cavities) of the brain, distending the brain from within. For reasons no one understands, hydrocephalic children occasionally end up like Denyse, significantly retarded but with unimpaired—indeed, overdeveloped—language skills. (Perhaps the ballooning ventricles crush much of the brain tissue necessary for everyday intelligence but leave intact some other portions that can develop language circuitry.) The various technical terms for the condition include "cocktail party conversation," "chatterbox syndrome," and "blathering."

Fluent grammatical language can in fact appear in many kinds of people with severe intellectual impairments, like schizophrenics, Alzheimer's patients, some autistic children, and some aphasics. One of the most fascinating syndromes recently came to light when the parents of a retarded girl with chatterbox syndrome in San Diego read an article about Chomsky's theories in a popular science magazine and called him at MIT, suggesting that their daughter might be of interest to him. Chomsky is a paper-and-pencil theoretician who wouldn't know Jabba the Hutt from the Cookie Monster, so he suggested that the parents bring their child to the laboratory of the psycholinguist Ursula Bellugi in La Jolla.

Bellugi, working with colleagues in molecular biology, neurology, and radiology, found that the child (whom they called Crystal), and a number of others they have subsequently tested, had a rare form of retardation called Williams syndrome. The syndrome seems to be associated with a defective gene on chromosome 11 involved in the regulation of calcium, and it acts in complex ways on the brain, skull, and internal organs during development, though no one knows why it has the effects it does. The children have an unusual appearance: they are short and slight, with narrow faces and broad foreheads, flat nasal bridges, sharp chins, star-shaped patterns in their irises, and full lips. They are sometimes called "elfin-faced" or "pixie people," but to me they look more like Mick Jagger. They are significantly retarded, with an IQ of about 50, and are incompetent at ordinary tasks like tying their shoes, finding their way, retrieving items from a cupboard, telling left from right, adding two numbers, drawing a bicycle, and suppressing their natural tendency to hug strangers. But like Denyse they are fluent, if somewhat prim, conversationalists. Here are two transcripts from Crystal when she was eighteen:

> And what an elephant is, it is one of the animals. And what the elephant does, it lives in the jungle. I can also live in the zoo. And what it has, it has long, gray ears, fan ears, ears that can blow in the wind. It has a long trunk that can pick up grass or pick up hay . . . If they're in a bad mood, it can be terrible . . . If the elephant gets mad, it could stomp; it could charge. Sometimes elephants can charge, like a bull can charge. They have big, long, tusks. They can damage a car . . . It could be dangerous. When they're in a pinch, when they're in a bad mood, it can be terrible. You don't want an elephant as a pet. You want a cat or a dog or a bird.

> This is a story about chocolates. Once upon a time, in Chocolate World there used to be a Chocolate Princess. She was such a yummy princess. She was on her chocolate throne and then some chocolate man came to see her. And the man bowed to her and he said these words to her. The man said to her, "Please, Princess Chocolate. I want you to see how I do my work. And it's hot outside in Chocolate World, and you might melt to the ground like melted butter. And if the sun changes to a different color, then the Chocolate World—and you—won't melt. You can be saved if the sun changes to a different color. And if it doesn't change to a different color, you and Chocolate World are doomed.

Laboratory tests confirm the impression of competence at grammar; the children understand complex sentences, and fix up ungrammatical sentences, at normal levels. And they have an especially charming quirk: they are fond of unusual words. Ask a normal child to name some animals, and you will get the standard inventory of pet store and barnyard: dog, cat, horse, cow, pig. Ask a Williams syndrome child, and you get a more interesting menagerie: unicorn, pteranodon, yak, ibex, water buffalo, sea lion, saber-tooth tiger, vulture, koala, dragon, and one that should be especially interesting to paleontologists, "brontosaurus rex." One eleven-year-old poured a glass of milk into the sink and said, "I'll have to evacuate it"; another handed Bellugi a drawing and announced, "Here, Doc, this is in remembrance of you."

People like Kirupano, Larry, the Hawaiian-born papaya grower, May-ela, Simon, Aunt Mae, Sarah, Mr. Ford, the K's, Denyse, and Crystal constitute a field guide to language users. They show that complex grammar is displayed across the full range of human habitats. You don't need to have left the Stone Age; you don't need to be middle class; you don't need to do well in school; you don't even need to be old enough for school. Your parents need not bathe you in language or even command a language. You don't need the intellectual where-withal to function in society, the skills to keep house and home together, or a particularly firm grip on reality. Indeed, you can possess all these advantages and still not be a competent language user, if you lack just the right genes or just the right bits of brain.

# 3

## Mentalese

*The year 1984 has come and gone, and it is losing its connotation of the* totalitarian nightmare of George Orwell's 1949 novel. But relief may be premature. In an appendix to *Nineteen Eighty-four,* Orwell wrote of an even more ominous date. In 1984, the infidel Winston Smith had to be converted with imprisonment, degradation, drugs, and torture; by 2050, there would be no Winston Smiths. For in that year the ultimate technology for thought control would be in place: the language Newspeak.

> The purpose of Newspeak was not only to provide a medium of expression for the world-view and mental habits proper to the devotees of Ingsoc [English Socialism], but to make all other modes of thought impossible. It was intended that when Newspeak had been adopted once and for all and Oldspeak forgotten, a heretical thought—that is, a thought diverging from the principles of Ingsoc—should be literally unthinkable, at least so far as thought is dependent on words. Its vocabulary was so constructed as to give exact and often very subtle expression to every meaning that a Party member could properly wish to express, while excluding all other meanings and also the pos-

sibility of arriving at them by indirect methods. This was done partly by the invention of new words, but chiefly by eliminating undesirable words and by stripping such words as remained of unorthodox meanings, and so far as possible of all secondary meanings whatever. To give a single example. The word *free* still existed in Newspeak, but it could only be used in such statements as "This dog is free from lice" or "This field is free from weeds." It could not be used in its old sense of "politically free" or "intellectually free," since political and intellectual freedom no longer existed even as concepts, and were therefore of necessity nameless.

    . . . A person growing up with Newspeak as his sole language would no more know that *equal* had once had the secondary meaning of "politically equal," or that *free* had once meant "intellectually free," than, for instance, a person who had never heard of chess would be aware of the secondary meanings attaching to *queen* and *rook*. There would be many crimes and errors which it would be beyond his power to commit, simply because they were nameless and therefore unimaginable.

But there is a straw of hope for human freedom: Orwell's caveat "at least so far as thought is dependent on words." Note his equivocation: at the end of the first paragraph, a concept is unimaginable and therefore nameless; at the end of the second, a concept is nameless and therefore unimaginable. *Is* thought dependent on words? Do people literally think in English, Cherokee, Kivunjo, or, by 2050, Newspeak? Or are our thoughts couched in some silent medium of the brain—a language of thought, or "mentalese"—and merely clothed in words whenever we need to communicate them to a listener? No question could be more central to understanding the language instinct.

    In much of our social and political discourse, people simply assume that words determine thoughts. Inspired by Orwell's essay "Politics and the English Language," pundits accuse governments of manipulating our minds with euphemisms like *pacification* (bombing), *revenue enhancement* (taxes), and *nonretention* (firing). Philosophers argue that since animals lack language, they must also lack consciousness—Wittgenstein wrote, "A dog could not have the thought 'perhaps it will rain tomorrow' "—and therefore they do not possess the rights of conscious beings. Some feminists blame sexist thinking on sexist language, like the use of *he* to refer to a generic person. Inevitably, reform movements have sprung up. Many replacements for *he* have been suggested over the years, including *E, hesh, po, tey, co, jhe, ve, xe, he'er, thon,* and *na.* The most extreme of these movements is General Semantics, begun in 1933 by the engineer Count Alfred Korzybski and popularized in long-time best-sellers by his disciples Stuart Chase and S. I. Hayakawa. (This is the same Hayakawa who later achieved notoriety as the protest-defying college president and snoozing U.S. senator.) General Semantics lays the blame for human folly on insidious "semantic damage" to thought perpetrated by the structure of language. Keeping a forty-year-old in prison for a theft he committed as a teenager assumes that the forty-year-old John and the eighteen-year-old John are "the same person," a cruel logical error that would be avoided if we referred to them not as *John* but as *John$_{1972}$* and *John$_{1994}$*, respectively. The verb *to be* is a particular source of illogic, because it identifies individuals with abstractions, as in *Mary is a woman,* and licenses evasions of responsibility, like Ronald Reagan's famous nonconfession *Mistakes were made.* One faction seeks to eradicate the verb altogether.

And supposedly there is a scientific basis for these assumptions: the famous Sapir-Whorf hypothesis of linguistic determinism, stating that people's thoughts are determined by the categories made available by their language, and its weaker version, linguistic relativity, stating that differences among languages cause differences in the thoughts of their speakers. People who remember little else from their college education can rattle off the factoids: the languages that carve the spectrum into color words at different places, the fundamentally different Hopi concept of time, the dozens of Eskimo words for snow. The implication is heavy: the foundational categories of reality are not "in" the world but are imposed by one's culture (and hence can be challenged, perhaps accounting for the perennial appeal of the hypothesis to undergraduate sensibilities).

But it is wrong, all wrong. The idea that thought is the same thing as language is an example of what can be called a conventional absurdity: a statement that goes against all common sense but that everyone believes because they dimly recall having heard it somewhere and because it is so pregnant with implications. (The "fact" that we use only five percent of our brains, that lemmings commit mass suicide, that the *Boy Scout Manual* annually outsells all other books, and that we can be coerced into buying by subliminal messages are other examples.) Think about it. We have all had the experience of uttering or writing a sentence, then stopping and realizing that it wasn't exactly what we meant to say. To have that feeling, there has to be a "what we meant to say" that is different from what we said. Sometimes it is not easy to find *any* words that properly convey a thought. When we hear or read, we usually remember the gist, not the exact words, so there has to be such a thing as a gist that is not the same as a bunch of words. And if thoughts depended on words, how could a new word ever be coined? How could a child learn a word to begin with? How could translation from one language to another be possible?

The discussions that assume that language determines thought carry on only by a collective suspension of disbelief. A dog, Bertrand Russell noted, may not be able to tell you that its parents were honest though poor, but can anyone really conclude from this that the dog is *unconscious*? (Out cold? A zombie?) A graduate student once argued with me using the following deliciously backwards logic: language must affect thought, because if it didn't, we would have no reason to fight sexist usage (apparently, the fact that it is offensive is not reason enough). As for government euphemism, it is contemptible not because it is a form of mind control but because it is a form of lying. (Orwell was quite clear about this in his masterpiece essay.) For example, "revenue enhancement" has a much broader meaning than "taxes," and listeners naturally assume that if a politician had meant "taxes" he would have said "taxes." Once a euphemism is pointed out, people are not so brainwashed that they have trouble understanding the deception. The National Council of Teachers of English annually lampoons government doublespeak in a widely reproduced press release, and calling attention to euphemism is a popular form of humor, like the speech from the irate pet store customer in *Monty Python's Flying Circus:*

> This parrot is no more. It has ceased to be. It's expired and gone to meet its maker. This is a late parrot. It's a stiff. Bereft of life, it rests in peace. If you hadn't nailed it to the perch, it would be pushing up the daisies. It's rung down the curtain and joined the choir invisible. This is an ex-parrot.

As we shall see in this chapter, there is no scientific evidence that languages dramatically shape their speakers' ways of thinking. But I want to do more than review the unintentionally comical history of attempts to prove that they do. The idea that language shapes thinking seemed plausible when scientists were in the dark about how thinking works or even how to study it. Now that cognitive scientists know how to think about thinking, there is less of a temptation to equate it with language just because words are more palpable than thoughts. By understanding *why* linguistic determinism is wrong, we will be in a better position to understand how language itself works when we turn to it in the next chapters.

The linguistic determinism hypothesis is closely linked to the names Edward Sapir and Benjamin Lee Whorf. Sapir, a brilliant linguist, was a student of the anthropologist Franz Boas. Boas and his students (who also include Ruth Benedict and Margaret Mead) were important intellectual figures in this century, because they argued that nonindustrial peoples were not primitive savages but had systems of language, knowledge, and culture as complex and valid in their world view as our own. In his study of Native American languages Sapir noted that speakers of different languages have to pay attention to different aspects of reality simply to put words together into grammatical sentences. For example, when English speakers decide whether or not to put *-ed* onto the end of a verb, they must pay attention to tense, the relative time of occurrence of the event they are referring to and the moment of speaking. Wintu speakers need not bother with tense, but when they decide which suffix to put on their verbs, they must pay attention to whether the knowledge they are conveying was learned through direct observation or by hearsay.

Sapir's interesting observation was soon taken much farther. Whorf was an inspector for the Hartford Fire Insurance Company and an amateur scholar of Native American languages, which led him to take courses from Sapir at Yale. In a much-quoted passage, he wrote:

> We dissect nature along lines laid down by our native languages. The categories and types that we isolate from the world of phenomena we do not find there because they stare every observer in the face; on the contrary, the world is presented in a kaleidoscopic flux of impressions which has to be organized by our minds—and this means largely by the linguistic systems in our minds. We cut nature up, organize it into concepts, and ascribe significances as we do, largely because we are parties to an agreement to organize it in this way—an agreement that holds throughout our speech community and is codified in the patterns of our language. The agreement is, of course, an implicit and unstated one, *but its terms are absolutely obligatory;* we cannot talk at all except by subscribing to the organization and classification of data which the agreement decrees.

What led Whorf to this radical position? He wrote that the idea first occurred to him in his work as a fire prevention engineer when he was struck by how language led workers to misconstrue dangerous situations. For example, one worker caused a serious explosion by tossing a cigarette into an "empty" drum that in fact was full of gasoline vapor. Another lit a blowtorch near a "pool of water" that was really a basin of decomposing tannery waste, which, far from being "watery," was releasing inflammable gases. Whorf's studies of American languages strengthened his conviction. For example, in Apache, *It is a dripping spring* must be expressed "As water, or springs, whiteness moves downward." "How utterly unlike our way of thinking!" he wrote.

But the more you examine Whorf's arguments, the less sense they make. Take the story about the worker and the "empty" drum. The seeds of disaster supposedly lay in the semantics of *empty*, which, Whorf claimed, means both "without its usual contents" and "null and void, empty, inert." The hapless worker, his conception of reality molded by his linguistic categories, did not distinguish between the "drained" and "inert" senses, hence, flick . . . boom! But wait. Gasoline vapor is invisible. A drum with nothing but vapor in it looks just like a drum with nothing in it at all. Surely this walking catastrophe was fooled by his eyes, not by the English language.

The example of whiteness moving downward is supposed to show that the Apache mind does not cut up events into distinct objects and actions. Whorf presented many such examples from Native American languages. The Apache equivalent of *The boat is grounded on the beach* is "It is on the beach pointwise as an event of canoe motion." *He invites people to a feast* becomes "He, or somebody, goes for eaters of cooked food." *He cleans a gun with a ramrod* is translated as "He directs a hollow moving dry spot by movement of tool." All this, to be sure, is utterly unlike our way of talking. But do we know that it is utterly unlike our way of thinking?

As soon as Whorf's articles appeared, the psycholinguists Eric Lenneberg and Roger Brown pointed out two non sequiturs in his argument. First, Whorf did not actually study any Apaches; it is not clear that he ever met one. His assertions about Apache psychology are based entirely on Apache grammar—making his argument circular. Apaches speak differently, so they must think differently. How do we know that they think differently? Just listen to the way they speak!

Second, Whorf rendered the sentences as clumsy, word-for-word translations, designed to make the literal meanings seem as odd as possible. But looking at the actual glosses that Whorf provided, I could, with equal grammatical justification, render the first sentence as the mundane "Clear stuff—water—is falling." Turning the tables, I could take the English sentence "He walks" and render it "As solitary masculinity, leggedness proceeds." Brown illustrates how strange the German mind must be, according to Whorf's logic, by reproducing

Mark Twain's own translation of a speech he delivered in flawless German to the Vienna Press Club:

> I am indeed the truest friend of the German language—and not only now, but from long since—yes, before twenty years already. . . . I would only some changes effect. I would only the language method—the luxurious, elaborate construction compress, the eternal parenthesis suppress, do away with, annihilate; the introduction of more than thirteen subjects in one sentence forbid; the verb so far to the front pull that one it without a telescope discover can. With one word, my gentlemen, I would your beloved language simplify so that, my gentlemen, when you her for prayer need, One her yonder-up understands.
>
> . . . I might gladly the separable verb also a little bit reform. I might none do let what Schiller did: he has the whole history of the Thirty Years' War between the two members of a separate verb inpushed. That has even Germany itself aroused, and one has Schiller the permission refused the History of the Hundred Years' War to compose—God be it thanked! After all these reforms established be will, will the German language the noblest and the prettiest on the world be.

Among Whorf's "kaleidoscopic flux of impressions," color is surely the most eye-catching. He noted that we see objects in different hues, depending on the wavelengths of the light they reflect, but that physicists tell us that wavelength is a continuous dimension with nothing delineating red, yellow, green, blue, and so on. Languages differ in their inventory of color words: Latin lacks generic "gray" and "brown"; Navajo collapses blue and green into one word; Russian has distinct words for dark blue and sky blue; Shona speakers use one word for the yellower greens and the greener yellows, and a different one for the bluer greens and the nonpurplish blues. You can fill in the rest of the argument. It is language that puts the frets in the spectrum; Julius Caesar would not know shale from Shinola.

But although physicists see no basis for color boundaries, physiologists do. Eyes do not register wavelength the way a thermometer registers temperature. They contain three kinds of cones, each with a different pigment, and the cones are wired to neurons in a way that makes the neurons respond best to red patches against a green background or vice versa, blue against yellow, black against white. No matter how influential language might be, it would seem preposterous to a physiologist that it could reach down into the retina and rewire the ganglion cells.

Indeed, humans the world over (and babies and monkeys, for that matter) color their perceptual worlds using the same palette, and this constrains the vocabularies they develop. Although languages may disagree about the wrappers in the sixty-four crayon box—the burnt umbers, the turquoises, the fuchsias—they agree much more on the wrappers in the eight-crayon box—the fire-engine reds, grass greens, lemon yellows. Speakers of different languages unanimously pick these shades as the best examples of their color words, as long as the language has a color word in that general part of the spectrum. And where languages do differ in their color words, they differ predictably, not according to the idiosyncratic taste of some word-coiner. Languages are organized a bit like the Crayola product line, the fancier ones adding colors to the more basic ones. If a language has only two color words, they are for black and white (usually encompassing dark and light, respectively). If it has three, they are for black, white, and red; if four, black, white, red, and either yellow or green. Five adds in both yellow and green; six, blue; seven, brown; more than seven, purple, pink, orange, or gray. But the clinching experiment was carried out in the New Guinea highlands with the Grand Valley Dani, a people speaking one of the black-and-white languages. The psychologist Eleanor Rosch found that the Dani were quicker at learning a new color category that was based on fire-engine red than a category based on an off-red. The way we see colors determines how we learn words for them, not vice versa.

The fundamentally different Hopi concept of time is one of the more startling claims about how minds can vary. Whorf wrote that the Hopi language contains "no words, grammatical forms, constructions, or expressions that refer directly to what we call 'time,' or to past, or future, or to enduring or lasting." He suggested, too, that the Hopi had "no general notion or intuition of TIME as a smooth flowing continuum in which everything in the universe proceeds at an equal rate, out of a future, through a present, into a past." According to Whorf, they did not conceptualize events as being like points, or lengths of time like days as countable things. Rather, they seemed to focus on change and process itself, and on psychological distinctions between presently known, mythical, and conjecturally distant. The Hopi also had little interest in "exact sequences, dating, calendars, chronology."

What, then, are we to make of the following sentence translated from Hopi?

> Then indeed, the following day, quite early in the morning at
> the hour when people pray to the sun, around that time then
> he woke up the girl again.

Perhaps the Hopi are not as oblivious to time as Whorf made them out to be. In his extensive study of the Hopi, the anthropologist Ekkehart Malotki, who reported this sentence, also showed that Hopi speech contains tense, metaphors for time, units of time (including days, numbers of days, parts of the day, yesterday and tomorrow, days of the week, weeks, months, lunar phases, seasons, and the year), ways to quantify units of time, and words like "ancient," "quick," "long time," and "finished." Their culture keeps records with sophisticated methods of dating, including a horizon-based sun calendar, exact ceremonial day sequences, knotted calendar strings, notched calendar sticks, and several devices for timekeeping using the principle of the sundial. No one is really sure how Whorf came up with his outlandish claims, but his limited, badly analyzed sample of Hopi speech and his long-time leanings toward mysticism must have contributed.

Speaking of anthropological canards, no discussion of language and thought would be complete without the Great Eskimo Vocabulary Hoax. Contrary to popular belief, the Eskimos do not have more words for snow than do speakers of English. They do not have four hundred words for snow, as it has been claimed in print, or two hundred, or one hundred, or forty-eight, or even nine. One dictionary puts the figure at two. Counting generously, experts can come up with about a dozen, but by such standards English would not be far behind, with *snow, sleet, slush, blizzard, avalanche, hail, hardpack, powder, flurry, dusting,* and a coinage of Boston's WBZ-TV meteorologist Bruce Schwoegler, *snizzling.*

Where did the myth come from? Not from anyone who has actually studied the Yupik and Inuit-Inupiaq families of polysynthetic languages spoken from Siberia to Greenland. The anthropologist Laura Martin has documented how the story grew like an urban legend, exaggerated with each retelling. In 1911 Boas casually mentioned that Eskimos used four unrelated word roots for snow. Whorf embellished the count to seven and implied that there were more. His article was widely reprinted, then cited in textbooks and popular books on language, which led to successively inflated estimates in other textbooks, articles, and newspaper columns of Amazing Facts.

The linguist Geoffrey Pullum, who popularized Martin's article in his essay "The Great Eskimo Vocabulary Hoax," speculates about why the story got so out of control: "The alleged lexical extravagance of the Eskimos comports so well with the many other facets of their polysynthetic perversity: rubbing noses; lending their wives to strangers; eating raw seal blubber; throwing Grandma out to be eaten by polar bears." It is an ironic twist. Linguistic relativity came out of the Boas school, as part of a campaign to show that nonliterate cultures were as complex and sophisticated as European ones. But the supposedly mind-broadening anecdotes owe their appeal to a patronizing willingness to treat other cultures' psychologies as weird and exotic compared to our own. As Pullum notes,

> Among the many depressing things about this credulous transmission and elaboration of a false claim is that even if there *were* a large number of roots for different snow types in some Arctic language, this would *not,* objectively, be intellectually interesting; it would be a most mundane and unremarkable fact.

Horsebreeders have various names for breeds, sizes, and ages of horses; botanists have names for leaf shapes; interior decorators have names for shades of mauve; printers have many different names for fonts (Carlson, Garamond, Helvetica, Times Roman, and so on), naturally enough. . . . Would anyone think of writing about printers the same kind of slop we find written about Eskimos in bad linguistics textbooks? Take [the following] random textbook . . . , with its earnest assertion "It is quite obvious that in the culture of the Eskimos . . . snow is of great enough importance to split up the conceptual sphere that corresponds to one word and one thought in English into several distinct classes . . ." Imagine reading: "It is quite obvious that in the culture of printers . . . fonts are of great enough importance to split up the conceptual sphere that corresponds to one word and one thought among non-printers into several distinct classes . . ." Utterly boring, even if true. Only the link to those legendary, promiscuous, blubber-gnawing hunters of the ice-packs could permit something this trite to be presented to us for contemplation.

If the anthropological anecdotes are bunk, what about controlled studies? The thirty-five years of research from the psychology laboratory is distinguished by how little it has shown. Most of the experiments have tested banal "weak" versions of the Whorfian hypothesis, namely that words can have some effect on memory or categorization. Some of these experiments have actually worked, but that is hardly surprising. In a typical experiment, subjects have to commit paint chips to memory and are tested with a multiple-choice procedure. In some of these studies, the subjects show slightly better memory for colors that have readily available names in their language. But even colors without names are remembered fairly well, so the experiment does not show that the colors are remembered by verbal labels alone. All it shows is that subjects remembered the chips in two forms, a nonverbal visual image and a verbal label, presumably because two kinds of memory, each one fallible, are better than one. In another type of experiment subjects have to say which two out of three color chips go together; they often put the ones together that have the same name in their language. Again, no surprise. I can imagine the subjects thinking to themselves, "Now how on earth does this guy expect me to pick two chips to put together? He didn't give me any hints, and they're all pretty similar. Well, I'd probably call those two 'green' and that one 'blue,' and that seems as good a reason to put them together as any." In these experiments, language is, technically speaking, influencing a form of thought in some way, but so what? It is hardly an example of incommensurable world views, or of concepts that are nameless and therefore unimaginable, or of dissecting nature along lines laid down by our native languages according to terms that are absolutely obligatory.

The only really dramatic finding comes from the linguist and now Swarthmore College president Alfred Bloom in his book *The Linguistic Shaping of Thought*. English grammar, says Bloom, provides its speakers with the subjunctive construction: *If John were to go to the hospital, he would meet Mary.* The subjunctive is used to express "counterfactual" situations, events that are known to be false but entertained as hypotheticals. (Anyone familiar with Yiddish knows a better example, the ultimate riposte to someone reasoning from improbable premises: *Az di bobe volt gehat beytsim volt zi geven mayn zeyde,* "If my grandmother had balls, she'd be my grandfather.") Chinese, in contrast, lacks a subjunctive and any other simple grammatical construction that directly expresses a counterfactual. The thought must be expressed circuitously, something like "If John is going to the hospital . . . but he is not going to the hospital . . . but if he is going, he meets Mary."

Bloom wrote stories containing sequences of implications from a counterfactual premise and gave them to Chinese and American students. For example, one story said, in outline, "Bier was an eighteenth-century European philosopher. There was some contact between the West and China at that time, but very few works of Chinese philosophy had been translated. Bier could not read Chinese, but if he had been able to read Chinese, he would have discovered B; what would have most influenced him would have been C; once influenced by that Chinese perspective, Bier would then have done D," and so on. The subjects were then asked to check off whether B, C, and D actually occurred. The American students gave the correct answer, no, ninety-eight percent of the time; the Chinese students gave the correct answer only seven percent of the time! Bloom concluded that the Chinese language renders its speakers unable to entertain hypothetical false worlds without great mental effort. (As far as I know, no one has tested the converse prediction on speakers of Yiddish.)

The cognitive psychologists Terry Au, Yohtaro Takano, and Lisa Liu were not exactly enchanted by these tales of the concreteness of the Oriental mind. Each one identified serious flaws in Bloom's experiments. One problem was that his stories were written in stilted Chinese. Another was that some of the science stories turned out, upon careful rereading, to be genuinely ambiguous. Chinese college students tend to have more science training than American students, and thus they were *better* at detecting the ambiguities that Bloom himself missed. When these flaws were fixed, the differences vanished.

People can be forgiven for overrating language. Words make noise, or sit on a page, for all to hear and see. Thoughts are trapped inside the head of the thinker. To know what someone else is thinking, or to talk to each other about the nature of thinking, we have to use—what else, words! It is no wonder that many commentators have trouble even conceiving of thought without words—or is it that they just don't have the language to talk about it?

As a cognitive scientist I can afford to be smug about common sense being true (thought is different from language) and linguistic determinism being a conventional absurdity. For two sets of tools now make it easier to think clearly about the whole problem. One is a body of experimental studies that break the word barrier and assess many kinds of nonverbal thought. The other is a theory of how thinking might work that formulates the questions in a satisfyingly precise way.

We have already seen an example of thinking without language: Mr. Ford, the fully intelligent aphasic discussed in Chapter 2. (One could, however, argue that his thinking abilities had been constructed before his stroke on the scaffolding of the language he then possessed.) We have also met deaf children who lack a language and soon invent one. Even more pertinent are the deaf adults occasionally discovered who lack any form of language whatsoever—no sign language, no writing, no lip reading, no speech. In her recent book *A Man Without Words,* Susan Schaller tells the story of Ildefonso, a twenty-seven-year-old illegal immigrant from a small Mexican village whom she met while working as a sign language interpreter in Los Angeles. Ildefonso's animated eyes conveyed an unmistakable intelligence and curiosity, and Schaller became his volunteer teacher and companion. He soon showed her that he had a full grasp of number: he learned to do addition on paper in three minutes and had little trouble understanding the base-ten logic behind two-digit numbers. In an epiphany reminiscent of the story of Helen Keller, Ildefonso grasped the principle of naming when Schaller tried to teach him the sign for "cat." A dam burst, and he demanded to be shown the sign for all the objects he was familiar with. Soon he was able to convey to Schaller parts of his life story: how as a child he had begged his desperately poor parents to send him to school, the kinds of crops he had picked in different states, his evasions of immigration authorities. He led Schaller to other languageless adults in forgotten corners of society. Despite their isolation from the verbal world, they displayed many abstract forms of thinking, like rebuilding broken locks, handling money, playing card games, and entertaining each other with long pantomimed narratives.

Our knowledge of the mental life of Ildefonso and other languageless adults must remain impressionistic for ethical reasons: when they surface, the first priority is to teach them language, not to study how they manage without it. But there are other languageless beings who have been studied experimentally, and volumes have been written about how they reason about space, time, objects, number, rate, causality, and categories. Let me recount three ingenious examples. One involves babies, who cannot think in words because they have not yet learned any. One involves monkeys, who cannot think in words because they are incapable of learning them. The third involves human adults, who, whether or not they think in words, claim their best thinking is done without them.

The developmental psychologist Karen Wynn has recently shown that five-month-old babies can do a simple form of mental arithmetic. She used a technique common in infant perception research. Show a baby a bunch of objects long enough, and the baby gets bored and looks away; change the scene, and if the baby notices the difference, he or she will regain interest. The methodology has shown that babies as young as five days old are sensitive to number. In one experiment, an experimenter bores a baby with an object, then occludes the object with an opaque screen. When the screen is removed, if the same object is present, the babies look for a little while, then gets bored again. But if, through invisible subterfuge, two or three objects have ended up there, the surprised babies stare longer.

In Wynn's experiment, the babies were shown a rubber Mickey Mouse doll on a stage until their little eyes wandered. Then a screen came up, and a prancing hand visibly reached out from behind a curtain and placed a second Mickey Mouse behind the screen. When the screen was removed, if there were two Mickey Mouses visible (something the babies had never actually seen), the babies looked for only a few moments. But if there was only one doll, the babies were captivated—even though this was exactly the scene that had bored them before the screen was put in place. Wynn also tested a second group of babies, and this time, after the screen came up to obscure a *pair* of dolls, a hand visibly reached behind the screen and removed one of them. If the screen fell to reveal a single Mickey, the babies looked briefly; if it revealed the old scene with two, the babies had more trouble tearing themselves away. The babies must have been keeping track of how many dolls were behind the screen, updating their counts as dolls were added or subtracted. If the number inexplicably departed from what they expected, they scrutinized the scene, as if searching for some explanation.

Vervet monkeys live in stable groups of adult males and females and their offspring. The primatologists Dorothy Cheney and Robert

Seyfarth have noticed that extended families form alliances like the Montagues and Capulets. In a typical interaction they observed in Kenya, one juvenile monkey wrestled another to the ground screaming. Twenty minutes later the victim's sister approached the perpetrator's sister and without provocation bit her on the tail. For the retaliator to have identified the proper target, she would have had to solve the following analogy problem: A (victim) is to B (myself) as C (perpetrator) is to X, using the correct relationship "sister of" (or perhaps merely "relative of"; there were not enough vervets in the park for Cheney and Seyfarth to tell).

But do monkeys really know how their groupmates are related to each other, and, more impressively, do they realize that different pairs of individuals like brothers and sisters can be related in the same way? Cheney and Seyfarth hid a loudspeaker behind a bush and played tapes of a two-year-old monkey screaming. The females in the area reacted by looking at the mother of the infant who had been recorded—showing that they not only recognized the infant by its scream but recalled who its mother was. Similar abilities have been shown in the longtailed macaques that Verena Dasser coaxed into a laboratory adjoining a large outdoor enclosure. Three slides were projected: a mother at the center, one of her offspring on one side, and an unrelated juvenile of the same age and sex on the other. Each screen had a button under it. After the monkey had been trained to press a button under the offspring slide, it was tested on pictures of other mothers in the group, each one flanked by a picture of that mother's offspring and a picture of another juvenile. More than ninety percent of the time the monkey picked the offspring. In another test, the monkey was shown two slides, each showing a pair of monkeys, and was trained to press a button beneath the slide showing a particular mother and her juvenile daughter. When presented with slides of new monkeys in the group, the subject monkey always picked the mother-and-offspring pair, whether the offspring was male, female, infant, juvenile, or adult. Moreover, the monkeys appeared to be relying not only on physical resemblance between a given pair of monkeys, or on the sheer number of hours they had previously spent together, as the basis for recognizing they were kin, but on something more subtle in the history of their interaction. Cheney and Seyfarth, who work hard at keeping track of who is related to whom in what way in the groups of animals they study, note that monkeys would make excellent primatologists.

Many creative people insist that in their most inspired moments they think not in words but in mental images. Samuel Taylor Coleridge wrote that visual images of scenes and words once appeared involuntarily before him in a dreamlike state (perhaps opium-induced). He managed to copy the first forty lines onto paper, resulting in the poem we know as "Kubla Khan," before a knock on the door shattered the images and obliterated forever what would have been the rest of the poem. Many contemporary novelists, like Joan Didion, report that their acts of creation begin not with any notion of a character or a plot but with vivid mental pictures that dictate their choice of words. The modern sculptor James Surls plans his projects lying on a couch listening to music; he manipulates the sculptures in his mind's eye, he says, putting an arm on, taking an arm off, watching the images roll and tumble.

Physical scientists are even more adamant that their thinking is geometrical, not verbal. Michael Faraday, the originator of our modern conception of electric and magnetic fields, had no training in mathematics but arrived at his insights by visualizing lines of force as narrow tubes curving through space. James Clerk Maxwell formalized the concepts of electromagnetic fields in a set of mathematical equations and is considered the prime example of an abstract theoretician, but he set down the equations only after mentally playing with elaborate imaginary models of sheets and fluids. Nikola Tesla's idea for the electrical motor and generator, Friedrich Kekulé's discovery of the benzene ring that kicked off modern organic chemistry, Ernest Lawrence's conception of the cyclotron, James Watson and Francis Crick's discovery of the DNA double helix—all came to them in images. The most famous self-described visual thinker is Albert Einstein, who arrived at some of his insights by imagining himself riding a beam of light and looking back at a clock, or dropping a coin while standing in a plummeting elevator. He wrote:

> The psychical entities which seem to serve as elements in thought are certain signs and more or less clear images which can be "voluntarily" reproduced and combined. . . . This combinatory play seems to be the essential feature in productive thought—before there is any connection with logical construction in words or other kinds of signs which can be communicated to others. The above-mentioned elements are, in my case, of visual and some muscular type. Conventional words or other signs have to be sought for laboriously only in a secondary state, when the mentioned associative play is sufficiently established and can be reproduced at will.

Another creative scientist, the cognitive psychologist Roger Shepard, had his own moment of sudden visual inspiration, and it led to a classic laboratory demonstration of mental imagery in mere mortals. Early one morning, suspended between sleep and awakening in a state of lucid consciousness, Shepard experienced "a spontaneous kinetic image of three-dimensional structures majestically turning in space." Within moments and before fully awakening, Shepard had a clear idea for the design of an experiment. A simple variant of his idea was later carried out with his then-student Lynn Cooper. Cooper and Shepard flashed thousands of slides, each showing a single letter of the alphabet, to their long-suffering student volunteers. Sometimes the letter was upright, but sometimes it was tilted or mirror-reversed or both. As an example, here are the sixteen versions of the letter *F*:

```
  0   +45  +90  +135  180  -135  -90  -45
```

The subjects were asked to press one button if the letter was normal (that is, like one of the letters in the top row of the diagram), another if it was a mirror image (like one of the letters in the bottom row). To do the task, the subjects had to compare the letter in the slide against some memory record of what the normal version of the letter looks like right-side up. Obviously, the right-side-up slide (0 degrees) is the quickest, because it matches the letter in memory exactly, but for the other orientations, some mental transformation to the upright is necessary first. Many subjects reported that they, like the famous sculptors and scientists, "mentally rotated" an image of the letter to the upright. By looking at the reaction times, Shepard and Cooper showed that this introspection was accurate. The upright letters were fastest, followed by the 45 degree letters, the 90 degree letters, and the 135 degree letters, with the 180 degree (upside-down) letters the slowest. In other words, the farther the subjects had to mentally rotate the letter, the longer they took. From the data, Cooper and Shepard estimated that letters revolve in the mind at a rate of 56 RPM.

Note that if the subjects had been manipulating something resembling *verbal descriptions* of the letters, such as "an upright spine with one horizontal segment that extends rightwards from the top and another horizontal segment that extends rightwards from the middle," the results would have been very different. Among all the topsy-turvy letters, the upside-down versions (180 degrees) should be fastest: one simply switches all the "top"s to "bottom"s and vice versa, and the "left"s to "right"s and vice versa, and one has a new description of the shape as it would appear right-side up, suitable for matching against memory. Sideways letters (90 degrees) should be slower, because "top" gets changed either to "right" or to "left," depending on whether it lies clockwise (+90 degrees) or counterclockwise (−90 degrees) from the upright. Diagonal letters (45 and 135 degrees) should be slowest, because every word in the description has to be replaced: "top" has to be replaced with either "top right" or "top left," and so on. So the order of difficulty should be 0, 180, 90, 45, 135, not the majestic rotation of 0, 45, 90, 135, 180 that Cooper and Shepard saw in the data. Many other experiments have corroborated the idea that visual thinking uses not language but a mental graphics system, with operations that rotate, scan, zoom, pan, displace, and fill in patterns of contours.

* * *

What sense, then, can we make of the suggestion that images, numbers, kinship relations, or logic can be represented in the brain without being couched in words? In the first half of this century, philosophers had an answer: none. Reifying thoughts as things in the head was a logical error, they said. A picture or family tree or number in the head would require a little man, a homunculus, to look at it. And what would be inside *his* head—even smaller pictures, with an even smaller man looking at them? But the argument was unsound. It took Alan Turing, the brilliant British mathematician and philosopher, to make the idea of a mental representation scientifically respectable. Turing described a hypothetical machine that could be said to engage in reasoning. In fact this simple device, named a Turing machine in his honor, is powerful enough to solve any problem that any computer, past, present, or future, can solve. And it clearly uses an internal symbolic representation—a kind of mentalese—without requiring a little man or any occult processes. By looking at how a Turing machine works, we can get a grasp of what it would mean for a human mind to think in mentalese as opposed to English.

In essence, to reason is to deduce new pieces of knowledge from old ones. A simple example is the old chestnut from introductory logic: if you know that Socrates is a man and that all men are mortal, you can figure out that Socrates is mortal. But how could a hunk of matter like a brain accomplish this feat? The first key idea is a *representation*: a physical object whose parts and arrangement correspond piece for piece to some set of ideas or facts. For example, the pattern of ink on this page

```
Socrates isa man
```

is a representation of the idea that Socrates is a man. The shape of one group of ink marks, `Socrates`, is a symbol that stands for the concept of Socrates. The shape of another set of ink marks, `isa`, stands for the concept of being an instance of, and the shape of the third, `man`, stands for the concept of man. Now, it is crucial to keep one thing in mind. I have put these ink marks in the shape of English words as a courtesy to you, the reader, so that you can keep them straight as we work through the example. But all that really matters is that they have different shapes. I could have used a star of David, a smiley face, and the Mercedes Benz logo, as long as I used them consistently.

Similarly, the fact that the `Socrates` ink marks are to the left of the `isa` ink marks on the page, and the `man` ink marks are to the right, stands for the idea that Socrates is a man. If I change any part of the representation, like replacing `isa` with `isasonofa`, or flipping the positions of `Socrates` and `man`, we would have a representation of a different idea. Again, the left-to-right English order is just a mnemonic device for your convenience. I could have done it right-to-left or up-and-down, as long as I used that order consistently.

Keeping these conventions in mind, now imagine that the page has a second set of ink marks, representing the proposition that every man is mortal:

```
Socrates isa man
Every man ismortal
```

To get reasoning to happen, we now need a *processor*. A processor is not a little man (so one needn't worry about an infinite regress of homunculi inside homunculi) but something much stupider: a gadget with a fixed number of reflexes. A processor can react to different pieces of a representation and do something in response, including altering the representation or making new ones. For example, imagine a machine that can move around on a printed page. It has a cutout in the shape of the letter sequence `isa`, and a light sensor that can tell when the cutout is superimposed on a set of ink marks in the exact shape of the cutout. The sensor is hooked up to a little pocket copier, which can duplicate any set of ink marks, either by printing identical ink marks somewhere else on the page or by burning them into a new cutout.

Now imagine that this sensor-copier-creeper machine is wired up with four reflexes. First, it rolls down the page, and whenever it detects some `isa` ink marks, it moves to the left, and copies the ink marks it finds there onto the bottom left corner of the page. Let loose on our page, it would create the following:

```
Socrates isa man
Every man ismortal




Socrates
```

Its second reflex, also in response to finding an `isa`, is to get itself to the right of that `isa` and copy any ink marks it finds there into the holes of a new cutout. In our case, this forces the processor to make a cutout in the shape of man. Its third reflex is to scan down the page checking for ink marks shaped like `Every`, and if it finds some, seeing if the ink marks to the right align with its new cutout. In our example, it finds one: the man in the middle of the second line. Its fourth reflex, upon finding such a match, is to move to the right and copy the ink marks it finds there onto the bottom center of the page. In our example, those are the ink marks `ismortal`. If you are following me, you'll see that our page now looks like this:

```
Socrates isa man
Every man ismortal


Socrates    ismortal
```

A primitive kind of reasoning has taken place. Crucially, although the gadget and the page it sits on collectively display a kind of intelligence, there is nothing in either of them that is itself intelligent. Gadget and page are just a bunch of ink marks, cutouts, photocells, lasers, and wires. What makes the whole device smart is the exact *correspondence* between the logician's rule "If X is a Y and all Y's are Z, then X is Z" and the way the device scans, moves, and prints. Logically speaking, "X is a Y" means that what is true of Y is also true of X, and mechanically speaking, X isa Y causes what is printed next to the Y to be also printed next to the X. The machine, blindly following the laws of physics, just responds to the shape of the ink marks isa (without understanding what it means to us) and copies other ink marks in a way that ends up mimicking the operation of the logical rule. What makes it "intelligent" is that the sequence of sensing and moving and copying results in its printing a representation of a conclusion that is true if and only if the page contains representations of premises that are true. If one gives the device as much paper as it needs, Turing showed, the machine can do anything that any computer can do—and perhaps, he conjectured, anything that any physically embodied mind can do.

Now, this example uses ink marks on paper as its representation and a copying-creeping-sensing machine as its processor. But the representation can be in any physical medium at all, as long as the patterns are used consistently. In the brain, there might be three groups of neurons, one used to represent the individual that the proposition is about (Socrates, Aristotle, Rod Stewart, and so on), one to represent the logical relationship in the proposition (is a, is not, is like, and so on), and one to represent the class or type that the individual is being categorized as (men, dogs, chickens, and so on). Each concept would correspond to the firing of a particular neuron; for example, in the first group of neurons, the fifth neuron might fire to represent Socrates and the seventeenth might fire to represent Aristotle; in the third group, the eighth neuron might fire to represent men, the twelfth neuron might fire to represent dogs. The processor might be a network of other neurons feeding into these groups, connected together in such a way that it reproduces the firing pattern in one group of neurons in some other group (for example, if the eighth neuron is firing in group 3, the processor network would turn on the eighth neuron in some fourth group, elsewhere in the brain). Or the whole thing could be done in silicon chips. But in all three cases the principles are the same. The way the elements in the processor are wired up would cause them to sense and copy pieces of a representation, and to produce new representations, in a way that mimics the rules of reasoning. With many thousands of representations and a set of somewhat more sophisticated processors (perhaps different kinds of representations and processors for different kinds of thinking), you might have a genuinely intelligent brain or computer. Add an eye that can detect certain contours in the world and turn on representations that symbolize them, and muscles that can act on the world whenever certain representations symbolizing goals are turned on, and you have a behaving organism (or add a TV camera and set of levers and wheels, and you have a robot).

This, in a nutshell, is the theory of thinking called "the physical symbol system hypothesis" or the "computational" or "representational" theory of mind. It is as fundamental to cognitive science as the cell doctrine is to biology and plate tectonics is to geology. Cognitive psychologists and neuroscientists are trying to figure out what kinds of representations and processors the brain has. But there are ground rules that must be followed at all times: no little men inside, and no peeking. The representations that one posits in the mind have to be arrangements of symbols, and the processor has to be a device with a fixed set of reflexes, period. The combination, acting all by itself, has to produce the intelligent conclusions. The theorist is forbidden to peer inside and "read" the symbols, "make sense" of them, and poke around to nudge the device in smart directions like some deus ex machina.

Now we are in a position to pose the Whorfian question in a precise way. Remember that a representation does not have to look like English or any other language; it just has to use symbols to represent concepts, and arrangements of symbols to represent the logical relations among them, according to some consistent scheme. But though internal representations in an English speaker's mind don't *have* to look like English, they *could*, in principle, look like English—or like whatever language the person happens to speak. So here is the question: Do they in fact? For example, if we know that Socrates is a man, is it because we have neural patterns that correspond one-to-one to the English words *Socrates, is, a,* and *man,* and groups of neurons in the brain that correspond to the subject of an English sentence, the verb, and the object, laid out in that order? Or do we use some other code for representing concepts and their relations in our heads, a language of thought or mentalese that is not the same as any of the world's languages? We can answer this question by seeing whether English sentences embody the information that a processor would need to perform valid sequences of reasoning—without requiring any fully intelligent homunculus inside doing the "understanding."

The answer is a clear no. English (or any other language people speak) is hopelessly unsuited to serve as our internal medium of computation. Consider some of the problems.

The first is ambiguity. These headlines actually appeared in newspapers:

Child's Stool Great for Use in Garden
Stud Tires Out
Stiff Opposition Expected to Casketless Funeral Plan
Drunk Gets Nine Months in Violin Case
Iraqi Head Seeks Arms

Queen Mary Having Bottom Scraped
Columnist Gets Urologist in Trouble with His Peers

Each headline contains a word that is ambiguous. But surely the thought underlying the word is *not* ambiguous; the writers of the headlines surely knew which of the two senses of the words *stool, stud,* and *stiff* they themselves had in mind. And if there can be two thoughts corresponding to one word, thoughts can't be words.

The second problem with English is its lack of logical explicitness. Consider the following example, devised by the computer scientist Drew McDermott:

Ralph is an elephant.
Elephants live in Africa.
Elephants have tusks.

Our inference-making device, with some minor modifications to handle the English grammar of the sentences, would deduce "Ralph lives in Africa" and "Ralph has tusks." This sounds fine but isn't. Intelligent you, the reader, knows that the Africa that Ralph lives in is the same Africa that all the other elephants live in, but that Ralph's tusks are his own. But the symbol-copier-creeper-sensor that is supposed to be a model of you *doesn't* know that, because the distinction is nowhere to be found in any of the statements. If you object that this is just common sense, you would be right—but it's common sense that we're trying to account for, and English sentences do not embody the information that a processor needs to carry out common sense.

A third problem is called "co-reference." Say you start talking about an individual by referring to him as *the tall blond man with one black shoe.* The second time you refer to him in the conversation you are likely to call him *the man;* the third time, just *him.* But the three expressions do not refer to three people or even to three ways of thinking about a single person; the second and third are just ways of saving breath. Something in the brain must treat them as the same thing; English isn't doing it.

A fourth, related problem comes from those aspects of language that can only be interpreted in the context of a conversation or text—what linguists call "deixis." Consider articles like *a* and *the*. What is the difference between *killed a policeman* and *killed the policeman*? Only that in the second sentence, it is assumed that some specific policeman was mentioned earlier or is salient in the context. Thus in isolation the two phrases are synonymous, but in the following contexts (the first from an actual newspaper article) their meanings are completely different:

> A policeman's 14-year-old son, apparently enraged after being disciplined for a bad grade, opened fire from his house, *killing a policeman* and wounding three people before he was shot dead.
>
> A policeman's 14-year-old son, apparently enraged after being disciplined for a bad grade, opened fire from his house, *killing the policeman* and wounding three people before he was shot dead.

Outside of a particular conversation or text, then, the words *a* and *the* are quite meaningless. They have no place in one's permanent mental database. Other conversation-specific words like *here, there, this, that, now, then, I, me, my, her, we,* and *you* pose the same problems, as the following old joke illustrates:

> First guy: I didn't sleep with my wife before we were married, did you?
>
> Second guy: I don't know. What was her maiden name?

A fifth problem is synonymy. The sentences

> Sam sprayed paint onto the wall.
> Sam sprayed the wall with paint.
> Paint was sprayed onto the wall by Sam.
> The wall was sprayed with paint by Sam.

refer to the same event and therefore license many of the same inferences. For example, in all four cases, one may conclude that the wall has paint on it. But they are four distinct arrangements of words. You know that they mean the same thing, but no simple processor, crawling over them as marks, would know that. Something else that is not one of those arrangements of words must be representing the single event that you know is common to all four. For example, the event might be represented as something like

(Sam spray paint$_i$) cause (paint$_i$ go to (on wall))

—which, assuming we don't take the English words seriously, is not too far from one of the leading proposals about what mentalese looks like.

These examples (and there are many more) illustrate a single important point. The representations underlying thinking, on the one hand, and the sentences in a language, on the other, are in many ways at cross-purposes. Any particular thought in our head embraces a vast amount of information. But when it comes to communicating a thought to someone else, attention spans are short and mouths are slow. To get information into a listener's head in a reasonable amount of time, a speaker can encode only a fraction of the message into words and must count on the listener to fill in the rest. But *inside a single head,* the demands are different. Air time is not a limited resource: different parts of the brain are connected to one another directly with thick cables that can transfer huge amounts of information quickly. Nothing can be left to the imagination, though, because the internal representations *are* the imagination.

We end up with the following picture. People do not think in English or Chinese or Apache; they think in a language of thought. This language of thought probably looks a bit like all these languages; presumably it has symbols for concepts, and arrangements of symbols that correspond to who did what to whom, as in the paint-spraying representation shown above. But compared with any given language, mentalese must be richer in some ways and simpler in others. It must be richer, for example, in that several concept symbols must correspond to a given English word like *stool* or *stud.* There must be extra paraphernalia that differentiate logically distinct kinds of concepts, like

Ralph's tusks versus tusks in general, and that link different symbols that refer to the same thing, like *the tall blond man with one black shoe* and *the man*. On the other hand, mentalese must be simpler than spoken languages; conversation-specific words and constructions (like *a* and *the*) are absent, and information about pronouncing words, or even ordering them, is unnecessary. Now, it could be that English speakers think in some kind of simplified and annotated quasi-English, with the design I have just described, and that Apache speakers think in a simplified and annotated quasi-Apache. But to get these languages of thought to subserve reasoning properly, they would have to look much more like each other than either one does to its spoken counterpart, and it is likely that they are the same: a universal mentalese.

Knowing a language, then, is knowing how to translate mentalese into strings of words and vice versa. People without a language would still have mentalese, and babies and many nonhuman animals presumably have simpler dialects. Indeed, if babies did not have a mentalese to translate to and from English, it is not clear how learning English could take place, or even what learning English would mean.

So where does all this leave Newspeak? Here are my predictions for the year 2050. First, since mental life goes on independently of particular languages, concepts of freedom and equality will be thinkable even if they are nameless. Second, since there are far more concepts than there are words, and listeners must always charitably fill in what the speaker leaves unsaid, existing words will quickly gain new senses, perhaps even regain their original senses. Third, since children are not content to reproduce any old input from adults but create a complex grammar that can go beyond it, they would creolize Newspeak into a natural language, possibly in a single generation. The twenty-first-century toddler may be Winston Smith's revenge.

# 4

# How Language Works

*Journalists say that when a dog bites a man that is not news, but when a man bites a dog that is news.* This is the essence of the language instinct: language conveys news. The streams of words called "sentences" are not just memory prods, reminding you of man and man's best friend and letting you fill in the rest; they tell you who in fact did what to whom. Thus we get more from most stretches of language than Woody Allen got from *War and Peace,* which he read in two hours after taking speed-reading lessons: "It was about some Russians." Language allows us to know how octopuses make love and how to remove cherry stains and why Tad was heartbroken, and whether the Red Sox will win the World Series without a good relief pitcher and how to build an atom bomb in your basement and how Catherine the Great died, among other things.

When scientists see some apparent magic trick in nature, like bats homing in on insects in pitch blackness or salmon returning to breed in their natal stream, they look for the engineering principles behind it. For bats, the trick turned out to be sonar; for salmon, it was locking in to a faint scent trail. What is the trick behind the ability of *Homo sapiens* to convey that man bites dog?

In fact there is not one trick but two, and they are associated with the names of two European scholars who wrote in the nineteenth century. The first principle, articulated by the Swiss linguist Ferdinand de Saussure, is "the arbitrariness of the sign," the wholly conventional pairing of a sound with a meaning. The word *dog* does not look like a dog, walk like a dog, or woof like a dog, but it means "dog" just the same. It does so because every English speaker has undergone an identical act of rote learning in childhood that links the sound to the meaning. For the price of this standardized memorization, the members of a language community receive an enormous benefit: the ability to convey a concept from mind to mind virtually instantaneously. Sometimes the gunshot marriage between sound and meaning can be amusing. As Richard Lederer points out in *Crazy English,* we drive on a parkway but park in a driveway, there is no ham in hamburger or bread in sweetbreads, and blueberries are blue but cranberries are not cran. But think about the "sane" alternative of depicting a concept so that receivers can apprehend the meaning in the form. The process is so challenging to the ingenuity, so comically unreliable, that we have made it into party games like Pictionary and charades.

The second trick behind the language instinct is captured in a phrase from Wilhelm Von Humboldt that presaged Chomsky: language "makes infinite use of finite media." We know the difference between the forgettable *Dog bites man* and the newsworthy *Man bites dog* because of the order in which *dog, man,* and *bites* are combined. That is, we use a code to translate between orders of words and combinations of thoughts. That code, or set of rules, is called a generative grammar; as I have mentioned, it should not be confused with the pedagogical and stylistic grammars we encountered in school.

The principle underlying grammar is unusual in the natural world. A grammar is an example of a "discrete combinatorial system." A finite number of discrete elements (in this case, words) are sampled, combined, and permuted to create larger structures (in this case, sentences) with properties that are quite distinct from those of their elements. For example, the meaning of *Man bites dog* is different from the meaning of any of the three words inside it, and different from the meaning of the same words combined in the reverse order. In a discrete combinatorial system like language, there can be an unlimited number of completely distinct combinations with an infinite range of properties. Another noteworthy discrete combinatorial system in the natural world is the genetic code in DNA, where four kinds of nucleotides are combined into sixty-four kinds of codons, and the codons can be strung into an unlimited number of different genes. Many biologists have capitalized on the close parallel between the principles of grammatical combination and the principles of genetic combination. In the technical language of genetics, sequences of DNA are said to contain "letters" and "punctuation"; may be "palindromic," "meaningless," or "synonymous"; are "transcribed" and "translated"; and are even stored in "libraries." The immunologist Niels Jerne entitled his Nobel Prize address "The Generative Grammar of the Immune System."

Most of the complicated systems we see in the world, in contrast, are *blending systems,* like geology, paint mixing, cooking, sound, light, and weather. In a blending system the properties of the combination lie *between* the properties of its elements, and the properties of the elements are lost in the average or mixture. For example, combining red paint and white paint results in pink paint. Thus the range of properties that can be found in a blending system are highly circumscribed, and the only way to differentiate large numbers of combinations is to discriminate tinier and tinier differences. It may not be a coincidence that the two systems in the universe that most impress us with their open-ended complex design—life and mind—are based on discrete combinatorial systems. Many biologists believe that if inheritance were not discrete, evolution as we know it could not have taken place.

The way language works, then, is that each person's brain contains a lexicon of words and the concepts they stand for (a mental dictionary) and a set of rules that combine the words to convey relationships among concepts (a mental grammar). We will explore the world of words in the next chapter; this one is devoted to the design of grammar.

The fact that grammar is a discrete combinational system has two important consequences. The first is the sheer vastness of language.

Go into the Library of Congress and pick a sentence at random from any volume, and chances are you would fail to find an exact repetition no matter how long you continued to search. Estimates of the number of sentences that an ordinary person is capable of producing are breathtaking. If a speaker is interrupted at a random point in a sentence, there are on average about ten different words that could be inserted at that point to continue the sentence in a grammatical and meaningful way. (At some points in a sentence, only one word can be inserted, and at others, there is a choice from among thousands; ten is the average.) Let's assume that a person is capable of producing sentences up to twenty words long. Therefore the number of sentences that a speaker can deal with in principle is at least $10^{20}$ (a one with twenty zeros after it, or a hundred million trillion). At a rate of five seconds a sentence, a person would need a childhood of about a hundred trillion years (with no time for eating or sleeping) to memorize them all. In fact, a twenty-word limitation is far too severe. The following comprehensible sentence from George Bernard Shaw, for example, is 110 words long:

> Stranger still, though Jacques-Dalcroze, like all these great teachers, is the completest of tyrants, knowing what is right and that he must and will have the lesson just so or else break his heart (not somebody else's, observe), yet his school is so fascinating that every woman who sees it exclaims: "Oh why was I not taught like this!" and elderly gentlemen excitedly enroll themselves as students and distract classes of infants by their desperate endeavours to beat two in a bar with one hand and three with the other, and start off on earnest walks around the room, taking two steps backward whenever M. Dalcroze calls out "Hop!"

Indeed, if you put aside the fact that the days of our age are threescore and ten, each of us is capable of uttering an *infinite* number of different sentences. By the same logic that shows that there are an infinite number of integers—if you ever think you have the largest integer, just add 1 to it and you will have another—there must be an infinite number of sentences. The *Guinness Book of World Records* once claimed to recognize the longest English sentence: a 1,300-word stretch in William Faulkner's novel *Absalom, Absalom!*, that begins:

> They both bore it as though in deliberate flagellant exal-
> tation . . .

I am tempted to achieve immortality by submitting the following record-breaker:

> Faulkner wrote, "They both bore it as though in deliberate
> flagellant exaltation . . ."

But it would be only the proverbial fifteen minutes of fame, for soon I could be bested by:

> Pinker wrote that Faulkner wrote, "They both bore it as
> though in deliberate flagellant exaltation . . ."

And that record, too, would fall when someone submitted:

> Who cares that Pinker wrote that Faulkner wrote, "They both
> bore it as though in deliberate flagellant exaltation . . ."?

And so on, ad infinitum. The infinite use of finite media distinguishes the human brain from virtually all the artificial language devices we commonly come across, like pull-string dolls, cars that nag you to close the door, and cheery voice-mail instructions ("Press the pound key for more options"), all of which use a fixed list of prefabricated sentences.

The second consequence of the design of grammar is that it is a code that is *autonomous* from cognition. A grammar specifies how words may combine to express meanings; that specification is independent of the particular meanings we typically convey or expect others to convey to us. Thus we all sense that some strings of words that can be given common-sense interpretations do not conform to the grammatical code of English. Here are some strings that we can easily interpret but that we sense are not properly formed:

Welcome to Chinese Restaurant. Please try your Nice Chi-
   nese Food with Chopsticks: the traditional and typical of
   Chinese glorious history and cultual.
It's a flying finches, they are.
The child seems sleeping.
Is raining.
Sally poured the glass with water.
Who did a book about impress you?

Skid crash hospital.
Drum vapor worker cigarette flick boom.
This sentence no verb.
This sentence has contains two verbs.
This sentence has cabbage six words.
This is not a complete. This either.

These sentences are "ungrammatical," not in the sense of split
infinitives, dangling participles, and the other hobgoblins of the
schoolmarm, but in the sense that every ordinary speaker of the casual
vernacular has a gut feeling that something is wrong with them,
despite their interpretability. Ungrammaticality is simply a consequence of our having a fixed code for interpreting sentences. For some
strings a meaning can be guessed, but we lack confidence that the
speaker has used the same code in producing the sentence as we used
in interpreting it. For similar reasons, computers, which are less for-
giving of ungrammatical input than human listeners, express their dis-
pleasure in all-too-familiar dialogues like this one:

```
> PRINT (x + 1
*****SYNTAX ERROR*****
```

The opposite can happen as well. Sentences can make no sense
but can still be recognized as grammatical. The classic example is a
sentence from Chomsky, his only entry in *Bartlett's Familiar Quotations*:

Colorless green ideas sleep furiously.

The sentence was contrived to show that syntax and sense can
be independent of each other, but the point was made long before

Chomsky; the genre of nonsense verse and prose, popular in the nineteenth century, depends on it. Here is an example from Edward Lear,
the acknowledged master of nonsense:

> It's a fact the whole world knows,
> That Pobbles are happier without their toes.

Mark Twain once parodied the romantic description of nature written
more for its mellifluousness than its content:

> It was a crisp and spicy morning in early October. The lilacs
> and laburnums, lit with the glory-fires of autumn, hung burn-
> ing and flashing in the upper air, a fairy bridge provided by
> kind Nature for the wingless wild things that have their homes
> in the tree-tops and would visit together; the larch and the
> pomegranate flung their purple and yellow flames in brilliant
> broad splashes along the slanting sweep of the woodland; the
> sensuous fragrance of innumerable deciduous flowers rose
> upon the swooning atmosphere; far in the empty sky a solitary
> esophagus slept upon motionless wing; everywhere brooded
> stillness, serenity, and the peace of God.

And almost everyone knows the poem in Lewis Carroll's *Through the
Looking-Glass* that ends:

> And, as in uffish thought he stood,
>   The Jabberwock, with eyes of flame,
> Came whiffling through the tulgey wood,
>   And burbled as it came!
>
> One, two! One, two! And through and through
>   The vorpal blade went snicker-snack!
> He left it dead, and with its head
>   He went galumphing back.
>
> "And hast thou slain the Jabberwock?
>   Come to my arms, my beamish boy!
> O frabjous day! Callooh! Callay!"
>   He chortled in his joy.

'Twas brillig, and the slithy toves
   Did gyre and gimble in the wabe:
All mimsy were the borogoves,
   And the mome raths outgrabe.

As Alice said, "Somehow it seems to fill my head with ideas—only I don't exactly know what they are!" But though common sense and common knowledge are of no help in understanding these passages, English speakers recognize that they are grammatical, and their mental rules allow them to extract precise, though abstract, frameworks of meaning. Alice deduced, "*Somebody* killed *something*: that's clear, at any rate—." And after reading Chomsky's entry in *Bartlett's,* anyone can answer questions like "What slept? How? Did one thing sleep, or several? What kind of ideas were they?"

How might the combinatorial grammar underlying human language work? The most straightforward way to combine words in order is explained in Michael Frayn's novel *The Tin Men.* The protagonist, Goldwasser, is an engineer working at an institute for automation. He must devise a computer system that generates the standard kinds of stories found in the daily papers, like "Paralyzed Girl Determined to Dance Again." Here he is hand-testing a program that composes stories about royal occasions:

He opened the filing cabinet and picked out the first card in the set. *Traditionally,* it read. Now there was a random choice between cards reading *coronations, engagements, funerals, weddings, comings of age, births, deaths,* or *the churching of women.* The day before he had picked *funerals,* and been directed on to a card reading with simple perfection *are occasions for mourning.* Today he closed his eyes, drew *weddings,* and was signposted on to *are occasions for rejoicing.*

*The wedding of X and Y* followed in logical sequence, and brought him a choice between *is no exception* and *is a case in point.* Either way there followed *indeed.* Indeed, whichever occasion one had started off with, whether coronations, deaths, or births, Goldwasser saw with intense mathematical pleasure, one now reached this same elegant bottleneck. He paused on *indeed*, then drew in quick succession *it is a particularly happy occasion*, *rarely*, and *can there have been a more popular young couple*.

From the next selection, Goldwasser drew *X has won himself/herself a special place in the nation's affections*, which forced him to go on to *and the British people have cleverly taken Y to their hearts already*.

Goldwasser was surprised, and a little disturbed, to realise that the word "fitting" had still not come up. But he drew it with the next card—*it is especially fitting that*.

This gave him *the bride/bridegroom should be*, and an open choice between *of such a noble and illustrious line*, *a commoner in these democratic times*, *from a nation with which this country has long enjoyed a particularly close and cordial relationship*, and *from a nation with which this country's relations have not in the past been always happy*.

Feeling that he had done particularly well with "fitting" last time, Goldwasser now deliberately selected it again. *It is also fitting that*, read the card, to be quickly followed by *we should remember*, and *X and Y are not mere symbols—they are a lively young man and a very lovely young woman*.

Goldwasser shut his eyes to draw the next card. It turned out to read *in these days when*. He pondered whether to select *it is fashionable to scoff at the traditional morality of marriage and family life* or *it is no longer fashionable to scoff at the traditional morality of marriage and family life*. The latter had more of the form's authentic baroque splendor, he decided.

Let's call this a word-chain device (the technical name is a "finite-state" or "Markov" model). A word-chain device is a bunch of lists of words (or prefabricated phrases) and a set of directors for going from list to list. A processor builds a sentence by selecting a word from one list, then a word from another list, and so on. (To recognize a sentence spoken by another person, one just checks the words against each list in order.) Word-chain systems are commonly used in satires like Frayn's, usually as do-it-yourself recipes for composing examples of a kind of verbiage. For example, here is a Social Science Jargon Generator, which the reader may operate by picking a word at random from the first column, then a word from the second, then one from the third, and stringing them together to form an impressive-sounding term like *inductive aggregating interdependence*.

| | | |
|---|---|---|
| dialectical | participatory | interdependence |
| defunctionalized | degenerative | diffusion |
| positivistic | aggregating | periodicity |
| predicative | appropriative | synthesis |
| multilateral | simulated | sufficiency |
| quantitative | homogeneous | equivalence |
| divergent | transfigurative | expectancy |
| synchronous | diversifying | plasticity |
| differentiated | cooperative | epigenesis |
| inductive | progressive | constructivism |
| integrated | complementary | deformation |
| distributive | eliminative | solidification |

Recently I saw a word-chain device that generates breathless book jacket blurbs, and another for Bob Dylan song lyrics.

A word-chain device is the simplest example of a discrete combinatorial system, since it is capable of creating an unlimited number of distinct combinations from a finite set of elements. Parodies notwithstanding, a word-chain device can generate infinite sets of grammatical English sentences. For example, the extremely simple scheme

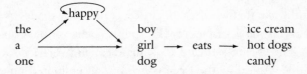

assembles many sentences, such as *A girl eats ice cream* and *The happy dog eats candy*. It can assemble an infinite number because of the loop at the top that can take the device from the *happy* list back to itself any number of times: *The happy dog eats ice cream, The happy happy dog eats ice cream,* and so on.

When an engineer has to build a system to combine words in particular orders, a word-chain device is the first thing that comes to mind. The recorded voice that gives you a phone number when you dial directory assistance is a good example. A human speaker is recorded uttering the ten digits, each in seven different sing-song patterns (one for the first position in a phone number, one for the second position, and so on). With just these seventy recordings, ten million phone numbers can be assembled; with another thirty recordings for three-digit area codes, ten billion numbers are possible (in practice, many are never used because of restrictions like the absence of 0 and 1 from the beginning of a phone number). In fact there have been serious efforts to model the English language as a very large word chain. To make it as realistic as possible, the transitions from one word list to another can reflect the actual probabilities that those kinds of words follow one another in English (for example, the word *that* is much more likely to be followed by *is* than by *indicates*). Huge databases of these "transition probabilities" have been compiled by having a computer analyze bodies of English text or by asking volunteers to name the words that first come to mind after a given word or series of words. Some psychologists have suggested that human language is based on a huge word chain stored in the brain. The idea is congenial to stimulus-response theories: a stimulus elicits a spoken word as a response, then the speaker perceives his or her own response, which serves as the next stimulus, eliciting one out of several words as the next response, and so on.

But the fact that word-chain devices seem ready-made for parodies like Frayn's raises suspicions. The point of the various parodies is that the genre being satirized is so mindless and cliché-ridden that a simple mechanical method can churn out an unlimited number of examples that can almost pass for the real thing. The humor works because of the discrepancy between the two: we all assume that people, even sociologists and reporters, are not really word-chain devices; they only seem that way.

The modern study of grammar began when Chomsky showed that word-chain devices are not just a bit suspicious; they are deeply, fundamentally, the wrong way to think about how human language works. They are discrete combinatorial systems, but they are the wrong kind. There are three problems, and each one illuminates some aspect of how language really does work.

First, a sentence of English is a completely different thing from a string of words chained together according to the transition probabilities of English. Remember Chomsky's sentence *Colorless green ideas sleep furiously.* He contrived it not only to show that nonsense can be grammatical but also to show that improbable word sequences can be grammatical. In English texts the probability that the word *colorless* is followed by the word *green* is surely zero. So is the probability that *green* is followed by *ideas, ideas* by *sleep,* and *sleep* by *furiously.* Nonetheless, the string is a well-formed sentence of English. Conversely, when one actually assembles word chains using probability tables, the resulting word strings are very far from being well-formed sentences. For example, say you take estimates of the set of words most likely to come after every four-word sequence, and use those estimates to grow a string word by word, always looking at the four most recent words to determine the next one. The string will be eerily Englishy, but not English, like *House to ask for is to earn out living by working towards a goal for his team in old New-York was a wonderful place wasn't it even pleasant to talk about and laugh hard when he tells lies he should not tell me the reason why you are is evident.*

The discrepancy between English sentences and Englishy word chains has two lessons. When people learn a language, they are learning how to put words in order, but not by recording which word follows which other word. They do it by recording which word *category*—noun, verb, and so on—follows which other category. That is, we can recognize *colorless green ideas* because it has the same order of adjectives and nouns that we learned from more familiar sequences like *strapless black dresses.* The second lesson is that the nouns and verbs and adjectives are not just hitched end to end in one long chain; there is some overarching blueprint or plan for the sentence that puts each word in a specific slot.

If a word-chain device is designated with sufficient cleverness, it can deal with these problems. But Chomsky had a definitive refutation of the very idea that a human language is a word chain. He proved that certain sets of English sentences could not, even in principle, be produced by a word-chain device, no matter how big or how faithful to probability tables the device is. Consider sentences like the following:

> Either the girl eats ice cream, or the girl eats candy.
> If the girl eats ice cream, then the boy eats hot dogs.

At first glance it seems easy to accommodate these sentences:

But the device does not work. *Either* must be followed later in a sentence by *or*; no one says *Either the girl eats ice cream, then the girl eats candy.* Similarly, *if* requires *then*; no one says *If the girl eats ice cream, or the girl likes candy.* But to satisfy the desire of a word early in a sentence for some other word late in the sentence, the device has to remember the early word while it is churning out all the words in between. And that is the problem: a word-chain device is an amnesiac, remembering only which word list it has just chosen from, nothing earlier. By the time it reaches the *or/then* list, it has no means of remembering whether it said *if* or *either* way back at the beginning. From our vantage point, peering down at the entire road map, we can remember which choice the device made at the first fork in the road, but the device itself, creeping antlike from list to list, has no way of remembering.

Now, you might think it would be a simple matter to redesign the device so that it does not have to remember early choices at late points in the sentence. For example, one could join up *either* and *or* and all the possible word sequences in between into one giant sequence, and *if* and *then* and all the sequences in between as a second giant sequence, before returning to a third copy of the sequence— yielding a chain so long I have to print it sideways (see page 88). There is something immediately disturbing about this solution: there are three identical subnetworks. Clearly, whatever people can say between an *either* and an *or,* they can say between an *if* and a *then,* and also after the *or* or the *then.* But this ability should come naturally out of the design of whatever the device is in people's heads that allows them to speak. It shouldn't depend on the designer's carefully writing down three identical sets of instructions (or, more plausibly, on the child's having to learn the structure of the English sentence three different times, once between *if* and *then,* once between *either* and *or,* and once after a *then* or an *or*).

But Chomsky showed that the problem is even deeper. Each of these sentences can be embedded in any of the others, including itself:

> If either the girl eats ice cream or the girl eats candy, then
>     the boy eats hot dogs.
> Either if the girl eats ice cream then the boy eats ice cream,
>     or if the girl eats ice cream then the boy eats candy.

For the first sentence, the device has to remember *if* and *either* so that it can continue later with *or* and *then,* in that order. For the second sentence, it has to remember *either* and *if* so that it can complete the sentence with *then* and *or.* And so on. Since there's no limit in principle to the number of *if*'s and *either*'s that can begin a sentence, each requiring its own order of *then*'s and *or*'s to complete it, it does no good to spell out each memory sequence as its own chain of lists; you'd need an infinite number of chains, which won't fit inside a finite brain.

This argument may strike you as scholastic. No real person ever begins a sentence with *Either either if either if if,* so who cares whether a putative model of that person can complete it with *then . . . then . . .*

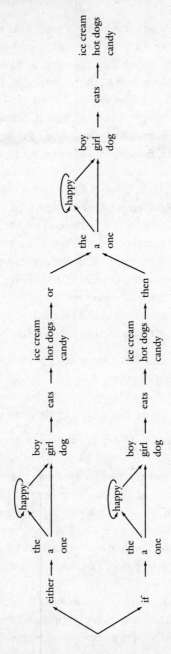

*or . . . then . . . or . . . or?* But Chomsky was just adopting the esthetic of the mathematician, using the interaction between *either-or* and *if-then* as the simplest possible example of a property of language—its use of "long-distance dependencies" between an early word and a later one—to prove mathematically that word-chain devices cannot handle these dependencies.

The dependencies, in fact, abound in languages, and mere mortals use them all the time, over long distances, often handling several at once—just what a word-chain device cannot do. For example, there is an old grammarian's saw about how a sentence can end in five prepositions. Daddy trudges upstairs to Junior's bedroom to read him a bedtime story. Junior spots the book, scowls, and asks, "Daddy, what did you bring that book that I don't want to be read to out of up for?" By the point at which he utters *read,* Junior has committed himself to holding four dependencies in mind: *to be read* demands *to, that book that* requires *out of, bring* requires *up,* and *what* requires *for.* An even better, real-life example comes from a letter to *TV Guide*:

> How Ann Salisbury can claim that Pam Dawber's anger at not receiving her fair share of acclaim for *Mork and Mindy*'s success derives from a fragile ego escapes me.

At the point just after the word *not,* the letter-writer had to keep four grammatical commitments in mind: (1) *not* requires *-ing* (her anger at *not* receiv*ing* acclaim); (2) *at* requires some kind of noun or gerund (her anger *at* not *receiving acclaim*); (3) the singular subject *Pam Dawber's anger* requires the verb fourteen words downstream to agree with it in number (Dawber's *anger . . . derives* from); (4) the singular subject beginning with *How* requires the verb twenty-seven words downstream to agree with it in number (*How . . . escapes* me). Similarly, a reader must keep these dependencies in mind while interpreting the sentence. Now, technically speaking, one could rig up a word-chain model to handle even these sentences, as long as there is some actual limit on the number of dependencies that the speaker need keep in mind (four, say). But the degree of redundancy in the device would be absurd; for each of the thousands of *combinations* of dependencies, an identical chain must be duplicated inside the device. In trying to fit such a superchain in a person's memory, one quickly runs out of brain.

The difference between the artificial combinatorial system we see in word-chain devices and the natural one we see in the human brain is summed up in a line from the Joyce Kilmer poem: "Only God can make a tree." A sentence is not a chain but a tree. In a human grammar, words are grouped into phrases, like twigs joined in a branch. The phrase is given a name—a mental symbol—and little phrases can be joined into bigger ones.

Take the sentence *The happy boy eats ice cream.* It begins with three words that hang together as a unit, the noun phrase *the happy boy.* In English a noun phrase (NP) is composed of a noun (N), sometimes preceded by an article or "determinator" (abbreviated "det") and any number of adjectives (A). All this can be captured in a rule that defines what English noun phrases look like in general. In the standard notation of linguistics, an arrow means "consists of," parentheses mean "optional," and an asterisk means "as many of them as you want," but I provide the rule just to show that all of its information can be captured precisely in a few symbols; you can ignore the notation and just look at the translation into ordinary words below it:

NP → (det) A* N
"A noun phrase consists of an optional determiner, followed
    by any number of adjectives, followed by a noun."

The rule defines an upside-down tree branch:

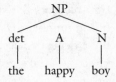

Here are two other rules, one defining the English sentence (S), the other defining the predicate or verb phrase (VP); both use the NP symbol as an ingredient:

S → NP VP
"A sentence consists of a noun phrase followed by a verb phrase."

VP → V NP
"A verb phrase consists of a verb followed by a noun phrase."

We now need a mental dictionary that specifies which words belong to which part-of-speech categories (noun, verb, adjective, preposition, determiner):

N → boy, girl, dog, cat, ice cream, candy, hot dogs
"Nouns may be drawn from the following list: *boy, girl, . . .*"

V → eats, likes, bites
"Verbs may be drawn from the following list: *eats, likes, bites.*"

A → happy, lucky, tall
"Adjectives may be drawn from the following list: *happy, lucky, tall.*"

det → a, the, one
"Determiners may be drawn from the following list: *a, the, one.*"

A set of rules like the ones I have listed—a "phrase structure grammar"—defines a sentence by linking the words to branches on an inverted tree:

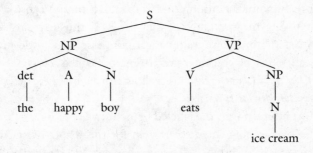

The invisible superstructure holding the words in place is a powerful invention that eliminates the problems of word-chain devices.

The key insight is that a tree is *modular,* like telephone jacks or garden hose couplers. A symbol like "NP" is like a connector or fitting of a certain shape. It allows one component (a phrase) to snap into any of several positions inside other components (larger phrases). Once a kind of phrase is defined by a rule and given its connector symbol, it never has to be defined again; the phrase can be plugged in anywhere there is a corresponding socket. For example, in the little grammar I have listed, the symbol "NP" is used both as the subject of a sentence (S → NP VP) and as the object of a verb phrase (VP → V NP). In a more realistic grammar, it would also be used as the object of a preposition (*near the boy*), in a possessor phrase (*the boy's bat*), as an indirect object (*give the boy a cookie*), and in several other positions. This plug-and-socket arrangement explains how people can use the same kind of phrase in many different positions in a sentence, including:

> [The happy happy boy] eats ice cream.
> I like [the happy happy boy].
> I gave [the happy happy boy] a cookie.
> [The happy happy boy]'s cat eats ice cream.

There is no need to learn that the adjective precedes the noun (rather than vice versa) for the subject, and then have to learn the same thing for the object, and again for the indirect object, and yet again for the possessor.

Note, too, that the promiscuous coupling of any phrase with any slot makes grammar autonomous from our common-sense expectations involving the meanings of the words. It thus explains why we can write and appreciate grammatical nonsense. Our little grammar defines all kinds of colorless green sentences, like *The happy happy candy likes the tall ice cream,* as well as conveying such newsworthy events as *The girl bites the dog.*

Most interestingly, the labeled branches of a phrase structure tree act as an overarching memory or plan for the whole sentence. This allows nested long-distance dependencies, like *if . . . then* and *either . . . or,* to be handled with ease. All you need is a rule defining a phrase that contains a copy of the very same kind of phrase, such as:

S → either S or S

"A sentence can consist of the word *either*, followed by a sentence, followed by the word *or*, followed by another sentence."

S → if S then S

"A sentence can consist of the word *if*, followed by a sentence, followed by the word *then*, followed by another sentence."

These rules embed one instance of a symbol inside another instance of the same symbol (here, a sentence inside a sentence), a neat trick—logicians call it "recursion"—for generating an infinite number of structures. The pieces of the bigger sentence are held together, in order, as a set of branches growing out of a common node. That node holds together each *either* with its *or*, each *if* with its *then*, as in the following diagram (the triangles are abbreviations for lots of underbrush that would only entangle us if shown in full):

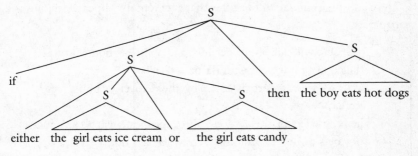

There is another reason to believe that a sentence is held together by a mental tree. So far I have been talking about stringing words into a grammatical order, ignoring what they mean. But grouping words into phrases is also necessary to connect grammatical sentences with their proper meanings, chunks of mentalese. We know that the sentence shown above is about a girl, not a boy, eating ice cream, and a boy, not a girl, eating hot dogs, and we know that the boy's snack is contingent on the girl's, not vice versa. That is because *girl* and *ice cream* are connected inside their own phrase, as are *boy* and *hot dogs*, as are the two sentences involving the girl. With a chaining device it's just one damn word after another, but with a phrase structure grammar the connectedness of words in the tree reflects the relatedness of ideas in mentalese. Phrase structure, then, is one solution to the engineering problem of taking an interconnected web of thoughts in the mind and encoding them as a string of words that must be uttered, one at a time, by the mouth.

One way to see how invisible phrase structure determines meaning is to recall one of the reasons mentioned in Chapter 3 that language and thought have to be different: a particular stretch of language can correspond to two distinct thoughts. I showed you examples like *Child's Stool Is Great for Use in Garden*, where the single word *stool* has two meanings, corresponding to two entries in the mental dictionary. But sometimes a whole sentence has two meanings, even if each individual word has only one meaning. In the movie *Animal Crackers*, Groucho Marx says, "I once shot an elephant in my pajamas. How he got into my pajamas I'll never know." Here are some similar ambiguities that accidentally appeared in newspapers:

> Yoko Ono will talk about her husband John Lennon who was killed in an interview with Barbara Walkers.
> Two cars were reported stolen by the Groveton police yesterday.
> The license fee for altered dogs with a certificate will be $3 and for pets owned by senior citizens who have not been altered the fee will be $1.50.
> Tonight's program discusses stress, exercise, nutrition, and sex with Celtic forward Scott Wedman, Dr. Ruth Westheimer, and Dick Cavett.
> We will sell gasoline to anyone in a glass container.
> For sale: Mixing bowl set designed to please a cook with round bottom for efficient beating.

The two meanings in each sentence come from the different ways in which the words can be joined up in a tree. For example, in *discuss sex with Dick Cavett*, the writer put the words together according to the tree below ("PP" means prepositional phrase): sex is what is to be discussed, and it is to be discussed with Dick Cavett.

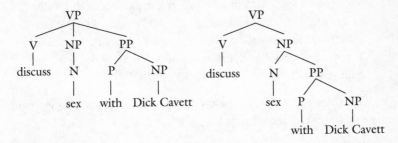

The alternative meaning comes from our analyzing the words according to the tree at the right: the words *sex with Dick Cavett* form a single branch of the tree, and sex with Dick Cavett is what is to be discussed.

Phrase structure, clearly, is the kind of stuff language is made of. But what I have shown you is just a toy. In the rest of this chapter I will try to explain the modern Chomskyan theory of how language works. Chomsky's writing are "classics" in Mark Twain's sense: something that everybody wants to have read and nobody wants to read. When I come across one of the countless popular books on mind, language, and human nature that refer to "Chomsky's deep structure of meaning common to all human languages" (wrong in two ways, we shall see), I know that Chomsky's books of the last twenty-five years are sitting on a high shelf in the author's study, their spines uncracked, their folios uncut. Many people want to have a go at speculating about the mind but have the same impatience about mastering the details of how language works that Eliza Doolittle showed to Henry Higgins in *Pygmalion* when she complained, "I don't want to talk grammar. I want to talk like a lady in a flower shop."

For nonspecialists the reaction is even more extreme. In Shakespeare's *The Second Part of King Henry VI,* the rebel Dick the Butcher speaks the well-known line "The first thing we do, let's kill all the lawyers." Less well known is the second thing Dick suggests they do:

behead Lord Say. Why? Here is the indictment presented by the mob's leader, Jack Cade:

> Thou hast most traitorously corrupted the youth of the realm in erecting a grammar school. . . . It will be proved to thy face that thou hast men about thee that usually talk of a noun and a verb, and such abominable words as no Christian ear can endure to hear.

And who can blame the grammarphobe, when a typical passage from one of Chomsky's technical works reads as follows?

> To summarize, we have been led to the following conclusions, on the assumption that the trace of a zero-level category must be properly governed. 1. VP is α-marked by I. 2. Only lexical categories are L-markers, so that VP is not L-marked by I. 3. α-government is restricted to sisterhood without the qualification (35). 4. Only the terminus of an $X^0$-chain can α-mark or Case-mark. 5. Head-to-head movement forms an A-chain. 6. SPEC-head agreement and chains involve the same indexing. 7. Chain coindexing holds of the links of an extended chain. 8. There is no accidental coindexing of I. 9. I-V coindexing is a form of head-head agreement; if it is restricted to aspectual verbs, then base-generated structures of the form (174) count as adjunction structures. 10. Possibly, a verb does not properly govern its α-marked complement.

All this is unfortunate. People, especially those who hold forth on the nature of mind, should be just plain curious about the code that the human species uses to speak and understand. In return, the scholars who study language for a living should see that such curiosity can be satisfied. Chomsky's theory need not be treated by either group as a set of cabalistic incantations that only the initiated can mutter. It is a set of discoveries about the design of language that can be appreciated intuitively if one first understands the problems to which the theory provides solutions. In fact, grasping grammatical theory provides an intellectual pleasure that is rare in the social sciences. When I entered high school in the late 1960s and electives were chosen for their "relevance," Latin underwent a steep decline in popularity (thanks to students like me, I confess). Our Latin teacher Mrs. Rillie, whose merry birthday parties for Rome failed to slow the decline, tried to persuade us that Latin grammar honed the mind with its demands for precision, logic, and consistency. (Nowadays, such arguments are more likely to come from the computer programming teachers.) Mrs. Rillie had a point, but Latin declensional paradigms are not the best way to convey the inherent beauty of grammar. The insights behind Universal Grammar are much more interesting, not only because they are more general and elegant but because they are about living minds rather than dead tongues.

Let's start with nouns and verbs. Your grammar teacher may have had you memorize some formula that equated parts of speech with kinds of meanings, like

> A NOUN's the name of any thing;
> As *school* or *garden, hoop* or *swing.*
> VERBS tell of something being done;
> To *read, count, sing, laugh, jump,* or *run.*

But as in most matters about language, she did not get it quite right. It is true that most names for persons, places, and things are nouns, but it is not true that most nouns are names for persons, places, or things. There are nouns with all kinds of meanings:

> the *destruction* of the city [an action]
> the *way* to San Jose [a path]
> *whiteness* moves downward [a quality]
> three *miles* along the path [a measurement in space]
> It takes three *hours* to solve the problem. [a measurement in time]
> Tell me the *answer.* ["what the answer is," a question]
> She is a *fool.* [a category or kind]
> a *meeting* [an event]

the *square root* of minus two [an abstract concept]
He finally kicked *the bucket*. [no meaning at all]

Likewise, though words for things being done, such as *count* and *jump,* are usually verbs, verbs can be other things, like mental states (*know, like*), possession (*own, have*), and abstract relations among ideas (*falsify, prove*).

Conversely, a single concept, like "being interested," can be expressed by different parts of speech:

her *interest* in fungi [noun]
Fungi are starting to *interest* her more and more. [verb]
She seems interested in fungi. Fungi seem *interesting* to her.
   [adjective]
*Interestingly,* the fungi grew an inch in an hour. [adverb]

A part of speech, then, is not a kind of meaning; it is a kind of token that obeys certain formal rules, like a chess piece or a poker chip. A noun, for example, is simply a word that does nouny things; it is the kind of word that comes after an article, can have an *'s* stuck onto it, and so on. There is a connection between concepts and part-of-speech categories, but it is a subtle and abstract one. When we construe an aspect of the world as something that can be identified and counted or measured and that can play a role in events, language often allows us to express that aspect as a noun, whether or not it is a physical object. For example, when we say *I have three reasons for leaving,* we are counting reasons as if they were objects (though of course we do not literally think that a reason can sit on a table or be kicked across a room). Similarly, when we construe some aspect of the world as an event or state involving several participants that affect one other, language often allows us to express that aspect as a verb. For example, when we say *The situation justified drastic measures,* we are talking about justification as if it were something the situation did, though again we know that justification is not something we can watch happening at a particular time and place. Nouns are *often* used for names of things, and verbs for something being done, but because the human mind can construe reality in a variety of ways, nouns and verbs are not limited to those uses.

Now what about the phrases that group words into branches? One of the most intriguing discoveries of modern linguistics is that there appears to be a common anatomy in all phrases in all the world's languages.

Take the English noun phrase. A noun phrase (NP) is named after one special word, a noun, that must be inside it. The noun phrase owes most of its properties to that one noun. For example, the NP *the cat in the hat* refers to a kind of cat, not a kind of hat; the meaning of the word *cat* is the core of the meaning of the whole phrase. Similarly, the phrase *fox in socks* refers to a fox, not socks, and the entire phrase is singular in number (that is, we say that the fox in socks *is* or *was* here, not *are* or *were* here), because the word *fox* is singular in number. This special noun is called the "head" of the phrase, and the information filed with that word in memory "percolates up" to the topmost node, where it is interpreted as characterizing the phrase as a whole. The same goes for verb phrases: *flying to Rio before the police catch him* is an example of flying, not an example of catching, so the verb *flying* is called its head. Here we have the first principle of building the meaning of a phrase out of the meaning of the words inside the phrase. What the entire phrase is "about" is what its head word is about.

The second principle allows phrases to refer not just to single things or actions in the world but to sets of players that interact with each other in a particular way, each with a specific role. For example, the sentence *Sergey gave the documents to the spy* is not just about any old act of giving. It choreographs three entities: Sergey (the giver), documents (the gift), and a spy (the recipient). These role-players are usually called "arguments," which has nothing to do with bickering; it's the term used in logic and mathematics for a participant in a relationship. A noun phrase, too, can assign roles to one or more players, as in *picture of John, governor of California,* and *sex with Dick Cavett,* each defining one role. The head and its role-players—other than the subject role, which is special—are joined together in a subphrase, smaller than an NP or a VP, that has the kind of non-mnemonic label that has made generative linguistics so uninviting, "N-bar" and "V-bar," named after the way they are written, $\overline{N}$ and $\overline{V}$:

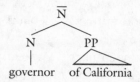

The third ingredient of a phrase is one or more modifiers (usually called "adjuncts"). A modifier is different from a role-player. Take the phrase *The man from Illinois*. Being a man from Illinois is not like being a governor of California. To be a governor, you have to be a governor of something; the Californianess plays a role in what it means for someone to be governor of California. In contrast, *from Illinois* is just a bit of information that we add on to help identify which man we are talking about; being from one state or another is not an inherent part of what it means to be a man. This distinction in meaning between role-players and modifiers ("arguments" and "adjuncts," in lingo) dictates the geometry of the phrase structure tree. The role-player stays next to the head noun inside the N-bar, but the modifier goes upstairs, though still inside the NP house:

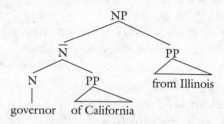

This restriction of the geometry of phrase structure trees is not just playing with notation; it is a hypothesis about how the rules of language are set up in our brains, governing the way we talk. It dictates that if a phrase contains both a role-player *and* a modifier, the role-player has to be closer to the head than the modifier is—there's no way the modifier could get between the head noun and the role-player without crossing branches in the tree (that is, sticking extraneous words in among the bits of the N-bar), which is illegal. Consider Ronald Reagan. He used to be the governor of California, but he was born in Tampico, Illinois. When he was in office, he could have been referred to as *the governor of California from Illinois* (role-player, then modifier). It would have sounded odd to refer to him as *the governor from Illinois of California* (modifier, then role-player). More pointedly, in 1964 Robert F. Kennedy's senatorial ambitions ran up against the inconvenient fact that both Massachusetts seats were already occupied (one by his younger brother Edward). So he simply took up residence in New York and ran for the U.S. Senate from there, soon becoming *the senator from New York from Massachusetts.* Not *the senator from Massachusetts from New York*—though that does come close to the joke that Bay Staters used to tell at the time, that they lived in the only state entitled to *three* senators.

Interestingly, what is true of N-bars and noun phrases is true of V-bars and verb phrases. Say that Sergey gave those documents to the spy in a hotel. The phrase *to the spy* is one of the role-players of the verb *give*—there is no such thing as giving without a getter. Therefore *to the spy* lives with the head verb inside the V-bar. But *in a hotel* is a modifier, a comment, an afterthought, and is kept outside the V-bar, in the VP. Thus the phrases are inherently ordered: we can say *gave the documents to the spy in a hotel,* but not *gave in a hotel the documents to the spy.* When a head is accompanied by just one phrase, however, that phrase can be either a role-player (inside the V-bar) or a modifier (outside the V-bar but inside the VP), and the actual order of the words is the same. Consider the following newspaper report:

> One witness told the commissioners that she had seen sexual intercourse taking place between two parked cars in front of her house.

The aggrieved woman had a modifier interpretation in mind for *between two parked cars,* but twisted readers give it a role-player interpretation.

The fourth and final component of a phrase is a special position reserved for subjects (which linguists call "SPEC," pronounced "speck," short for "specifier"; don't ask). The subject is a special role-player, usually the causal agent if there is one. For example, in the verb phrase *the guitarists destroy the hotel room,* the prhase *the guitarists* is the subject; it is the causal agent of the event consisting of the hotel room being destroyed. Actually, noun phrases can have subjects too, as in the parallel NP *the guitarists' destruction of the hotel room.* Here, then, is the full anatomy of a VP and of an NP:

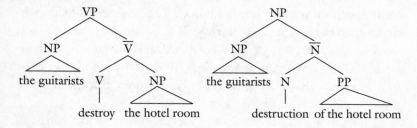

Now the story begins to get interesting. You must have noticed that noun phrases and verb phrases have a lot in common: (1) a head, which gives the phrase its name and determines what it is about, (2) some role-players, which are grouped with the head inside a subphrase (the N-bar or V-bar), (3) modifiers, which appear outside the N- or V-bar, and (4) a subject. The orderings inside a noun phrase and inside a verb phrase are the same: the noun comes before its role-players *(the destruction of the hotel room,* not *the of the hotel room destruction),* and the verb comes before *its* role-players *(to destroy the hotel room,* not *to the hotel room destroy).* The modifiers go to the right in both cases, the subject to the left. It seems as if there is a standard design to the two phrases.

In fact, the design pops up all over the place. Take, for example, the prepositional phrase (PP) *in the hotel.* It has a head, the preposition *in,* which means something like "interior region," and then a role, the thing whose interior region is being picked out, in this case a hotel. And the same goes for the adjective phrase (AP): in *afraid of the wolf,* the head adjective, *afraid,* occurs before its role-player, the source of the fear.

With this common design, there is no need to write out a long list of rules to capture what is inside a speaker's head. There may be just one pair of super-rules for the entire language, where the distinctions among nouns, verbs, prepositions, and adjectives are collapsed and all four are specified with a variable like "X." Since a phrase just inherits the properties of its head (*a tall man* is a kind of *man*), it's redundant to call a phrase headed by a noun a "noun phrase"—we could just call it an "X phrase," since the nounhood of the head noun, like the manhood of the head noun and all the other information in the head noun, percolates up to characterize the whole phrase. Here is what the super-rules look like (as before, focus on the summary of the rule, not the rule itself):

XP → (SPEC) $\overline{X}$ YP*
"A phrase consists of an optional subject, followed by an X-bar, followed by any number of modifiers."

$\overline{X}$ → X ZP*
"An X-bar consists of a head word, followed by any number of role-players."

Just plug in noun, verb, adjective, or preposition for X, Y, and Z, and you have the actual phrase structure rules that spell the phrases. This streamlined version of phrase structure is called "the X-bar theory."

This general blueprint for phrases extends even farther, to other languages. In English, the head of a phrase comes before its role-players. In many languages, it is the other way around—but it is the other way around across the board, across all the kinds of phrases in the language. For example, in Japanese, the verb comes *after* its object, not before: they say *Kenji sushi ate,* not *Kenji ate sushi.* The preposition comes after its noun phrase: *Kenji to,* not *to Kenji* (so they are actually called "postpositions"). The adjective comes after its complement: *Kenji than taller,* not *taller than Kenji.* Even the words marking questions are flipped: they say, roughly, *Kenji eat did?,* not

*Did Kenji eat?* Japanese and English are looking-glass versions of each other. And such consistency has been found in scores of languages: if a language has the verb before the object, as in English, it will also have prepositions; if it has the verb after the object, as in Japanese, it will have postpositions.

This is a remarkable discovery. It means that the super-rules suffice not only for all phrases in English but for all phrases in all languages, with one modification: removing the left-to-right order from each super-rule. The trees become mobiles. One of the rules would say:

$$\overline{X} \rightarrow \{ZP^\star, X\}$$

"An X-bar is composed of a head X and any number of role-players, in either order."

To get English, one appends a single bit of information saying that the order within an X-bar is "head-first." To get Japanese, that bit of information would say that the order is "head-last." Similarly, the other super-rule (the one for phrases) can be distilled so that left-to-right order boils away, and an ordered phrase in a particular language can be reconstituted by adding back either "X-bar-first" or "X-bar-last." The piece of information that makes one language different from another is called a parameter.

In fact, the super-rule is beginning to look less like an exact blueprint for a particular phrase and more like a general guideline or principle for what phrases must look like. The principle is usable only after you combine it with a language's particular setting for the order parameter. This general conception of grammar, first proposed by Chomsky, is called the "principles and parameters" theory.

Chomsky suggests that the unordered super-rules (principles) are universal and innate, and that when children learn a particular language, they do not have to learn a long list of rules, because they were born knowing the super-rules. All they have to learn is whether their particular language has the parameter value head-first, as in English, or head-last, as in Japanese. They can do that merely by noticing whether a verb comes before or after its object in any sentence in their parents' speech. If the verb comes before the object, as in *Eat your spinach!*, the child concludes that the language is head-first; if it comes after, as in *Your spinach eat!*, the child concludes that the language is head-last. Huge chunks of grammar are then available to the child, all at once, as if the child were merely flipping a switch to one of two possible positions. If this theory of language learning is true, it would help solve the mystery of how children's grammar explodes into adult-like complexity in so short a time. They are not acquiring dozens or hundreds of rules; they are just setting a few mental switches.

The principles and parameters of phrase structure specify only what kinds of ingredients may go into a phrase in what order. They do not spell out any particular phrase. Left to themselves, they would run amok and produce all kinds of mischief. Take a look at the following sentences, which all conform to the principles or super-rules. The ones I have marked with an asterisk do not sound right.

Melvin dined.
*Melvin dined the pizza.

Melvin devoured the pizza.
*Melvin devoured.

Melvin put the car in the garage.
*Melvin put.

*Melvin put the car.
*Melvin put in the garage.

Sheila alleged that Bill is a liar.
*Sheila alleged the claim.
*Sheila alleged.

It must be the verb's fault. Some verbs, like *dine*, refuse to appear in the company of a direct object noun phrase. Others, like *devour*, won't appear without one. This is true even though *dine* and *devour* are very close in meaning, both being ways of eating. You may dimly recall from grammar lessons that verbs like *dine* are called "intransitive" and verbs like *devour* are called "transitive." But verbs come in many flavors, not just these two. The verb *put* is not content unless it has both an object NP *(the car)* and a prepositional phrase *(in the garage)*. The verb *allege* requires an embedded sentence *(that Bill is a liar)* and nothing else.

Within a phrase, then, the verb is a little despot, dictating which of the slots made available by the super-rules are to be filled. These demands are stored in the verb's entry in the mental dictionary, more or less as follows:

> *dine:*
> verb
> means "to eat a meal in a refined setting"
> eater = subject

> *devour:*
> verb
> means "to eat something ravenously"
> eater = subject
> thing eaten = object

> *put:*
> verb
> means "to cause something to go to some place"
> putter = subject
> thing put = object
> place = prepositional object

> *allege:*
> verb
> means "to declare without proof"
> declarer = subject
> declaration = complement sentence

Each of these entries lists a definition (in mentalese) of some kind of event, followed by the players that have roles in the event. The entry indicates how each role-player may be plugged into the sentence—as a subject, an object, a prepositional object, an embedded sentence, and so on. For a sentence to feel grammatical, the verb's demands must be satisfied. *Melvin devoured* is bad because *devour*'s desire for a "thing eaten" role is left unfulfilled. *Melvin dined the pizza* is bad because *dine* didn't order *pizza* or any other object.

Because verbs have the power to dictate how a sentence conveys who did what to whom, one cannot sort out the roles in a sentence without looking up the verb. That is why your grammar teacher got it wrong when she told you that the subject of the sentence is the "doer of the action." The subject of the sentence is often the doer, but only when the verb says so; the verb can also assign it other roles:

> The big bad wolf *frightened* the three little pigs. [The subject is doing the frightening.]
> The three little pigs *feared* the big bad wolf. [The subject is being frightened.]
>
> My true love *gave* me a partridge in a pear tree. [The subject is doing the giving.]
> I *received* a partridge in a pear tree from my true love. [The subject is being given to.]
>
> Dr. Nussbaum *performed* plastic surgery. [The subject is operating on someone.]
> Cheryl *underwent* plastic surgery. [The subject is being operated on.]

In fact, many verbs have two distinct entries, each casting a different set of roles. This can give rise to a common kind of ambiguity, as in the old joke: "Call me a taxi." "OK, you're a taxi." In one of the Harlem Globetrotters' routines, the referee tells Meadowlark Lemon to shoot the ball. Lemon points his finger at the ball and shouts, "Bang!" The comedian Dick Gregory tells of walking up to a lunch counter in Mississippi during the days of racial segregation. The waitress said to him, "We don't serve colored people." "That's fine," he replied, "I don't eat colored people. I'd like a piece of chicken."

So how do we actually distinguish *Man bites dog* from *Dog bites man*? The dictionary entry for *bite* says "The biter is the subject; the bitten thing is the object." But how do we *find* subjects and objects in the tree? Grammar puts little tags on the noun phrases that can be matched up with the roles laid out in a verb's dictionary entry. These tags are called *cases*. In many languages, cases appear as prefixes or suffixes on the nouns. For example, in Latin, the nouns for man and dog, *homo* and *canis,* change their endings depending on who is biting whom:

> Canis hominem mordet. [not news]
> Homo canem mordet. [news]

Julius Caesar knew who bit whom because the noun corresponding to the bitee appeared with *-em* at the end. Indeed, this allowed Caesar to find the biter and bitee even when the order of the two was flipped, which Latin allows: *Hominem canis mordet* means the same thing as *Canis hominem mordet,* and *Canem homo mordet* means the same thing as *Homo canem mordet.* Thanks to case markers, verbs' dictionary entries can be relieved of the duty of keeping track of where their role-players actually appear in the sentence. A verb need only indicate that, say, the doer is a subject; whether the subject is in first or third or fourth position in the sentence is up to the rest of the grammar, and the interpretation is the same. Indeed, in what are called "scrambling" languages, case markers are exploited even further: the article, adjective, and noun inside a phrase are each tagged with a particular case marker, and the speaker can scramble the words of the phrase all over the sentence (say, put the adjective at the end for emphasis), knowing that the listener can mentally join them back up. This process, called agreement or concord, is a second engineering solution (aside from phrase structure itself) to the problem of encoding a tangle of interconnected thoughts into strings of words that appear one after the other.

Centuries ago, English, like Latin, had suffixes that marked case overtly. But the suffixes have all eroded, and overt case survives only in the personal pronouns—*I, he, she, we, they* are used for the subject role; *my, his, her, our, their* are used for the possessor role; *me, him, her, us, them* are used for all other roles. (The *who/whom* distinction could be added to this list, but it is on the way out; in the United States, *whom* is used consistently only by careful writers and pretentious speakers.) Interestingly, since we all know to say *He saw us* but never *Him saw we,* the syntax of case must still be alive and well in English. Though nouns appear physically unchanged no matter what role they play, they are tagged with silent cases. Alice realized this after spotting a mouse swimming nearby in her pool of tears:

> "Would it be of any use, now," thought Alice, "to speak to this mouse? Everything is so out-of-the-way down here, that I should think very likely it can talk: at any rate, there's no harm in trying." So she began. "O Mouse, do you know the way out of this pool? I am very tired of swimming about here, O Mouse!" (Alice thought this must be the right way of speaking to a mouse: she had never done such a thing before, but she remembered having seen, in her brother's Latin Grammar, "A Mouse—of a mouse—to a mouse—a mouse—O mouse!")

English speakers tag a noun phrase with a case by seeing what the noun is adjacent to, generally a verb or preposition (but for Alice's mouse, the archaic "vocative" case marker *O*). They use these case tags to match up each noun phrase with its verb-decreed role.

The requirement that noun phrases must get case tags explains why certain sentences are impossible even though the super-rules admit them. For example, a direct object role-player has to come right after the verb, before any other role-player: one says *Tell Mary that John is coming,* not *Tell that John is coming Mary.* The reason is that the NP *Mary* cannot just float around tagless but must be case-marked, by sitting adjacent to the verb. Curiously, while verbs and prepositions can mark case on their adjacent NP's, nouns and adjectives cannot: *governor California* and *afraid the wolf,* though interpretable, are ungrammatical. English demands that the meaningless preposition *of* precede the noun, as in *governor of California* and *afraid of the wolf,* for no reason other than to give it a case tag. The sentences we utter are kept under tight rein by verbs and prepositions—phrases cannot just show up anywhere they feel like in the VP but must have a job description and be wearing an identity badge at all times. Thus we cannot say things like *Last night I slept bad dreams a hangover snoring no pajamas sheets were wrinkled,* even though a listener could guess what that would mean. This marks a major difference between human languages and, for example, pidgins and the signing of chimpanzees, where any word can pretty much go anywhere.

Now, what about the most important phrase of all, the sentence? If a noun phrase is a phrase built around a noun, and a verb phrase is a phrase built around a verb, what is a sentence built around?

The critic Mary McCarthy once said of her rival Lillian Hellman, "Every word she writes is a lie, including 'and' and 'the.' " The insult relies on the fact that a sentence is the smallest thing that can be either true or false; a single word cannot be either (so McCarthy is alleging that Hellman's lying extends deeper than one would have thought possible). A sentence, then, must express some kind of meaning that does not clearly reside in its nouns and verbs but that embraces the entire combination and turns it into a proposition that can be true or false. Take, for example, the optimistic sentence *The Red Sox will win the World Series.* The word *will* does not apply to the Red Sox alone, nor to the World Series alone, nor to winning alone; it applies to an entire concept, the-Red-Sox-winning-the-World-Series. That concept is timeless and therefore truthless. It can refer equally well to some past glory, a hypothetical future one, even to the mere logical possibility, bereft of any hope that it will ever happen. But the word *will* pins the concept down to temporal coordinates, namely the stretch of time subsequent to the moment the sentence is uttered. If I declare "The Red Sox will win the World Series," I can be right or wrong (probably wrong, alas).

The word *will* is an example of an auxiliary, a word that expresses layers of meaning having to do with the truth of a proposition as the speaker conceives it. These layers also include negation (as in *won't* and *doesn't*), necessity (*must*), and possibility (*might* and *can*). Auxiliaries typically occur at the periphery of sentence trees, mirroring the fact that they assert something about the rest of the sentence taken as a whole. The auxiliary is the head of the sentence in exactly the same way that a noun is the head of the noun phrase. Since the auxiliary is also called INFL (for "inflection"), we can call the sentence an IP (an INFL phrase or auxiliary phrase). Its subject position is reserved for the subject of the entire sentence, reflecting the fact that a sentence is an assertion that some predicate (the VP) is true of its subject. Here, more or less, is what a sentence looks like in the current version of Chomsky's theory:

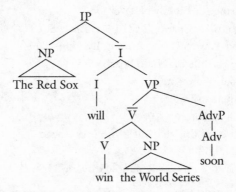

An auxiliary is an example of a "function word," a different kind of word from nouns, verbs, and adjectives, the "content" words. Function words include articles (*the, a, some*), pronouns (*he, she*), the possessive marker *'s*, meaningless prepositions like *of*, words that introduce complements like *that* and *to*, and conjunctions like *and* and *or*. Function words are bits of crystallized grammar; they delineate larger phrases into which NP's and VP's and AP's fit, thereby providing a scaffolding for the sentence. Accordingly, the mind treats function words differently from content words. People add new content words to the language all the time (like the noun *fax*, and the verb *to snarf*, meaning to retrieve a computer file), but the function words form a closed club that resists new members. That is why all the attempts to introduce gender-neutral pronouns like *hesh* and *thon* have failed. Recall, too, that patients with damage to the language areas of the brain have more trouble with function words like *or* and *be* than with content words like *oar* and *bee*. When words are expensive, as in telegrams and headlines, writers tend to leave the function words out, hoping that the reader can reconstruct them from the order of the content words. But because function words are the most reliable clues to the phrase structure of the sentence, telegraphic language is always a gamble. A reporter once sent Cary Grant the telegram, "How old Cary Grant?" He replied, "Old Cary Grant fine." Here are some headlines from a collection called *Squad Helps Dog Bite Victim,* put together by the staff of the *Columbia Journalism Review*:

New Housing for Elderly Not Yet Dead
New Missouri U. Chancellor Expects Little Sex
12 on Their Way to Cruise Among Dead in Plane Crash
N.J. Judge to Rule on Nude Beach
Chou Remains Cremated
Chinese Apeman Dated
Hershey Bars Protest
Reagan Wins on Budget, But More Lies Ahead
Deer Kill 130,000
Complaints About NBA Referees Growing Ugly

Function words also capture much of what makes one language grammatically different from another. Though all languages have function words, the properties of the words differ in ways that can have large effects on the structure of the sentences in the language. We have already seen one example: overt case and agreement markers in Latin allow noun phrases to be scrambled; silent ones in English force them to remain in place. Function words capture the grammatical look and feel of a language, as in these passages that use a language's function words but none of its content words:

DER JAMMERWOCH

Es brillig war. Die schlichte Toven
Wirrten und wimmelten in Waben.

LE JASEROQUE

Il brilgue: les tôves lubricilleux
Se gyrent en vrillant dans la guave.

The effect can also be seen in passages that take the function words from one language but the content words from another, like the following pseudo-German notice that used to be posted in many university computing centers in the English-speaking world:

ACHTUNG! ALLES LOOKENSPEEPERS!

Das computermachine ist nicht fuer gefingerpoken und mittengrabben. Ist easy schnappen der springenwerk, blowenfusen und poppencorken mit spitzensparken. Ist nicht fuer gewerken bei das dumpkopfen. Das rubbernecken sightseeren keepen das cottenpickenen hans in das pockets muss; relaxen und watchen das blinkenlichten.

Turnabout being fair play, computer operators in Germany have posted a translation into pseudo-English:

ATTENTION

This room is fulfilled mit special electronische equippment. Fingergrabbing and pressing the cnoeppkes from the computers is allowed for die experts only! So all the "lefthanders" stay away and do not disturben the brainstorming von here working intelligencies. Otherwise you will be out thrown and kicked andeswhere! Also: please keep still and only watchen astaunished the blinkenlights.

Anyone who goes to cocktail parties knows that one of Chomsky's main contributions to intellectual life is the concept of "deep structure," together with the "transformations" that map it onto "surface structure." When Chomsky introduced the terms in the behaviorist climate of the early 1960s, the reaction was sensational. Deep structure came to refer to everything that was hidden, profound, universal, or meaningful, and before long there was talk of the deep structure of visual perception, stories, myths, poems, paintings, musical compositions, and so on. Anticlimactically, I must now divulge that "deep structure" is a prosaic technical gadget in grammatical theory. It is not the meaning of a sentence, nor is it what is universal across all human languages. Though universal grammar and abstract phrase structures seem to be permanent features of grammatical theory, many linguists—including, in his most recent writings, Chomsky himself—think one can do without deep structure per se. To discourage all the hype incited by the word "deep," linguists now usually refer to it as "d-structure." The concept is actually quite simple.

Recall that for a sentence to be well formed, the verb must get what it wants: all the roles listed in the verb's dictionary entry must appear in their designated positions. But in many sentences, the verb does not seem to be getting what it wants. Remember that *put* requires a subject, an object, and a prepositional phrase; *He put the car* and *He put in the garage* sound incomplete. How, then, do we account for the following perfectly good sentences?

> The car was put in the garage.
> What did he put in the garage?
> Where did he put the car?

In the first sentence, *put* seems to be doing fine without an object, which is out of character. Indeed, now it rejects one: *The car was put the Toyota in the garage* is awful. In the second sentence, *put* also appears in public objectless. In the third, its obligatory prepositional phrase is missing. Does this mean we need to add new dictionary entries for *put,* allowing it to appear in some places without its object or its prepositional phrase? Obviously not, or *He put the car* and *He put in the garage* would slip back in.

In some sense, of course, the required phrases really are there—they're just not where we expect them. In the first sentence, a passive construction, the NP *the car,* playing the role of "thing put" which ordinarily would be the object, shows up in the subject position instead. In the second sentence, a *wh*-question (that is, a question formed with *who, what, where, when,* or *why*), the "thing put" role is expressed by the word *what* and shows up at the beginning. In the third sentence, the "place" role also shows up at the beginning instead of after the object, where it ordinarily belongs.

A simple way to account for the entire pattern is to say that every sentence has two phrase structures. The phrase structure we have been talking about so far, the one defined by the super-rules, is the deep structure. Deep structure is the interface between the mental dictionary and phrase structure. In the deep structure, all the role-players for *put* appear in their expected places. Then a transformational operation can "move" a phrase to a previously unfilled slot elsewhere in the tree. That is where we find the phrase in the actual sentence. This new tree is the surface structure (now called "s-structure," because as a mere "surface" representation it never used to get proper respect). Here are the deep structure and surface structure of a passive sentence:

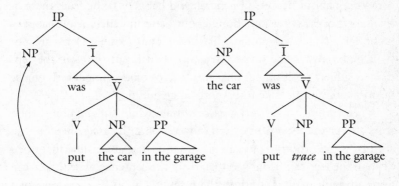

In the deep structure on the left, *the car* is where the verb wanted it; in the surface structure on the right, it is where we actually hear it. In the surface structure, the position from which the phrase was moved contains an inaudible symbol that was left behind by the movement transformation, called a "trace." The trace serves as a reminder of the role that the moved phrase is playing. It tells us that to find out what *the car* is doing in the putting event, we should look up the "object" slot in the entry for the verb *put*; that slot says "thing put." Thanks to the trace, the surface structure contains the information needed to recover the meaning of the sentence; the original deep structure, which was used only to plug in the right sets of words from the lexicon, plays no role.

Why do languages bother with separate deep structures and surface structures? Because it takes more than just keeping the verb happy—what deep structure does—to have a usable sentence. A given concept often has to play one kind of role, defined by the verb in the verb phrase, and simultaneously a separate role, independent of the verb, defined by some other layer of the tree. Consider the difference between *Beavers build dams* and its passive, *Dams are built by beavers*. Down in the verb phrase—the level of who did what to whom—the nouns are playing the same roles in both sentences. Beavers do the building, dams get built. But up at the sentence (IP) level—the level of subject-predicate relations, of what is being asserted to be true of what—they are playing different roles. The active sentence is saying something about beavers in general, and happens to be true; the passive sentence is saying something about dams in general, and happens to be false (since some dams, like the Grand Coulee Dam, are not built by beavers). The surface structure, which puts *dams* in the sentence's subject position but links it to a trace of its original verb phrase position, allows the cake to be both eaten and had.

The ability to move phrases around while still retaining their roles also gives the speaker of a rigid-word-order language like English a bit of wiggle room. For example, phrases that are ordinarily buried deep in the tree can be moved to early in the sentence, where they can hook up with material fresh in the listener's mind. For example, if a play-by-play announcer has been describing Nevin Markwart's progression down the ice, he could say *Markwart spears Gretzky!!!* But if it was Wayne Gretzky the announcer had been describing, he would say *Gretzky is speared by Markwart!!!!* Moreover, because a passive participle has the option of leaving the doer role, ordinarily the subject, unfilled in deep structure, it is useful when one wants to avoid mentioning that role altogether, as in Ronald Reagan's evasive concession *Mistakes were made*.

Hooking up players with different roles in different scenarios is something that grammar excels at. In a *wh*-question like

What did he put [*trace*] in the garage?

the noun phrase *what* gets to live a double life. Down in the who-did-what-to-whom realm of the verb phrase, the position of the trace indicates that the entity has the role of the thing being put; up in the what-is-being-asserted-of-what realm of the sentence, the word *what* indicates that the point of the sentence is to ask the listener to provide the identity of something. If a logician were to express the meaning behind the sentence, it would be something like "For which *x*, John put *x* in the garage." When these movement operations are combined with other components of syntax, as in *She was told by Bob to be examined by a doctor* or *Who did he say that Barry tried to convince to leave?* or *Tex is fun for anyone to tease,* the components interact to determine the meaning of the sentence in chains of deduction as intricate and precise as the workings of a fine Swiss watch.

Now that I have dissected syntax in front of you, I hope your reaction is more favorable than Eliza Doolittle's or Jack Cade's. At the very least I hope you are impressed at how syntax is a Darwinian "organ of extreme perfection and complication." Syntax is complex, but the complexity is there for a reason. For our thoughts are surely even more complex, and we are limited by a mouth that can pronounce a single word at a time. Science has begun to crack the beautifully designed code that our brains use to convey complex thoughts as words and their orderings.

The workings of syntax are important for another reason. Grammar offers a clear refutation of the empiricist doctrine that there is nothing in the mind that was not first in the senses. Traces, cases, X-bars, and the other paraphernalia of syntax are colorless, odorless, and tasteless, but they, or something like them, must be a part of our unconscious mental life. This should not be surprising to a thoughtful computer scientist. There is no way one can write a halfway intelligent program without defining variables and data structures that do not directly correspond to anything in the input or output. For example, a graphics program that had to store an image of a triangle inside a circle would not store the actual keystrokes that the user typed to draw the shapes, because the same shapes could have been drawn in a different order or with a different device like a mouse or a light pen. Nor would it store the list of dots that have to be lit up to display the shapes on a video screen, because the user might later want to move the circle around and leave the triangle in place, or make the circle bigger or smaller, and one long list of dots would not allow the program to know which dots belong to the circle and which to the triangle. Instead, the shapes would be stored in some more abstract format (like the coordinates of a few defining points for each shape), a format that mirrors neither the inputs nor the outputs to the program but that can be translated to and from them when the need arises.

Grammar, a form of mental software, must have evolved under similar design specifications. Though psychologists under the influence of empiricism often suggest that grammar mirrors commands to the speech muscles, melodies in speech sounds, or mental scripts for the ways that people and things tend to interact, I think all these suggestions miss the mark. Grammar is a protocol that has to interconnect the ear, the mouth, and the mind, three very different kinds of machine. It cannot be tailored to any of them but must have an abstract logic of its own.

The idea that the human mind is designed to use abstract variables and data structures used to be, and in some circles still is, a shocking and revolutionary claim, because the structures have no direct counterpart in the child's experience. Some of the organization of grammar would have to be there from the start, part of the language-learning mechanism that allows children to make sense out of the noises they hear from their parents. The details of syntax have figured prominently in the history of psychology, because they are a case where complexity in the mind is not caused by learning; learning is caused by complexity in the mind. And that was real news.

# 5

✥

# Words, Words, Words

*The word* glamour *comes from the word* grammar, *and since the* Chomskyan revolution the etymology has been fitting. Who could not be dazzled by the creative power of the mental grammar, by its ability to convey an infinite number of thoughts with a finite set of rules? There has been a book on mind and matter called *Grammatical Man*, and a Nobel Prize lecture comparing the machinery of life to a generative grammar. Chomsky has been interviewed in *Rolling Stone* and alluded to on *Saturday Night Live*. In Woody Allen's story "The Whore of Mensa," the patron asks, "Suppose I wanted Noam Chomsky explained to me by two girls?" "It'd cost you," she replies.

Unlike the mental grammar, the mental dictionary has had no cachet. It seems like nothing more than a humdrum list of words, each transcribed into the head by dull-witted rote memorization. In the preface to his *Dictionary*, Samuel Johnson wrote:

> It is the fate of those who dwell at the lower employments of life, to be rather driven by the fear of evil, than attracted by the prospect of good; to be exposed to censure, without hope of praise; to be disgraced by miscarriage, or punished for neglect, where success would have been without applause, and diligence without reward.

Among these unhappy mortals is the writer of dictionaries.

Johnson's own dictionary defines *lexicographer* as "a harmless drudge,
that busies himself in tracing the original, and detailing the significa-
tion of words."

In this chapter we will see that the stereotype is unfair. The world
of words is just as wondrous as the world of syntax, or even more so.
For not only are people as infinitely creative with words as they are
with phrases and sentences, but memorizing individual words
demands its own special virtuosity.

Recall the *wug*-test, passed by any preschooler: "Here is a wug.
Now there are two of them. There are two___." Before being so challenged, the child has neither heard anyone say, nor been rewarded for
saying, the word *wugs*. Therefore words are not simply retrieved from
a mental archive. People must have a mental rule for generating new
words from old ones, something like "To form the plural of a noun,
add the suffix -*s*." The engineering trick behind human language—its
being a discrete combinatorial system—is used in at least two different
places: sentences and phrases are built out of words by the rules of
syntax, and the words themselves are built out of smaller bits by
another set of rules, the rules of "morphology."

The creative powers of English morphology are pathetic compared to what we find in other languages. The English noun comes in
exactly two forms (*duck* and *ducks*), the verb in four (*quack, quacks,
quacked, quacking*). In modern Italian and Spanish every verb has
about fifty forms; in classical Greek, three hundred and fifty; in Turkish, two million! Many of the languages I have brought up, such as
Eskimo, Apache, Hopi, Kivunjo, and American Sign Language, are
known for this prodigious ability. How do they do it? Here is an
example from Kivunjo, the Bantu language that was said to make
English look like checkers compared to chess. The verb "Näïkì-
ḿlyïïà," meaning "He is eating it for her," is composed of eight parts:

- N-: A marker indicating that the word is the "focus" of that
  point in the conversation.

- -ä-: A subject agreement marker. It identifies the eater as falling into Class 1 of the sixteen gender classes, "human singular." (Remember that to a linguist "gender" means kind, not sex.) Other genders embrace nouns that pertain to several humans, thin or extended objects, objects that come in pairs or clusters, the pairs or clusters themselves, instruments, animals, body parts, diminutives (small or cute versions of things), abstract qualities, precise locations, and general locales.

- -ï-: Present tense. Other tenses in Bantu can refer to today, earlier today, yesterday, no earlier than yesterday, yesterday or earlier, in the remote past, habitually, ongoing, consecutively, hypothetically, in the future, at an indeterminate time, not yet, and sometimes.

- -kì-: An object agreement marker, in this case indicating that the thing eaten falls into gender Class 7.

- -ḿ-: A benefactive marker, indicating for whose benefit the action is taking place, in this case a member of gender Class 1.

- -lyì-: The verb, "to eat."

- -ï-: An "applicative" marker, indicating that the verb's cast of players has been augmented by one additional role, in this case the benefactive. (As an analogy, imagine that in English we had to add a suffix to the verb *bake* when it is used in *I baked her a cake* as opposed to the usual *I baked a cake*.)

- -à: A final vowel, which can indicate indicative versus subjunctive mood.

If you multiply out the number of possible combinations of the seven prefixes and suffixes, the product is about half a million, and that is the number of possible forms per verb in the language. In effect, Kivunjo and languages like it are building an entire sentence inside a single complex word, the verb.

But I have been a bit unfair to English. English is genuinely crude in its "inflectional" morphology, where one modifies a word to fit the sentence, like marking a noun for the plural with -s or a verb for past tense with -ed. But English holds its own in "derivational" morphology, where one creates a new word out of an old one. For example, the suffix -able, as in *learnable, teachable,* and *huggable,* converts a verb meaning "to do X" into an adjective meaning "capable of having X done to it." Most people are surprised to learn how many derivational suffixes there are in English. Here are the more common ones:

| | | | |
|---|---|---|---|
| -able | -ate | -ify | -ize |
| -age | -ed | -ion | -ly |
| -al | -en | -ish | -ment |
| -an | -er | -ism | -ness |
| -ant | -ful | -ist | -ory |
| -ance | -hood | -ity | -ous |
| -ary | -ic | -ive | -y |

In addition, English is free and easy with "compounding," which glues two words together to form a new one, like *toothbrush* and *mouse-eater.* Thanks to these processes, the number of possible words, even in morphologically impoverished English, is immense. The computational linguist Richard Sproat compiled all the distinct words used in the forty-four million words of text from Associated Press news stories beginning in mid-February 1988. Up through December 30, the list contained three hundred thousand distinct word forms, about as many as in a good unabridged dictionary. You might guess that this would exhaust the English words that would ever appear in such stories. But when Sproat looked at what came over the wire on December 31, he found no fewer than thirty-five new forms, including *instrumenting, counterprograms, armhole, part-Vulcan, fuzzier, groveled, boulderlike, mega-lizard, traumatological,* and *ex-critters.*

Even more impressive, the output of one morphological rule can be the input to another, or to itself: one can talk about the *unmicrowaveability* of some French fries or a *toothbrush-holder fastener box* in which to keep one's toothbrush-holder fasteners. This makes the number of possible words in a language bigger than immense; like the number of sentences, it is infinite. Putting aside fanciful coinages concocted for immortality in *Guinness,* a candidate for the longest word to date in English might be *floccinaucinihilipilification,* defined in the *Oxford English Dictionary* as "the categorizing of something as worthless or trivial." But that is a record meant to be broken:

> *floccinaucinihilipilificational:* pertaining to the categorizing of something as worthless or trivial
> *floccinaucinihilipilificationalize:* to cause something to pertain to the categorizing of something as worthless or trivial
> *floccinaucinihilipilificationalization:* the act of causing something to pertain to the categorizing of something as worthless or trivial
> *floccinaucinihilipilificationalizational:* pertaining to the act of causing something to pertain to the categorizing of something as worthless or trivial
> *floccinaucinihilipilificationalizationalize:* to cause something to pertain to the act of causing something to pertain . . .

Or, if you suffer from sesquipedaliaphobia, you can think of your *great-grandmother,* your *great-great-grandmother,* your *great-great-great-grandmother,* and so on, limited only in practice by the number of generations since Eve.

What's more, words, like sentences, are too delicately layered to be generated by a chaining device (a system that selects an item from one list, then moves on to some other list, then to another). When Ronald Reagan proposed the Strategic Defense Initiative, popularly known as Star Wars, he imagined a future in which an incoming Soviet missile would be shot down by an *anti-missile missile.* But critics pointed out that the Soviet Union could counterattack with an *anti-anti-missile-missile missile.* No problem, said his MIT-educated engineers; we'll just build an *anti-anti-anti-missile-missile-missile missile.* These high-tech weapons need a high-tech grammar—something that can keep track of all the *anti*'s at the beginning of the word so that it can complete the word with an equal number of *missile*'s, plus one, at the end. A word structure grammar (a phrase structure grammar for words) that can embed a word in between an *anti-* and its *missile* can achieve these objectives; a chaining device cannot, because it has forgotten the pieces that it laid down at the beginning of the long word by the time it gets to the end.

Like syntax, morphology is a cleverly designed system, and many of the seeming oddities of words are predictable products of its internal logic. Words have a delicate anatomy consisting of pieces, called morphemes, that fit together in certain ways. The word structure system is an extension of the X-bar phase structure system, in which big nounish things are built out of smaller nounish things, smaller nounish things are built out of still smaller nounish things, and so on. The biggest phrase involving nouns is the noun phrase; a noun phrase contains an N-bar; an N-bar contains a noun—the word. Jumping from syntax to morphology, we simply continue the dissection, analyzing the word into smaller and smaller nounish pieces.

Here is a picture of the structure of the word *dogs:*

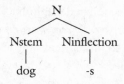

The top of this mini-tree is "N" for "noun"; this allows the docking maneuver in which the whole word can be plugged into the noun slot inside any noun phrase. Down inside the word, we have two parts: the bare word form *dog*, usually called the stem, and the plural inflection *-s*. The rule responsible for inflected words (the rule of *wug*-test fame) is simply

N → Nstem Ninflection
"A noun can consist of a noun stem followed by a noun inflection."

The rule nicely interfaces with the mental dictionary: *dog* would be listed as a noun stem meaning "dog," and *-s* would be listed as a noun inflection meaning "plural of."

This rule is the simplest, most stripped-down example of anything we would want to call a rule of grammar. In my laboratory we use it as an easily studied instance of mental grammar, allowing us to document in great detail the psychology of linguistic rules from infancy to old age in both normal and neurologically impaired people, in much the same way that biologists focus on the fruit fly *Drosophila* to study the machinery of genes. Though simple, the rule that glues an inflection to a stem is a surprisingly powerful computational operation. That is because it recognizes an abstract mental symbol, like "noun stem," instead of being associated with a particular list of words or a particular list of sounds or a particular list of meanings. We can use the rule to inflect any item in the mental dictionary that lists "noun stem" in its entry, without caring what the word means; we can convert not only *dog* to *dogs* but also *hour* to *hours* and *justification* to *justifications*. Likewise, the rule allows us to form plurals without caring what the word sounds like; we pluralize unusual-sounding words as in *the Gorbachevs, the Bachs,* and *the Mao Zedongs*. For the same reason, the rule is perfectly happy applying to brand-new nouns, like *faxes, dweebs, wugs,* and *zots*.

We apply the rule so effortlessly that perhaps the only way I can drum up some admiration for what it accomplishes is to compare humans with a certain kind of computer program that many computer scientists tout as the wave of the future. These programs, called "artificial neural networks," do not apply a rule like the one I have just shown you. An artificial neural network works by analogy, converting *wug* to *wugged* because it is vaguely similar to *hug–hugged, walk–walked,* and thousands of other verbs the network has been trained to recognize. But when the network is faced with a new verb that is unlike anything it has previously been trained on, it often mangles it, because the network does not have an abstract, all-embracing category "verb stem" to fall back on and add an affix to. Here are some comparisons between what people typically do and what artificial neural networks typically do when given a *wug*-test:

| VERB | TYPICAL PAST-TENSE FORM GIVEN BY PEOPLE | TYPICAL PAST-TENSE FORM GIVEN BY NEURAL NETWORKS |
|---|---|---|
| mail | mailed | membled |
| conflict | conflicted | conflafted |
| wink | winked | wok |
| quiver | quivered | quess |
| satisfy | satisfied | sedderded |
| smairf | smairfed | sprurice |
| trilb | trilbed | treelilt |
| smeej | smeejed | leefloag |
| frilg | frilged | freezled |

Stems can be built out of parts, too, in a second, deeper level of word assembly. In compounds like *Yugoslavia report, sushi-lover, broccoli-green,* and *toothbrush,*

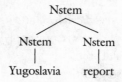

two stems are joined together to form a new stem, by the rule

Nstem → Nstem Nstem
"A noun stem can consist of a noun stem followed by another
    noun stem."

In English, a compound is often spelled with a hyphen or by running its two words together, but it can also be spelled with a space between the two components as if they were still separate words. This confused your grammar teacher into telling you that in *Yugoslavia report,* "Yugoslavia" is an adjective. To see that this can't be right, just try comparing it with a real adjective like *interesting.* You can say *This report seems interesting* but not *This report seems Yugoslavia!* There is a simple way to tell whether something is a compound word or a phrase: compounds generally have stress on the first word, phrases on the second. A *dark róom* (phrase) is any room that is dark, but a *dárk room* (compound word) is where photographers work, and a dark-room can be lit when the photographer is done. A *black bóard* (phrase) is necessarily a board that is black, but some *bláckboards* (compound word) are green or even white. Without pronunciation or punctuation as a guide, some word strings can be read either as a phrase or as a compound, like the following headlines:

Squad Helps Dog Bite Victim
Man Eating Piranha Mistakenly Sold as Pet Fish
Juvenile Court to Try Shooting Defendant

New stems can also be formed out of old ones by adding affixes (prefixes and suffixes), like the *-al, -ize,* and *-ation* I used recursively to get longer and longer words ad infinitum (as in *sensationalizatio-nalization*). For example, *-able* combines with any verb to create an adjective, as in *crunch–crunchable.* The suffix *-er* converts any verb to a noun, as in *crunch–cruncher,* and the suffix *-ness* converts any adjective into a noun, as in *crunchy–crunchiness.*

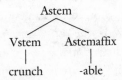

The rule forming them is

Astem → Stem Astemaffix
"An adjective stem can consist of a stem joined to a suffix."

and a suffix like *-able* would have a mental dictionary entry like the following:

*-able:*
    adjective stem affix
    means "capable of being X'd"
    attach me to a verb stem

Like inflections, stem affixes are promiscuous, mating with any stem that has the right category label, and so we have *crunchable, scrunchable, shmooshable, wuggable,* and so on. Their meanings are predictable: capable of being crunched, capable of being scrunched, capable of being shmooshed, even capable of being "wugged," whatever *wug* means. (Though I can think of an exception: in the sentence *I asked him what he thought of my review in his book, and his response was unprintable,* the word *unprintable* means something much more specific than "incapable of being printed.")

The scheme for computing the meaning of a stem out of the meaning of its parts is similar to the one used in syntax: one special element is the "head," and it determines what the conglomeration refers to. Just as the phrase *the cat in the hat* is a kind of cat, showing that *cat* is its head, a *Yugoslavia report* is a kind of report, and *shmooshability* is a kind of ability, so *report* and *-ability* must be the heads of those words. The head of an English word is simply its rightmost morpheme.

Continuing the dissection we can tease stems into even smaller parts. The smallest part of a word, the part that cannot be cut up into any smaller parts, is called its root. Roots can combine with special suffixes to form stems. For example, the root *Darwin* can be found inside the stem *Darwinian*. The stem *Darwinian* in turn can be fed into the suffixing rule to yield the new stem *Darwinianism*. From there, the inflectional rule could even give us the word *Darwinianisms,* embodying all three levels of word structure:

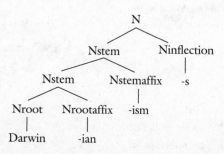

Interestingly, the pieces fit together in only certain ways. Thus *Darwinism*, a stem formed by the stem suffix *-ism*, cannot be a host for *-ian*, because *-ian* attaches only to roots; hence *Darwinismian* (which would mean "pertaining to Darwinism") sounds ridiculous. Similarly, *Darwinsian* ("pertaining to the two famous Darwins, Charles and Erasmus"), *Darwinsianism*, and *Darwinsism* are quite impossible, because whole inflected words cannot have any root or stem suffixes joined to them.

Down at the bottommost level of roots and root affixes, we have entered a strange world. Take *electricity*. It seems to contain two parts, *electric* and *-ity:*

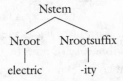

But are these words really assembled by a rule, gluing a dictionary entry for *-ity* onto the root *electric*, like this?

Nstem → Nroot Nrootsuffix
"A noun stem can be composed of a noun root and a suffix."

*-ity:*
    noun root suffix
    means "the state of being X"
    attach me to a noun root

Not this time. First, you can't get *electricity* simply by gluing together the word *electric* and the suffix *-ity*—that would sound like "electrick itty." The root that *-ity* is attached to has changed its pronunciation to "electriss." That residue, left behind when the suffix has been removed, is a root that cannot be pronounced in isolation.

Second, root-affix combinations have unpredictable meanings; the neat scheme for interpreting the meaning of the whole from the meaning of the parts breaks down. *Complexity* is the state of being complex, but *electricity* is not the state of being electric (you would never say that the electricity of this new can opener makes it convenient); it is the force powering something electric. Similarly, *instrumental* has nothing to do with instruments, *intoxicate* is not about toxic substances, one does not recite at a *recital,* and a five-speed *transmission* is not an act of transmitting.

Third, the supposed rule and affix do not apply to words freely, unlike the other rules and affixes we have looked at. For example, something can be *academic* or *acrobatic* or *aerodynamic* or *alcoholic,* but *academicity, acrobaticity, aerodynamicity,* and *alcoholicity* sound horrible (to pick just the first four words ending in *-ic* in my electronic dictionary).

So at the third and most microscopic level of word structure, roots and their affixes, we do not find bona fide rules that build words according to predictable formulas, *wug*-style. The stems seem to be stored in the mental dictionary with their own idiosyncratic meanings attached. Many of these complex stems originally were formed after the Renaissance, when scholars imported many words and suffixes into English from Latin and French, using some of the rules appropriate to those languages of learning. We have inherited the words, but not the rules. The reason to think that modern English speakers mentally analyze these words as trees at all, rather than as homogeneous strings of sound, is that we all sense that there is a natural break point between the *electric* and the *-ity.* We also recognize that there is an affinity between the word *electric* and the word *electricity,* and we recognize that any other word containing *-ity* must be a noun.

Our ability to appreciate a pattern inside a word, while knowing that the pattern is not the product of some potent rule, is the inspiration for a whole genre of wordplay. Self-conscious writers and speakers often extend Latinate root suffixes to new forms by analogy, such as *religiosity, criticality, systematicity, randomicity, insipidify, calumniate, conciliate, stereotypy, disaffiliate, gallonage,* and *Shavian.* The words have an air of heaviosity and seriosity about them, making the style an easy target for parody. A 1982 editorial cartoon by Jeff Mac-Nelly put the following resignation speech into the mouth of Alexander Haig, the malaprop-prone Secretary of State:

I decisioned the necessifaction of the resignatory action/ option due to the dangerosity of the trendflowing of foreign policy away from our originatious careful coursing towards consistensivity, purposity, steadfastnitude, and above all, clarity.

Another cartoon, by Tom Toles, showed a bearded academician explaining the reason verbal Scholastic Aptitude Test scores were at an all-time low:

Incomplete implementation of strategized programmatics designated to maximize acquisition of awareness and utilization of communications skills pursuant to standardized review and assessment of languaginal development.

In the culture of computer programmers and managers, this analogy-making is used for playful precision, not pomposity. *The New Hacker's Dictionary*, a compilation of hackish jargon, is a near-exhaustive catalogue of the not-quite-freely-extendible root affixes in English:

*ambimoustrous* adj. Capable of operating a mouse with either hand.

*barfulous* adj. Something that would make anyone barf.

*bogosity* n. The degree to which something is bogus.

*bogotify* v. To render something bogus.

*bozotic* adj. Having the quality of Bozo the Clown.

*cuspy* adj. Functionally elegant.

*depeditate* v. To cut the feet off of (e.g., while printing the bottom of a page).

*dimwittery* n. Example of a dim-witted statement.

*geekdom* n. State of being a techno-nerd.

*marketroid* n. Member of a company's marketing department.

*mumblage* n. The topic of one's mumbling.

*pessimal* adj. Opposite of "optimal."

*wedgitude* n. The state of being wedged (stuck; incapable of proceeding without help).

*wizardly* adj. Pertaining to expert programmers.

Down at the level of word roots, we also find messy patterns in irregular plurals like *mouse–mice* and *man–men* and in irregular past-tense forms like *drink–drank* and *seek–sought*. Irregular forms tend to come in families, like *drink–drank, sink–sank, shrink–shrank, stink–stank, sing–sang, ring–rang, spring–sprang, swim–swam,* and *sit–sat,* or *blow–blew, know–knew, grow–grew, throw–threw, fly–flew,* and *slay–slew*. This is because thousands of years ago Proto-Indo-European, the language ancestral to English and most other European languages, had rules that replaced one vowel with another to form the past tense, just as we now have a rule that adds *-ed*. The irregular or "strong" verbs in modern English are mere fossils of these rules; the rules themselves are dead and gone. Most verbs that would seem eligible to belong to the irregular families are arbitrarily excluded, as we see in the following doggerel:

> Sally Salter, she was a young teacher who taught,
> And her friend, Charley Church, was a preacher who praught;
> Though his enemies called him a screecher, who scraught.
>
> His heart, when he saw her, kept sinking, and sunk;
> And his eye, meeting hers, began winking, and wunk;
> While she in her turn, fell to thinking, and thunk.
>
> In secret he wanted to speak, and he spoke,
> To seek with his lips what his heart long had soke,
> So he managed to let the truth leak, and it loke.
>
> The kiss he was dying to steal, then he stole;
> At the feet where he wanted to kneel, then he knole;
> And he said, "I feel better than ever I fole."

People must simply be memorizing each past-tense form separately. But as this poem shows, they can be sensitive to the patterns among them and can even extend the patterns to new words for humorous effect, as in Haigspeak and hackspeak. Many of us have been tempted by the cuteness of *sneeze–snoze, squeeze–squoze, take–took–tooken,* and *shit-shat,* which are based on analogies with *freeze–froze, break–broke–broken,* and *sit–sat*. In *Crazy English* Richard

Lederer wrote an essay called "Foxen in the Henhice," featuring irregular plurals gone mad: *booth–beeth, harmonica–harmonicae, mother–methren, drum–dra, Kleenex–Kleenices,* and *bathtub–bath-tubim.* Hackers speak of *faxen, VAXen, boxen, meece,* and *Macinteesh. Newsweek* magazine once referred to the white-caped, rhinestone-studded Las Vegas entertainers as *Elvii.* In the *Peanuts* comic strip, Linus's teacher Miss Othmar once had the class glue eggshells into model *igli.* Maggie Sullivan wrote an article in the *New York Times* calling for "strengthening" the English language by conjugating more verbs as if they were strong:

> *Subdue, subdid, subdone:* Nothing could have subdone him the way her violet eyes subdid him.
> *Seesaw, sawsaw, seensaw:* While the children sawsaw, the old man thought of long ago when he had seensaw.
> *Pay, pew, pain:* He had pain for not choosing a wife more care-fully.
> *Ensnare, ensnore, ensnorn:* In the 60's and 70's, Sominex ads ensnore many who had never been ensnorn by ads before.
> *Commemoreat, commemorate, commemoreaten:* At the banquet to commemoreat Herbert Hoover, spirits were high, and by the end of the evening many other Republicans had been commemoreaten.

In Boston there is an old joke about a woman who landed at Logan Airport and asked the taxi driver, "Can you take me someplace where I can get scrod?" He replied, "Gee, that's the first time I've heard it in the pluperfect subjunctive."

Occasionally a playful or cool-sounding form will catch on and spread through the language community, as *catch–caught* did several hundred years ago on the analogy of *teach–taught* and as *sneak–snuck* is doing today on the analogy of *stick–stuck.* (I am told that *has tooken* is the preferred form among today's mall rats.) This process can be seen clearly when we compare dialects, which retain the products of their own earlier fads. The curmudgeonly columnist H. L. Mencken was also a respectable amateur linguist, and he documented many past-tense forms found in American regional dialects, like *heat–het* (similar to *bleed–bled*), *drag–drug* (*dig–dug*), and *help–holp* (*tell–told*). Dizzy Dean, the St. Louis Cardinals pitcher and CBS announcer, was notorious for saying "He slood into second base," common in his native Arkansas. For four decades English teachers across the nation engaged in a letter-writing campaign to CBS demanding that he be removed, much to his delight. One of his replies, during the Great Depression, was "A lot of folks that ain't sayin' 'ain't' ain't eatin'." Once he baited them with the following play-by-play:

> The pitcher wound up and flang the ball at the batter. The batter swang and missed. The pitcher flang the ball again and this time the batter connected. He hit a high fly right to the center fielder. The center fielder was all set to catch the ball, but at the last minute his eyes were blound by the sun and he dropped it!

But successful adoptions of such creative extensions are rare; irregulars remain mostly as isolated oddballs.

Irregularity in grammar seems like the epitome of human eccentricity and quirkiness. Irregular forms are explicitly abolished in "rationally designed" languages like Esperanto, Orwell's Newspeak, and Planetary League Auxiliary Speech in Robert Heinlein's science fiction novel *Time for the Stars.* Perhaps in defiance of such regimentation, a woman in search of a nonconformist soulmate recently wrote this personal ad in the *New York Review of Books*:

> **Are you an irregular verb** who believes nouns have more power than adjectives? Unpretentious, professional DWF, 5 yr. European resident, sometime violinist, slim, attractive, with married children. . . . Seeking sensitive, sanguine, youthful man, mid 50's– 60's, health-conscious, intellectually

adventurous, who values truth, loyalty,
and openness.

A general statement of irregularity and the human condition comes
from the novelist Marguerite Yourcenar: "Grammar, with its mixture
of logical rule and arbitrary usage, proposes to a young mind a fore-
taste of what will be offered to him later on by law and ethics, those
sciences of human conduct, and by all the systems wherein man has
codified his instinctive experience."

For all its symbolism about the freewheeling human spirit,
though, irregularlity is tightly encapsulated in the word-building system; the system as a whole is quite cuspy. Irregular forms are roots,
which are found inside stems, which are found inside words, some of
which can be formed by regular inflection. This layering not only predicts many of the possible and impossible words of English (for example, why *Darwinianism* sounds better than *Darwinismian*); it
provides a neat explanation for many trivia questions about seemingly
illogical usage, such as: Why in baseball is a batter said to have *flied
out*—why has no mere mortal ever *flown out* to center field? Why is
the hockey team in Toronto called the *Maple Leafs* and not the *Maple
Leaves?* Why do many people say *Walkmans,* rather than *Walkmen,* as
the plural of *Walkman?* Why would it sound odd for someone to say
that all of his daughter's friends are *low-lives?*

Consult any style manual or how-to book on grammar, and it
will give one or two explanations as to why the irregular is tossed
aside—both wrong. One is that the books are closed on irregular
words in English; any new form added to the language must be regular. Not true: if I coin new words like *to re-sing* or *to out-sing,* their
pasts are *re-sang* and *out-sang,* not *re-singed* and *out-singed.* Similarly,
I recently read that there are peasants who run around with small
tanks in China's oil fields, scavenging oil from unguarded wells; the
article calls them *oil-mice,* not *oil-mouses.* The second explanation is
that when a word acquires a new, nonliteral sense, like baseball's *fly
out,* that sense requires a regular form. The oil-mice clearly falsify that
explanation, as do the many other metaphors based on irregular

nouns, which steadfastly keep their irregularity: *sawteeth* (not *saw-tooths*), *Freud's intellectual children* (not *childs*), *snowmen* (not *snow-mans*), and so on. Likewise, when the verb *to blow* developed slang meanings like *to blow him away* (assassinate) and *to blow it off* (dismiss casually), the past-tense forms remained irregular: *blew him away* and *blew off the exam*, not *blowed him away* and *blowed off the exam*.

The real rationale for *flied out* and *Walkmans* comes from the algorithm for interpreting the meanings of complex words from the meanings of the simple words they are built out of. Recall that when a big word is built out of smaller words, the big word gets all its properties from one special word sitting inside it at the extreme right: the head. The head of the verb *to overshoot* is the verb *to shoot,* so *overshooting* is a kind of *shooting,* and it is a verb, because *shoot* is a verb. Similarly, a *workman* is a singular noun, because *man,* its head, is a singular noun, and it refers to a kind of man, not a kind of work. Here is what the word structures look like:

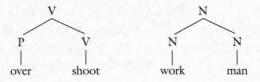

Crucially, the percolation conduit from the head to the top node applies to *all* the information stored with the head word: not just its nounhood or verbhood, and not just its meaning, but any irregular form that is stored with it, too. For example, part of the mental dictionary entry for *shoot* would say "I have my own irregular past-tense form, *shot.*" This bit of information percolates up and applies to the complex word, just like any other piece of information. The past tense of *overshoot* is thus *overshot* (not *overshooted* ). Likewise, the word *man* bears the tag "My plural is *men.*" Since *man* is the head of *workman,* the tag percolates up to the N symbol standing for *workman,* and so the plural of *workman* is *workmen.* This is also why we get *out-sang, oil-mice, sawteeth,* and *blew him away.*

Now we can answer the trivia questions. The source of quirkiness in words like *fly out* and *Walkmans* is their *headlessness*. A headless word is an exceptional item that, for one reason or another, differs in some property from its rightmost element, the one it would be based on if it were like ordinary words. A simple example of a headless word is a *low-life*—not a kind of life at all but a kind of person, namely one who leads a low life. In the word *low-life*, then, the normal percolation pipeline must be blocked. Now, a pipeline inside a word cannot be blocked for just one kind of information; if it is blocked for one thing, nothing passes through. If *low-life* does not get its meaning from *life*, it cannot get its plural from *life* either. The irregular form associated with *life*, namely *lives*, is trapped in the dictionary, with no way to bubble up to the whole word *low-life*. The all-purpose regular rule, "Add the *-s* suffix," steps in by default, and we get *low-lifes*. By similar unconscious reasoning, speakers arrive at *saber-tooths* (a kind of tiger, not a kind of tooth), *tenderfoots* (novice cub scouts, who are not a kind of foot but a kind of youngster that has tender feet), *flatfoots* (also not a kind of foot but a slang term for policemen), and *still lifes* (not a kind of life but a kind of painting).

Since the Sony Walkman was introduced, no one has been sure whether two of them should be *Walkmen* or *Walkmans*. (The nonsexist alternative *Walkperson* would leave us on the hook, because we would be faced with a choice between *Walkpersons* and *Walkpeople*.) The temptation to say *Walkmans* comes from the word's being headless: a Walkman is not a kind of man, so it must not be getting its meaning from the word *man* inside it, and by the logic of headlessness it shouldn't receive a plural form from *man*, either. But it is hard to be comfortable with any kind of plural, because the relation between *Walkman* and *man* feels utterly obscure. It feels obscure because the word was not put together by any recognizable scheme. It is an example of the pseudo-English that is popular in Japan in signs and product names. (For example, one popular soft drink is called Sweat, and T-shirts have enigmatic inscriptions like CIRCUIT BEAVER, NURSE MENTALITY, and BONERACTIVE WEAR.) The Sony Corporation has an official answer to the question of how to refer to more than one Walkman. Fearing that their trademark, if converted to a noun, may become as generic as *aspirin* or *kleenex,* they sidestep the grammatical issues by insisting upon *Walkman Personal Stereos.*

What about flying out? To the baseball cognoscenti, it is not directly based on the familiar verb *to fly* ("to proceed through the air") but on the noun *a fly* ("a ball hit on a conspicuously parabolic trajectory"). To *fly out* means "to make an out by hitting a fly that gets caught." The noun *a fly,* of course, itself came from the verb *to fly.* The word-within-a-word-within-a-word structure can be seen in this bamboo-like tree:

Since the whole word, represented by its topmost label, is a verb, but the element it is made out of one level down is a noun, *to fly out,* like *low-life,* must be headless—if the noun *fly* were its head, *fly out* would have to be a noun, too, which it is not. Lacking a head and its associated data pipeline, the irregular forms of the original verb *to fly,* namely *flew* and *flown,* are trapped at the bottommost level and cannot bubble up to attach to the whole word. The regular *-ed* rule rushes in in its usual role as the last resort, and thus we say that Wade Boggs *flied out.* What kills the irregularity of *to fly out,* then, is not its specialized meaning, but its being a verb based on a word that is not a verb. By the same logic, we say *They ringed the city with artillery* ("formed a ring around it"), not *They rang the city with artillery,* and *He grandstanded to the crowd* ("played to the grandstand"), not *He grandstood to the crowd.*

This principle works every time. Remember Sally Ride, the astronaut? She received a lot of publicity because she was America's first woman in space. But recently Mae Jemison did her one better. Not only is Jemison America's first *black* woman in space, but she appeared in *People* magazine in 1993 in their list of the fifty most beautiful people in the world. Publicity-wise, she has out-Sally-Rided Sally Ride (not *has out-Sally-Ridden Sally Ride*). For many years New York State's most infamous prison was Sing Sing. But since the riot at the Attica Correctional Facility in 1971, Attica has become even more infamous: it has out-Sing-Singed Sing Sing (not *has out-Sing-Sung Sing Sing*).

As for the Maple Leafs, the noun being pluralized is not *leaf,* the unit of foliage, but a noun based on the *name* Maple Leaf, Canada's national symbol. A name is not the same thing as a noun. (For example, whereas a noun may be preceded by an article like *the,* a name may not be: you cannot refer to someone as *the Donald,* unless you are Ivana Trump, whose first language is Czech.) Therefore, the noun *a Maple Leaf* (referring to, say, the goalie) must be headless, because it is a noun based on a word that is not a noun. And a noun that does not get its nounhood from one of its components cannot get an irregular plural from that component either; hence it defaults to the regular form *Maple Leafs.* This explanation also answers a question that kept bothering David Letterman throughout one of his recent *Late Night* shows: why is the new major league baseball team in Miami called the Florida Marlins rather than the Florida Marlin, given that those fish are referred to in the plural as *marlin*? Indeed, the explanation applies to all nouns based on names:

> I'm sick of dealing with all the *Mickey Mouses* in this administration. [not *Mickey Mice*]
>
> Hollywood has been relying on movies based on comic book heroes and their sequels, like the three *Supermans* and the two *Batmans.* [not *Supermen* and *Batmen*]
>
> Why has the second half of the twentieth century produced no *Thomas Manns*? [not *Thomas Menn*]
>
> We're having Julia Child and her husband over for dinner tonight. You know, *the Childs* are great cooks. [not *the Children*]

Irregular forms, then, live at the bottom of word structure trees, where roots and stems from the mental dictionary are inserted. The developmental psycholinguist Peter Gordon has capitalized on this effect in an ingenious experiment that shows how children's minds seem to be designed with the logic of word structure built in.

Gordon focused on a seeming oddity first noticed by the linguist Paul Kiparsky: compounds can be formed out of irregular plurals but not out of regular plurals. For example, a house infested with mice can be described as *mice-infested*, but it sounds awkward to describe a house infested with rats as *rats-infested*. We say that it is *rat-infested*, even though by definition one rat does not make an infestation. Similarly, there has been much talk about *men-bashing* but no talk about *gays-bashing* (only *gay-bashing*), and there are *teethmarks*, but no *clawsmarks*. Once there was a song about a *purple-people-eater*, but it would be ungrammatical to sing about a *purple-babies-eater*. Since the licit irregular plurals and the illicit regular plurals have similar meanings, it must be the grammar of irregularity that makes the difference.

The theory of word structure explains the effect easily. Irregular plurals, because they are quirky, have to be stored in the mental dictionary as roots or stems; they cannot be generated by a rule. Because of this storage, they can be fed into the compounding rule that joins an existing stem to another existing stem to yield a new stem. But regular plurals are not stems stored in the mental dictionary; they are complex words that are assembled on the fly by inflectional rules whenever they are needed. They are put together too late in the root-to-stem-to-word assembly process to be available to the compounding rule, whose inputs can only come out of the dictionary.

Gordon found that three- to five-year-old children obey this restriction fastidiously. Showing the children a puppet, he first asked them, "Here is a monster who likes to eat mud. What do you call him?" He then gave them the answer, *a mud-eater*, to get them started. Children like to play along, and the more gruesome the meal, the more eagerly they fill in the blank, often to the dismay of their onlooking parents. The crucial parts came next. A "monster who likes to eat mice," the children said, was a *mice-eater*. But a "monster who likes to eat rats" was never called a *rats-eater*, only a *rat-eater*. (Even the children who made the error *mouses* in their spontaneous speech never called the puppet a *mouses-eater*.) The children, in other words, respected the subtle restrictions on combining plurals and compounds inherent in the word structure rules. This suggests that the rules take the same form in the unconscious mind of the child as they do in the unconscious mind of the adult.

But the most interesting discovery came when Gordon examined how children might have acquired this constraint. Perhaps, he reasoned, they learned it from their parents by listening for whether the plurals that occur inside the parents' compounds are irregular, regular, or both, and then duplicate whatever kinds of compounds they hear. This would be impossible, he discovered. Motherese just doesn't have any compounds containing plurals. Most compounds are like *toothbrush*, with singular nouns inside them; compounds like *mice-infested*, though grammatically possible, are seldom used. The children produced *mice-eater* but never *rats-eater*, even though they had no evidence from adult speech that this is how languages work. We have another demonstration of knowledge despite "poverty of the input," and it suggests that another basic aspect of grammar may be innate. Just as Crain and Nakayama's Jabba experiment showed that in syntax children automatically distinguish between word strings and phrase structures, Gordon's mice-eater experiment shows that in morphology children automatically distinguish between roots stored in the mental dictionary and inflected words created by a rule.

A word, in a word, is complicated. But then what in the world *is* a word? We have just seen that "words" can be built out of parts by morphological rules. But then what makes them different from phrases or sentences? Shouldn't we reserve the word "word" for a thing that has to be rote-memorized, the arbitrary Saussurean sign that exemplifies the first of the two principles of how language works (the other being the discrete combinatorial system)? The puzzlement comes from the fact that the everyday word "word" is not scientifically precise. It can refer to two things.

The concept of a word that I have used so far in this chapter is a linguistic object that, even if built out of parts by the rules of morphology, behaves as the indivisible, smallest unit with respect to the rules of syntax—a "syntactic atom," in *atom*'s original sense of something that cannot be split. The rules of syntax can look inside a sentence or phrase and cut and paste the smaller phrases inside it. For example, the rule for producing questions can look inside the sentence *This monster eats mice* and move the phrase corresponding to *mice* to the front, yielding *What did this monster eat?* But the rules of syntax halt at the boundary between a phrase and a word; even if the word is built out of parts, the rules cannot look "inside" the word and fiddle with those parts. For example, the question rule cannot look inside the word *mice-eater* in the sentence *This monster is a mice-eater* and move the morpheme corresponding to *mice* to the front; the resulting question is virtually unintelligible: *What is this monster an -eater?* (Answer: mice.) Similarly, the rules of syntax can stick an adverb inside a phrase, as in *This monster eats mice quickly.* But they cannot stick an adverb inside a word, as in *This monster is a mice-quickly-eater.* For these reasons, we say that words, even if they are generated out of parts by one set of rules, are not the same thing as phrases, which are generated out of parts by a different set of rules. Thus one precise sense of our everyday term "word" refers to the units of language that are the products of morphological rules, and which are unsplittable by syntactic rules.

The second, very different sense of "word" refers to a rote-memorized chunk: a string of linguistic stuff that is arbitrarily associated with a particular meaning, one item from the long list we call the mental dictionary. The grammarians Anna Maria Di Sciullo and Edwin Williams coined the term "listeme," the unit of a memorized list, to refer to this sense of "word" (their term is a play on "morpheme," the unit of morphology, and "phoneme," the unit of sound). Note that a listeme need not coincide with the first precise sense of "word," a syntactic atom. A listeme can be a tree branch any size, as long as it cannot be produced mechanically by rules and therefore has to be memorized. Take idioms. There is no way to predict the meaning of *kick the bucket, buy the farm, spill the beans, bite the bullet, screw the pooch, give up the ghost, hit the fan,* or *go bananas* from the meanings of their components using the usual rules of heads and role-players. *Kicking the bucket* is not a kind of kicking, and buckets have nothing to do with it. The meanings of these phrase-sized units have to be memorized as listemes, just as if they were simple word-sized units, and so they are really "words" in this second sense. Di Sciullo and Williams, speaking as grammatical chauvinists, describe the mental dictionary (lexicon) as follows: "If conceived of as the set of listemes, the lexicon is incredibly boring by its very nature. . . . The lexicon is like a prison—it contains only the lawless, and the only thing that its inmates have in common is their lawlessness."

In the rest of this chapter I turn to the second sense of "word," the listeme. It will be a kind of prison reform: I want to show that the lexicon, though a repository of lawless listemes, is deserving of respect and appreciation. What seems to a grammarian like an act of brute force incarceration—a child hears a parent use a word and thenceforth retains that word in memory—is actually an inspiring feat.

One extraordinary feature of the lexicon is the sheer capacity for memorization that goes into building it. How many words do you think an average person knows? If you are like most writers who have offered an opinion based on the number of words they hear or read, you might guess a few hundred for the uneducated, a few thousand for the literate, and as many as 15,000 for gifted wordsmiths like Shakespeare (that is how many distinct words are found in his collected plays and sonnets).

The real answer is very different. People can recognize vastly more words than they have occasion to use in some fixed period of time or space. To estimate the size of a person's vocabulary—in the sense of memorized listemes, not morphological products, of course, because the latter are infinite—psychologists use the following method. Start with the largest unabridged dictionary available; the smaller the dictionary, the more words a person might know but not get credit for. Funk & Wagnall's *New Standard Unabridged Dictionary,* to take an example, has 450,000 entries, a healthy number, but too many to test exhaustively. (At thirty seconds a word, eight hours a day, it would take more than a year to test a single person.) Instead, draw a sample—say, the third entry from the top of the first column on every eighth left-hand page. Entries often have many meanings, such as "*hard*: (1) firm; (2) difficult; (3) harsh; (4) toilsome . . ." and so on, but counting them would require making arbitrary decisions about how to lump or split the meanings. Thus it is practical only to estimate how many words a person has learned at least one meaning for, not how many meanings a person has learned altogether. The testee is presented with each word in the sample, and asked to choose the closest synonym from a set of alternatives. After a correction for guessing, the proportion correct is multiplied by the size of the dictionary, and that is an estimate of the person's vocabulary size.

Actually, another correction must be applied first. Dictionaries are consumer products, not scientific instruments, and for advertising purposes their editors often inflate the number of entries. ("Authoritative. Comprehensive. Over 1.7 million words of text and 160,000 definitions. Includes a 16-page full-color atlas.") They do it by including compounds and affixed forms whose meanings are predictable from the meanings of their roots and the rules of morphology, and thus are not true listemes. For example, my desk dictionary includes, together with *sail*, the derivatives *sailplane, sailer, sailless, sailing-boat,* and *sailcloth,* whose meanings I could deduce even if I had never heard them before.

The most sophisticated estimate comes from the psychologists William Nagy and Richard Anderson. They began with a list of 227,553 different words. Of these, 45,453 were simple roots and stems. Of the remaining 182,100 derivatives and compounds, they estimated that all but 42,080 could be understood in context by someone who knew their components. Thus there were a total of 44,453 + 42,080 = 88,533 listeme words. By sampling from this list and testing the sample, Nagy and Anderson estimated that an average American high school graduate knows 45,000 words—three times as many as Shakespeare managed to use! Actually, this is an underestimate, because proper names, numbers, foreign words, acronyms, and many common undecomposable compounds were excluded. There is no need to follow the rules of Scrabble in estimating vocabulary size; these forms are all listemes, and a person should be given credit for them. If they had been included, the average high school graduate would probably be credited with something like 60,000 words (a tetrabard?), and superior students, because they read more, would probably merit a figure twice as high, an octobard.

Is 60,000 words a lot or a little? It helps to think of how quickly they must have been learned. Word learning generally begins around the age of twelve months. Therefore, high school graduates, who have been at it for about seventeen years, must have been learning an average of ten new words a day continuously since their first birthdays, or about a new word every ninety waking minutes. Using similar techniques, we can estimate that an average six-year-old commands about 13,000 words (notwithstanding those dull, dull *Dick and Jane* reading primers, which were based on ridiculously lowball estimates). A bit of arithmetic shows that preliterate children, who are limited to ambient speech, must be lexical vacuum cleaners, inhaling a new word every two waking hours, day in, day out. Remember that we are talking about listemes, each involving an arbitrary pairing. Think about having to memorize a new batting average or treaty date or phone number every ninety minutes of your waking life since you took your first steps. The brain seems to be reserving an especially capacious storage space and an especially rapid transcribing mechanism for the mental dictionary. Indeed, naturalistic studies by the psychologist Susan Carey have shown that if you casually slip a new color word like *olive* into a conversation with a three-year-old, the child will probably remember something about it five weeks later.

Now think of what goes into each act of memorization. A word is the quintessential symbol. Its power comes from the fact that every member of a linguistic community uses it interchangeably in speaking and understanding. If you use a word, then as long as it is not too obscure I can take it for granted that if I later utter it to a third party, he will understand my use of it the same way I understood yours. I do not have to try the word back on you to see how you react, or test it out on every third party and see how they react, or wait for you to use it with third parties. This sounds more obvious than it is. After all, if I observe that a bear snarls before it attacks, I cannot expect to scare a mosquito by snarling at it; if I bang a pot and the bear flees, I cannot expect the bear to bang a pot to scare hunters. Even within our species, learning a word from another person is not just a case of imitating that person's behavior. Actions are tied to particular kinds of actors and targets of the action in ways that words are not. If a girl learns to flirt by watching her older sister, she does not flirt with the sister or with their parents but only with the kind of person that she observes to be directly affected by the sister's behavior. Words, in contrast, are a universal currency within a community. In order to learn to use a word upon merely hearing it used by others, babies must tacitly assume that a word is not merely a person's characteristic behavior in affecting the behavior of others, but a shared bidirectional symbol, available to convert meaning to sound by any person when the person speaks, and sound to meaning by any person when the person listens, according to the same code.

Since a word is a pure symbol, the relation between its sound and its meaning is utterly arbitrary. As Shakespeare (using a mere tenth of a percent of his written lexicon and a far tinier fraction of his mental one) put it,

> What's in a name? that which we call a rose
> By any other name would smell as sweet.

Because of that arbitrariness, there is no hope that mnemonic tricks might lighten the memorization burden, at least for words that are not built out of other words. Babies should not, and apparently do not, expect *cattle* to mean something similar to *battle*, or *singing* to be like *stinging*, or *coats* to resemble *goats*. Onomatopoeia, where it is found, is of no help, because it is almost as conventional as any other word sound. In English, pigs go "oink"; in Japanese, they go "boo-boo." Even in sign languages the mimetic abilities of the hands are put aside and their configurations are treated as arbitrary symbols. Residues of resemblance between a sign and its referent can occasionally be discerned, but like onomatopoeia they are so much in the eye or ear of the beholder that they are of little use in learning. In American Sign Language the sign for "tree" is a motion of a hand as if it was a branch waving in the wind; in Chinese Sign Language "tree" is indicated by the motion of sketching a tree trunk.

The psychologist Laura Ann Petitto has a startling demonstration that the arbitrariness of the relation between a symbol and its meaning is deeply entrenched in the child's mind. Shortly before they turn two, English-speaking children learn the pronouns *you* and *me*. Often they reverse them, using *you* to refer to themselves. The error is forgivable. *You* and *me* are "deictic" pronouns, whose referent shifts with the speaker: *you* refers to you when I use it but to me when you use it. So children may need some time to get that down. After all, Jessica hears her mother refer to her, Jessica, using *you;* why should she not think that *you* means "Jessica"?

Now, in ASL the sign for "me" is a point to one's chest; the sign for "you" is a point to one's partner. What could be more transparent? One would expect that using "you" and "me" in ASL would be as foolproof as knowing how to point, which all babies, deaf and hearing, do before their first birthday. But for the deaf children Petitto studied, pointing is not pointing. The children used the sign of pointing to their conversational partners to mean "me" at exactly the age at which hearing children use the spoken sound *you* to mean "me." The children were treating the gesture as a pure linguistic symbol; the fact that it pointed somewhere did not register as being relevant. This attitude is appropriate in learning sign languages; in ASL, the pointing hand-shape is like a meaningless consonant or vowel, found as a component of many other signs, like "candy" and "ugly."

There is one more reason we should stand in awe of the simple act of learning a word. The logician W. V. O. Quine asks us to imagine a linguist studying a newly discovered tribe. A rabbit scurries by, and a native shouts, "Gavagai!" What does *gavagai* mean? Logically speaking, it needn't be "rabbit." It could refer to that particular rabbit (Flopsy, for example). It could mean any furry thing, any mammal, or any member of that species of rabbit (say, *Oryctolagus cuniculus*), or any member of that variety of that species (say, chinchilla rabbit). It could mean scurrying rabbit, scurrying thing, rabbit plus the ground it scurries upon, or scurrying in general. It could mean footprint-maker, or habitat for rabbit-fleas. It could mean the top half of a rabbit, or rabbit-meat-on-the-hoof, or possessor of at least one rabbit's foot. It could mean anything that is either a rabbit or a Buick. It could mean collection of undetached rabbit parts, or "Lo! Rabbithood again!," or "It rabbiteth," analogous to "It raineth."

The problem is the same when the child is the linguist and the parents are the natives. Somehow a baby must intuit the correct meaning of a word and avoid the mind-boggling number of logically impeccable alternatives. It is an example of a more general problem that Quine calls "the scandal of induction," which applies to scientists and children alike: how can they be so successful at observing a finite set of events and making some correct generalization about all future events of that sort, rejecting an infinite number of false generalizations that are also consistent with the original observations?

We all get away with induction because we are not open-minded logicians but happily blinkered humans, innately constrained to make only certain kinds of guesses—the probably correct kinds—about how the world and its occupants work. Let's say the word-learning baby has a brain that carves the world into discrete, bounded, cohesive objects and into the actions they undergo, and that the baby forms mental categories that lump together objects that are of the same kind. Let's also say that babies are designed to expect a language to contain words for kinds of objects and words for kinds of actions—nouns and verbs, more or less. Then the undetached rabbit parts, rabbit-trod ground, intermittent rabbiting, and other accurate descriptions of the scene will, fortunately, not occur to them as possible meanings of *gavagai*.

But could there really be a preordained harmony between the child's mind and the parent's? Many thinkers, from the woolliest mystics to the sharpest logicians, united only in their assault on common sense, have claimed that the distinction between an object and an action is not in the world or even in our minds, initially, but is imposed on us by our language's distinction between nouns and verbs. And if it is the word that delineates the thing and the act, it cannot be the concepts of thing and act that allow for the learning of the word.

I think common sense wins this one. In an important sense, there really are things and kinds of things and actions out there in the world, and our mind is designed to find them and to label them with words. That important sense is Darwin's. It's a jungle out there, and the organism designed to make successful predictions about what is going to happen next will leave behind more babies designed just like it. Slicing space-time into objects and actions is an eminently sensible way to make predictions given the way the world is put together. Conceiving of an extent of solid matter as a thing—that is, giving a single mentalese name to all of its parts—invites the prediction that those parts will continue to occupy some region of space and will move as a unit. And for many portions of the world, that prediction is correct. Look away, and the rabbit still exists; lift the rabbit by the scruff of the neck, and the rabbit's foot and the rabbit ears come along for the ride.

What about kinds of things, or categories? Isn't it true that no two individuals are exactly alike? Yes, but they are not arbitary collections of properties, either. Things that have long furry ears and tails like pom-poms also tend to eat carrots, scurry into burrows, and breed like, well, rabbits. Lumping objects into categories—giving them a category label in mentalese—allows one, when viewing an entity, to infer some of the properties one cannot directly observe, using the properties one *can* observe. If Flopsy has long furry ears, he is a "rabbit"; if he is a rabbit, he might scurry into a burrow and quickly make more rabbits.

Moreover, it pays to give objects several labels in mentalese, designating different-sized categories like "cottontail rabbit," "rabbit," "mammal," "animal," and "living thing." There a tradeoff involved in choosing one category over another. It takes less effort to determine that Peter Cottontail is an animal than that he is a cottontail (for example, an animallike motion will suffice for us to recognize that he is an animal, leaving it open whether or not he is a cottontail). But we can predict more new things about Peter if we know he is a cottontail than if we merely know he is an animal. If he is a cottontail, he likes carrots and inhabits open country or woodland clearings; if he is merely an animal, he could eat anything and live anywhere, for all one knows. The middle-sized or "basic-level" category "rabbit" represents a compromise between how easy it is to label something and how much good the label does you.

Finally, why separate the rabbit from the scurry? Presumably because there are predictable consequences of rabbithood that cut across whether it is scurrying, eating, or sleeping: make a loud sound, and in all cases it will be down a hole lickety-split. The consequences of making a loud noise in the presence of lionhood, whether eating or sleeping, are predictably different, and that is a difference that makes a difference. Likewise, scurrying has certain consequences regardless of who is doing it; whether it be rabbit or lion, a scurrier does not remain in the same place for long. With sleeping, a silent approach will generally work to keep a sleeper—rabbit or lion—motionless. Therefore a powerful prognosticator should have separate sets of mental labels for kinds of objects and kinds of actions. That way, it does not have to learn separately what happens when a rabbit scurries, what happens when a lion scurries, what happens when a rabbit sleeps, what happens when a iion sleeps, what happens when a gazelle scurries, what happens when a gazelle sleeps, and on and on; knowing about rabbits and lions and gazelles in general, and scurrying and sleeping in general, will suffice. With $m$ objects and $n$ actions, a knower needn't go through $m \times n$ learning experiences; it can get away with $m + n$ of them.

So even a wordless thinker does well to chop continuously flowing experience into things, kinds of things, and actions (not to mention places, paths, events, states, kinds of stuff, properties, and other types of concepts). Indeed, experimental studies of baby cognition have shown that infants have the concept of an object before they learn any words for objects, just as we would expect. Well before their first birthday, when first words appear, babies seem to keep track of the bits of stuff that we would call objects: they show surprise if the parts of an object suddenly go their own ways, or if the object magically appears or disappears, passes through another solid object, or hovers in the air without visible means of support.

Attaching words to these concepts, of course, allows one to share one's hard-won discoveries and insights about the world with the less experienced or the less observant. Figuring out which word to attach to which concept is the *gavagai* problem, and if infants start out with concepts corresponding to the kinds of meanings that languages use, the problem is partly solved. Laboratory studies confirm that young children assume that certain kinds of concepts get certain types of words, and other kinds of concepts cannot be the meaning of a word at all. The developmental psychologists Ellen Markman and Jeanne Hutchinson gave two- and three-year-old children a set of pictures, and for each picture asked them to "find another one that is the same as this." Children are intrigued by objects that interact, and when faced with these instructions they tend to select pictures that make groups of role-players like a blue jay and a nest or a dog and a bone. But when Markman and Hutchinson told them to "find another *dax* that is the same as this *dax*," the children's criterion shifted. A word must label a *kind* of thing, they seemed to be reasoning, so they put together a bird with another type of bird, a dog with another type of dog. For a child, a *dax* simply cannot mean "a dog or its bone," interesting though the combination may be.

Of course, more than one word can be applied to a thing: Peter Cottontail is not only a *rabbit* but an *animal* and a *cottontail*. Children have a bias to interpret nouns as middle-level kinds of objects like "rabbit," but they also must overcome that bias, to learn other types of words like *animal*. Children seem to manage this by being in sync with a striking feature of language. Though most common words have many meanings, few meanings have more than one word. That is, homonyms are plentiful, synonyms rare. (Virtually all supposed synonyms have some difference in meaning, however small. For example, *skinny* and *slim* differ in their connotation of desirability; *police-*

*man* and *cop* differ in formality.) No one really knows why languages are so stingy with words and profligate with meanings, but children seem to expect it (or perhaps it is this expectation that causes it!), and that helps them further with the *gavagai* problem. If a child already knows a word for a kind of thing, then when another word is used for it, he or she does not take the easy but wrong way and treat it as a synonym. Instead, the child tries out some other possible concept. For example, Markman found that if you show a child a pair of pewter tongs and call it *biff*, the child interprets *biff* as meaning tongs in general, showing the usual bias for middle-level objects, so when asked for "more biffs," the child picks out a pair of plastic tongs. But if you show the child a pewter cup and call it *biff*, the child does not interpret *biff* as meaning "cup," because most children already know a word that means "cup," namely, *cup*. Loathing synonyms, the children guess that *biff* must mean something else, and the stuff the cup is made of is the next most readily available concept. When asked for more *biffs*, the child chooses a pewter spoon or pewter tongs.

Many other ingenious studies have shown how children home in on the correct meanings for different kinds of words. Once children know some syntax, they can use it to sort out different kinds of meaning. For example, the psychologist Roger Brown showed children a picture of hands kneading a mass of little squares in a bowl. If he asked them, "Can you see any sibbing?," the children pointed to the hands. If instead he asked them, "Can you see a sib?," they point to the bowl. And if he asked, "Can you see any sib?," they point to the stuff inside the bowl. Other experiments have uncovered great sophistication in children's understanding of how classes of words fit into sentence structures and how they relate to concepts and kinds.

So what's in a name? The answer, we have seen, is, a great deal. In the sense of a morphological product, a name is an intricate structure, elegantly assembled by layers of rules and lawful even at its quirkiest. And in the sense of a listeme, a name is a pure symbol, part of a cast of thousands, rapidly acquired because of a harmony between the mind of the child, the mind of the adult, and the texture of reality.

# 6

⁂

# The Sounds of Silence

*When I was a student I worked in a laboratory at McGill University* that studied auditory perception. Using a computer, I would synthesize trains of overlapping tones and determine whether they sounded like one rich sound or two pure ones. One Monday morning I had an odd experience: the tones suddenly turned into a chorus of screaming munchkins. Like this: (beep boop-boop) (beep boop-boop) (beep boop-boop) HUMPTY-DUMPTY-HUMPTY-DUMPTY-HUMPTY-DUMPTY (beep boop-boop) (beep boop-boop) HUMPTY-DUMPTY-HUMPTY-DUMPTY-HUMPTY-HUMPTY-DUMPTY-DUMPTY (beep boop-boop) (beep boop-boop) (beep boop-boop) HUMPTY-DUMPTY (beep boop-boop) HUMPTY-HUMPTY-HUMPTY-DUMPTY (beep boop-boop). I checked the oscilloscope: two streams of tones, as programmed. The effect had to be perceptual. With a bit of effort I could go back and forth, hearing the sound as either beeps or munchkins. When a fellow student entered, I recounted my discovery, mentioning that I couldn't wait to tell Professor Bregman, who directed the laboratory. She offered some advice: don't tell anyone, except perhaps Professor Poser (who directed the psychopathology program).

Years later I discovered what I had discovered. The psychologists

Robert Remez, David Pisoni, and their colleagues, braver men than I am, published an article in *Science* on "sine-wave speech." They synthesized three simultaneous wavering tones. Physically, the sound was nothing at all like speech, but the tones followed the same contours as the bands of energy in the sentence. "Where were you a year ago?" Volunteers described what they heard as "science fiction sounds" or "computer bleeps." A second group of volunteers was told that the sounds had been generated by a bad speech synthesizer. They were able to make out many of the words, and a quarter of them could write down the sentence perfectly. The brain can hear speech content in sounds that have only the remotest resemblance to speech. Indeed, sine-wave speech is how mynah birds fool us. They have a valve on each bronchial tube and can control them independently, producing two wavering tones which we hear as speech.

Our brains can flip between hearing something as a bleep and hearing it as a word because phonetic perception is like a sixth sense. When we listen to speech the actual sounds go in one ear and out the other; what we perceive is *language*. Our experience of words and syllables, of the "b"-ness of *b* and the "ee"-ness of *ee,* is as separable from our experience of pitch and loudness as lyrics are from a score. Sometimes, as in sine-wave speech, the senses of hearing and phonetics compete over which gets to interpret a sound, and our perception jumps back and forth. Sometimes the two senses simultaneously interpret a single sound. If one takes a tape recording of *da,* electronically removes the initial chirplike portion that distinguishes the *da* from *ga* and *ka,* and plays the chirp to one ear and the residue to the other, what people hear is a chirp in one ear and *da* in the other—a single clip of sound is perceived simultaneously as *d*-ness and a chirp. And sometimes phonetic perception can transcend the auditory channel. If you watch an English-subtitled movie in a language you know poorly, after a few minutes you may feel as if you are actually understanding the speech. In the laboratory, researchers can dub a speech sound like *ga* onto a close-up video of a mouth articulating *va, ba, tha,* or *da.* Viewers literally *hear* a consonant like the one they see the mouth making—an astonishing illusion with the pleasing name "McGurk effect," after one of its discoverers.

Actually, one does not need electronic wizardry to create a speech illusion. All speech is an illusion. We hear speech as a string of separate words, but unlike the tree falling in the forest with no one to hear it, a word boundary with no one to hear it has no sound. In the speech sound wave, one word runs into the next seamlessly; there are no little silences between spoken words the way there are white spaces between written words. We simply hallucinate word boundaries when we reach the edge of a stretch of sound that matches some entry in our mental dictionary. This becomes apparent when we listen to speech in a foreign language: it is impossible to tell where one word ends and the next begins. The seamlessness of speech is also apparent in "oronyms," strings of sound that can be carved into words in two different ways:

> The good can decay many ways.
> The good candy came anyways.

> The stuffy nose can lead to problems.
> The stuff he knows can lead to problems.

> Some others I've seen.
> Some mothers I've seen.

Oronyms are often used in songs and nursery rhymes:

> I scream,
> You scream,
> We all scream
> For ice cream.

> Mairzey doats and dozey doats
> And little lamsey divey,
> A kiddley-divey do,
> Wouldn't you?

> Fuzzy Wuzzy was a bear,
> Fuzzy Wuzzy had no hair.

> Fuzzy Wuzzy wasn't fuzzy,
> Was he?

> In fir tar is,
> In oak none is.
> In mud eel is,
> In clay none is.
> Goats eat ivy.
> Mares eat oats.

And some are discovered inadvertently by teachers reading their students' term papers and homework assignments:

> Jose can you see by the donzerly light? [Oh say can you see
>     by the dawn's early light?]
> It's a doggy-dog world. [dog-eat-dog]
> Eugene O'Neill won a Pullet Surprise. [Pulitzer Prize]
> My mother comes from Pencil Vanea. [Pennsylvania]
> He was a notor republic. [notary public]
> They played the Bohemian Rap City. [Bohemian Rhapsody]

Even the sequence of sounds we think we hear within a word is an illusion. If you were to cut up a tape of someone's saying *cat*, you would not get pieces that sounded like *k*, *a*, and *t* (the units called "phonemes" that correspond roughly to the letters of the alphabet). And if you spliced the pieces together in the reverse order, they would be unintelligible, not *tack*. As we shall see, information about each component of a word is smeared over the entire word.

Speech perception is another one of the biological miracles making up the language instinct. There are obvious advantages to using the mouth and ear as a channel of communication, and we do not find any hearing community opting for sign language, though it is just as expressive. Speech does not require good lighting, face-to-face contact, or monopolizing the hands and eyes, and it can be shouted over long distances or whispered to conceal the message. But to take advantage of the medium of sound, speech has to overcome the problem that the ear is a narrow informational bottleneck. When engineers first tried to develop reading machines for the blind in the 1940s, they devised a set of noises that corresponded to the letters of the alphabet. Even with heroic training, people could not recognize the sounds at a rate faster than good Morse code operators, about three units a second. Real speech, somehow, is perceived an order of magnitude faster: ten to fifteen phonemes per second for casual speech, twenty to thirty per second for the man in the late-night Veg-O-Matic ads, and as many as forty to fifty per second for artificially sped-up speech. Given how the human auditory system works, this is almost unbelievable. When a sound like a click is repeated at a rate of twenty times a second or faster, we no longer hear it as a sequence of separate sounds but as a low buzz. If we can hear forty-five phonemes per second, the phonemes cannot possibly be consecutive bits of sound; each moment of sound must have several phonemes packed into it that our brains somehow unpack. As a result, speech is by far the fastest way of getting information into the head through the ear.

No human-made system can match a human in decoding speech. It is not for lack of need or trying. A speech recognizer would be a boon to quadriplegics and other disabled people, to professionals who have to get information into a computer while their eyes or hands are busy, to people who never learned to type, to users of telephone services, and to the growing number of typists who are victims of repetitive-motion syndromes. So it is not surprising that engineers have been working for more than forty years to get computers to recognize the spoken word. The engineers have been frustrated by a tradeoff. If a system has to be able to listen to many different people, it can recognize only a tiny number of words. For example, telephone companies are beginning to install directory assistance systems that can recognize anyone saying the word *yes,* or, in the more advanced systems, the ten English digits (which, fortunately for the engineers, have very different sounds). But if a system has to recognize a large number of words, it has to be trained to the voice of a single speaker. No system today can duplicate a person's ability to recognize both many words and many speakers. Perhaps the state of the art is a system called Dragon-Dictate, which runs on a personal computer and can recognize 30,000 words. But it has severe limitations. It has to be trained extensively on the voice of the user. You . . . have . . . to . . . talk . . . to . . . it . . . like . . . this, with quarter-second pauses between the words (so it operates at about one-fifth the rate of ordinary speech). If you have to use a word that is not in its dictionary, like a name, you have to spell it out using the "Alpha, Bravo, Charlie" alphabet. And the program still garbles words about fifteen percent of the time, more than once per sentence. It is an impressive product but no match for even a mediocre stenographer.

The physical and neural machinery of speech is a solution to two problems in the design of the human communication system. A person might know 60,000 words, but a person's mouth cannot make 60,000 different noises (at least, not ones that the ear can easily discriminate). So language has exploited the principle of the discrete combinatorial system again. Sentences and phrases are built out of words, words are built out of morphemes, and morphemes, in turn, are built out of phonemes. Unlike words and morphemes, though, phonemes do not contribute bits of meaning to the whole. The meaning of *dog* is not predictable from the meaning of *d*, the meaning of *o*, the meaning of *g*, and their order. Phonemes are a different kind of linguistic object. They connect outward to speech, not inward to mentalese: a phoneme corresponds to an act of making a sound. A division into independent discrete combinatorial systems, one combining meaningless sounds into meaningful morphemes, the others combining meaningful morphemes into meaningful words, phrases, and sentences, is a fundamental design feature of human language, which the linguist Charles Hockett has called "duality of patterning."

But the phonological module of the language instinct has to do more than spell out the morphemes. The rules of language are discrete combinatorial systems: phonemes snap cleanly into morphemes, morphemes into words, words into phrases. They do not blend or melt or coalesce: *Dog bites man* differs from *Man bites dog,* and believing in God is different from believing in Dog. But to get these structures out of one head and into another, they must be converted to audible signals. The audible signals people can produce are not a series of crisp beeps like on a touch-tone phone. Speech is a river of breath, bent into hisses and hums by the soft flesh of the mouth and throat. The problems Mother Nature faced are digital-to-analog conversion when the talker encodes strings of discrete symbols into a continuous stream of sound, and analog-to-digital conversion when the listener decodes continuous speech back into discrete symbols.

The sounds of language, then, are put together in several steps. A finite inventory of phonemes is sampled and permuted to define words, and the resulting strings of phonemes are then massaged to make them easier to pronounce and understand before they are actually articulated. I will trace out these steps for you and show you how they shape some of our everyday encounters with speech: poetry and song, slips of the ear, accents, speech recognition machines, and crazy English spelling.

One easy way to understand speech sounds is to track a glob of air through the vocal tract into the world, starting in the lungs.

When we talk, we depart from our usual rhythmic breathing and take in quick breaths of air, then release them steadily, using the muscles of the ribs to counteract the elastic recoil force of the lungs. (If we did not, our speech would sound like the pathetic whine of a released balloon.) Syntax overrides carbon dioxide: we suppress the delicately tuned feedback loop that controls our breathing rate to regulate oxygen intake, and instead we time our exhalations to the length of the phrase or sentence we intend to utter. This can lead to mild hyperventilation or hypoxia, which is why public speaking is so exhausting and why it is difficult to carry on a conversation with a jogging partner.

The air leaves the lungs through the trachea (windpipe), which opens into the larynx (the voice-box, visible on the outside as the Adam's apple). The larynx is a valve consisting of an opening (the glottis) covered by two flaps of retractable muscular tissue called the vocal folds (they are also called "vocal cords" because of an early anatomist's error; they are not cords at all). The vocal folds can close off the glottis tightly, sealing the lungs. This is useful when we want to stiffen our upper body, which is a floppy bag of air. Get up from your chair without using your arms; you will feel your larynx tighten. The larynx is also closed off in physiological functions like coughing and defecation. The grunt of the weightlifter or tennis player is a reminder that we use the same organ to seal the lungs and to produce sound.

The vocal folds can also be partly stretched over the glottis to produce a buzz as the air rushes past. This happens because the high-pressure air pushes the vocal folds open, at which point they spring back and get sucked together, closing the glottis until air pressure builds up and pushes them open again, starting a new cycle. Breath is thus broken into a series of puffs of air, which we perceive as a buzz, called "voicing." You can hear and feel the buzz by making the sounds *sssssss*, which lacks voicing, and *zzzzzzzz*, which has it.

The frequency of the vocal folds' opening and closing determines the pitch of the voice. By changing the tension and position of the vocal folds, we can control the frequency and hence the pitch. This is most obvious in humming or singing, but we also change pitch continuously over the course of a sentence, a process called intonation. Normal intonation is what makes natural speech sound different from the speech of robots in old science fiction movies and of the Coneheads on *Saturday Night Live*. Intonation is also controlled in sarcasm, emphasis, and an emotional tone of voice such as anger or cheeriness. In "tone languages" like Chinese, rising or falling tones distinguish certain vowels from others.

Though voicing creates a sound wave with a dominant frequency of vibration, it is not like a tuning fork or a test of the Emergency Broadcasting System, a pure tone with that frequency alone. Voicing is a rich, buzzy sound with many "harmonics." A male voice is a wave with vibrations not only at 100 cycles per second but also at 200 cps, 300 cps, 400 cps, 500 cps, 600 cps, 700 cps, and so on, all the way up to 4000 cps and beyond. A female voice has vibrations at 200 cps, 400 cps, 600 cps, and so on. The richness of the sound source is crucial—it is the raw material that the rest of the vocal tract sculpts into vowels and consonants.

If for some reason we cannot produce a hum from the larynx, any rich source of sound will do. When we whisper, we spread the vocal folds, causing the air stream to break apart chaotically at the edges of the folds and creating a turbulence or noise that sounds like hissing or radio static. A hissing noise is not a neatly repeating wave consisting of a sequence of harmonics, as we find in the periodic sound of a speaking voice, but a jagged, spiky wave consisting of a hodgepodge of constantly changing frequencies. This mixture, though, is all that the rest of the vocal tract needs for intelligible whispering. Some laryngectomy patients are taught "esophageal speech," or controlled burping, which provides the necessary noise. Others place a vibrator against their necks. In the 1970s the guitarist Peter Frampton funneled the amplified sound of his electric guitar through a tube into his mouth, allowing him to articulate his twangings. The effect was good for a couple of hit records before he sank into rock-and-roll oblivion.

The richly vibrating air then runs through a gantlet of chambers before leaving the head: the throat or "pharynx" behind the tongue, the mouth region between the tongue and palate, the opening between the lips, and an alternative route to the external world through the nose. Each chamber has a particular length and shape, which affects the sound passing through by the phenomenon called "resonance." Sounds of different frequencies have different wavelengths (the distance between the crests of the sound wave); higher pitches have shorter wavelengths. A sound wave moving down the length of a tube bounces back when it reaches the opening at the other end. If the length of the tube is a certain fraction of the wavelength of the sound, each reflected wave will reinforce the next incoming one; if it is of a different length, they will interfere with one another. (This is similar to how you get the best effect pushing a child on a swing if you synchronize each push with the top of the arc.) Thus a tube of a particular length amplifies some sound frequencies and filters out others. You can hear the effect when you fill a bottle. The noise of the sloshing water gets filtered by the chamber of air between the surface and the opening: the more water, the smaller the chamber, the higher the resonant frequency of the chamber, and the tinnier the gurgle.

What we hear as different vowels are the different combinations of amplifications and filtering of the sound coming up from the larynx. These combinations are produced by moving five speech organs around in the mouth to change the shapes and lengths of the resonant cavities that the sound passes through. For example, *ee* is defined by two resonances, one from 200 to 350 cps produced mainly by the throat cavity, and the other from 2100 to 3000 cps produced mainly by the mouth cavity. The range of frequencies that a chamber filters is independent of the particular mixture of frequencies that enters it, so we can hear an *ee* as an *ee* whether it is spoken, whispered, sung high, sung low, burped, or twanged.

The tongue is the most important of the speech organs, making language truly the "gift of tongues." Actually, the tongue is three organs in one: the hump or body, the tip, and the root (the muscles that anchor it to the jaw). Pronounce the vowels in *bet* and *butt* repeatedly, *e-uh, e-uh, e-uh*. You should feel the body of your tongue moving forwards and backwards (if you put a finger between your teeth, you can feel it with the finger). When your tongue is in the front of your mouth, it lengthens the air chamber behind it in your throat and shortens the one in front of it in your mouth, altering one of the resonances: for the *bet* vowel, the mouth amplifies sounds near 600 and 1800 cps; for the *butt* vowel, it amplifies sounds near 600 and 1200. Now pronounce the vowels in *beet* and *bat* alternately. The body of your tongue will jump up and down, at right angles to the *bet-butt* motion; you can even feel your jaw move to help it. This, too, alters the shapes of the throat and mouth chambers, and hence their resonances. The brain interprets the different patterns of amplification and filtering as different vowels.

The link between the postures of the tongue and the vowels it sculpts gives rise to a quaint curiosity of English and many other languages called phonetic symbolism. When the tongue is high and at the front of the mouth, it makes a small resonant cavity there that amplifies some higher frequencies, and the resulting vowels like *ee* and *i* (as in *bit*) remind people of little things. When the tongue is low and to the back, it makes a large resonant cavity that amplifies some lower frequencies, and the resulting vowels like *a* in *father* and *o* in *core* and in *cot* remind people of large things. Thus mice are t*ee*ny and squ*ea*k, but elephants are hum*o*ngous and r*oa*r. Audio speakers have small tw*ee*ters for the high sounds and large w*oo*fers for the low ones. English speakers correctly guess that in Chinese *ch'ing* means light and *ch'ung* means heavy. (In controlled studies with large numbers of foreign words, the hit rate is statistically above chance, though just barely.) When I questioned our local computer wizard about what she meant when she said she was going to *frob* my workstation, she gave me this tutorial on hackerese. When you get a brand-new graphic equalizer for your stereo and aimlessly slide the knobs up and down to hear the effects, that is *frobbing*. When you move the knobs by medium-sized amounts to get the sound to your general liking, that is *twiddling*. When you make the final small adjustments to get it perfect, that is *tweaking*. The *ob, id,* and *eak* sounds perfectly follow the large-to-small continuum of phonetic symbolism.

And at the risk of sounding like Andy Rooney on *Sixty Minutes,* have you ever wondered why we say *fiddle-faddle* and not *faddle-fiddle*? Why is it *ping-pong* and *pitter-patter* rather than *pong-ping* and *patter-pitter*? Why *dribs and drabs,* rather than vice versa? Why can't a kitchen be *span and spic*? Whence *riff-raff, mish-mash, flim-flam, chit-chat, tit for tat, knick-knack, zig-zag, sing-song, ding-dong, King Kong, criss-cross, shilly-shally, see-saw, hee-haw, flip-flop, hippity-hop, tick-tock, tic-tac-toe, eeny-meeny-miney-moe, bric-a-brac, clickety-clack, hickory-dickory-dock, kit and kaboodle,* and *bibbity-bobbity-boo*? The answer is that the vowels for which the tongue is high and in the front always come before the vowels for which the tongue is low and in the back. No one knows why they are aligned in this order, but it seems to be a kind of syllogism from two other oddities. The first is that words that connote me-here-now tend to have higher and fronter vowels than verbs that connote distance from "me": *me* versus *you, here* versus *there, this* versus *that.* The second is that words that connote me-here-now tend to come before words that connote literal or metaphorical distance from "me" (or a prototypical generic speaker): *here and there* (not *there and here*), *this and that, now and then, father and son, man and machine, friend or foe, the Harvard-Yale game* (among Harvard students), *the Yale-Harvard game* (among Yalies), *Serbo-Croatian* (among Serbs), *Croat-Serbian* (among Croats). The syllogism seems to be: "me" = high front vowel; me first; therefore, high front vowel first. It is as if the mind just cannot bring itself to flip a coin in ordering words; if meaning does not determine the order, sound is brought to bear, and the rationale is based on how the tongue produces the vowels.

Let's look at the other speech organs. Pay attention to your lips when you alternate between the vowels in *boot* and *book*. For *boot*, you round the lips and protrude them. This adds an air chamber, with its own resonances, to the front of the vocal tract, amplifying and filtering other sets of frequencies and thus defining other vowel contrasts. Because of the acoustic effects of the lips, when we talk to a happy person over the phone, we can literally hear the smile.

Remember your grade-school teacher telling you that the vowel sounds in *bat, bet, bit, bottle,* and *butt* were "short," and the vowel sounds in *bait, beet, bite, boat,* and *boot* were "long"? And you didn't know what she was talking about? Well, forget it; her information is five hundred years out of date. Older stages of English differentiated words by whether their vowels were pronounced quickly or were drawn out, a bit like the modern distinction between *bad* meaning "bad" and *baaaad* meaning "good." But in the fifteenth century English pronunciation underwent a convulsion called the Great Vowel Shift. The vowels that had simply been pronounced longer now became "tense": by advancing the tongue root (the muscles attaching the tongue to the jaw), the tongue becomes tense and humped rather than lax and flat, and the hump narrows the air chamber in the mouth above it, changing the resonances. Also, some tense vowels in modern English, like in *bite* and *brow*, are "diphthongs," two vowels pronounced in quick succession as if they were one: ba-eet, bra-oh.

You can hear the effects of the fifth speech organ by drawing out the vowel in *Sam* and *sat*, postponing the final consonant indefinitely. In most dialects of English, the vowels will be different: the vowel in *Sam* will have a twangy, nasal sound. That is because the soft palate or velum (the fleshy flap at the back of the hard palate) is opened, allowing air to flow out through the nose as well as through the mouth. The nose is another resonant chamber, and when vibrating air flows through it, yet another set of frequencies gets amplified and filtered. English does not differentiate words by whether their vowels are nasal or not, but many languages, like French, Polish, and Portuguese, do. English speakers who open their soft palate even when pronouncing *sat* are said to have a "nasal" voice. When you have a cold and your nose is blocked, opening the soft palate makes no difference, and your voice is the opposite of nasal.

So far we have just discussed the vowels—sounds where the air has clear passage from the larynx to the world. When some barrier is put in the way, one gets a consonant. Pronounce *sssss*. The tip of your tongue—the sixth speech organ—is brought up almost against the gum ridge, leaving a small opening. When you force a stream of air through the opening, the air breaks apart turbulently, creating noise. Depending on the size of the opening and the length of the resonant cavities in front of it, the noise will have some of its frequencies louder than others, and the peak and range of frequencies define the sound we hear as *s*. This noise-making comes from the friction of moving air, so this kind of sound is called a fricative. When rushing air is squeezed between the tongue and palate, we get *sh;* between the tongue and teeth, *th;* and between the lower lip and teeth, *f.* The body of the tongue, or the vocal folds of the larynx, can also be positioned to create turbulence, defining the various "ch" sounds in languages like German, Hebrew, and Arabic (*Bach, Chanukah,* and so on).

Now pronounce a *t.* The tip of the tongue gets in the way of the airstream, but this time it does not merely impede the flow; it stops it entirely. When the pressure builds up, you release the tip of the tongue, allowing the air to pop out (flutists use this motion to demarcate musical notes). Other "stop" consonants can be formed by the lips *(p),* by the body of the tongue pressed against the palate *(k),* and by the larynx (in the "glottal" consonants in *uh-oh*). What a listener hears when you produce a stop consonant is the following. First, nothing, as the air is dammed up behind the stoppage: stop consonants are the sounds of silence. Then, a brief burst of noise as the air is released; its frequency depends on the size of the opening and the resonant cavities in front of it. Finally, a smoothly changing resonance, as voicing fades in while the tongue is gliding into the position of whatever vowel comes next. As we shall see, this hop-skip-and-jump makes life miserable for speech engineers.

Finally, pronounce *m*. Your lips are sealed, just like for *p*. But this time the air does not back up silently; you can say *mmmmm* until you are out of breath. That is because you have also opened your soft palate, allowing all of the air to escape through your nose. The voicing sound is now amplified at the resonant frequencies of the nose and of the part of the mouth behind the blockage. Releasing the lips causes a sliding resonance similar in shape to what we heard for the release in *p,* except without the silence, noise burst, and fade-in. The sound *n* works similarly to *m*, except that the blockage is created by the tip of the tongue, the same organ used for *d* and *s*. So does the *ng* in *sing,* except that the body of the tongue does the job.

Why do we say *razzle-dazzle* instead of *dazzle-razzle?* Why *super-duper, helter-skelter, harum-scarum, hocus-pocus, willy-nilly, hully-gully, roly-poly, holy moly, herky-jerky, walkie-talkie, namby-pamby, mumbo-jumbo, loosey-goosey, wing-ding, wham-bam, hobnob, razza-matazz,* and *rub-a-dub-dub?* I thought you'd never ask. Consonants differ in "obstruency"—the degree to which they impede the flow of air, ranging from merely making it resonate, to forcing it noisily past an obstruction, to stopping it up altogether. The word beginning with the less obstruent consonant always comes before the word beginning with the more obstruent consonant. Why ask why?

Now that you have completed a guided tour up the vocal tract, you can understand how the vast majority of sounds in the world's languages are created and heard. The trick is that a speech sound is not a single gesture by a single organ. Every speech sound is a *combination* of gestures, each exerting its own pattern of sculpting of the sound wave, all executed more or less simultaneously—that is one of the reasons speech can be so rapid. As you may have noticed, a sound can be nasal or not, and produced by the tongue body, the tongue tip, or the lips, in all six possible combinations:

|  | Nasal<br>(Soft Palate Open) | Not Nasal<br>(Soft Palate Closed) |
|---|:---:|:---:|
| Lips | *m* | *p* |
| Tongue tip | *n* | *t* |
| Tongue body | *ng* | *k* |

Similarly, voicing combines in all possible ways with the choice of speech organ:

|  | Voicing<br>(Larynx Hums) | No Voicing<br>(Lrynx Doesn't Hum) |
|---|:---:|:---:|
| Lips | *b* | *p* |
| Tongue tip | *d* | *t* |
| Tongue body | *g* | *k* |

Speech sounds thus nicely fill the rows and columns and layers of a multidimensional matrix. First, one of the six speech organs is chosen as the major articulator: the larynx, soft palate, tongue body, tongue tip, tongue root, or lips. Second, a manner of moving that articulator is selected: fricative, stop, or vowel. Third, configurations of the other speech organs can be specified: for the soft palate, nasal or not; for the larynx, voiced or not; for the tongue root, tense or lax; for the lips, rounded or unrounded. Each manner or configuration is a symbol for a set of commands to the speech muscles, and such symbols are called features. To articulate a phoneme, the commands must be executed with precise timing, the most complicated gymnastics we are called upon to perform.

English multiplies out enough of these combinations to define 40 phonemes, a bit above the average for the world's languages. Other languages range from 11 (Polynesian) to 141 (Khoisan or · "Bushman"). The total inventory of phonemes across the world numbers in the thousands, but they are all defined as combinations of the six speech organs and their shapes and motions. Other mouth sounds are not used in any language: scraping teeth, clucking the tongue against the floor of the mouth, making raspberries, and squawking like Donald Duck, for instance. Even the unusual Khoisan and Bantu clicks (similar to the sound of *tsk-tsk* and made famous by the Xhosa pop singer Miriam Makeba) are not miscellanous phonemes added to those languages. Clicking is a manner-of-articulation feature, like stop or fricative, and it combines with all the other features to define a new layer of rows and columns in the language's table of phonemes. There are clicks produced by the lips, tongue tip, and tongue body, any of which can be nasalized or not, voiced or not, and so on, as many as 48 click sounds in all!

An inventory of phonemes is one of the things that gives a language its characteristic sound pattern. For example, Japanese is famous for not distinguishing *r* from *l*. When I arrived in Japan on November 4, 1992, the linguist Masaaki Yamanashi greeted me with a twinkle and said, "In Japan, we have been very interested in Clinton's erection."

We can often recognize a language's sound pattern even in a speech stream that contains no real words, as with the Swedish chef on *The Muppets* or John Belushi's samurai dry cleaner. The linguist Sarah G. Thomason has found that people who claim to be channeling back to past lives or speaking in tongues are really producing gibberish that conforms to a sound pattern vaguely reminiscent of the claimed language. For example, one hypnotized channeler, who claimed to be a nineteenth-century Bulgarian talking to her mother about soldiers laying waste to the countryside, produced generic pseudo-Slavic gobbledygook like this:

> Ovishta reshta rovishta. Vishna beretishti? Ushna barishta dashto. Na darishnoshto. Korapshnoshashit darishtoy. Aobashni bedetpa.

And of course, when the words in one language are pronounced with the sound pattern of another, we call it a foreign accent, as in the following excerpt from a fractured fairy tale by Bob Belviso:

GIACCHE ENNE BINNESTAUCCHE

Uans appona taim uase disse boi. Neimmese Giacche. Naise boi. Live uite ise mamma. Mainde da cao.

Uane dei, di spaghetti ise olle ronne aute. Dei goine feinte fromme no fudde. Mamma soi orais, "Oreie Giacche, teicche da cao enne traide erra forre bocchese spaghetti enne somme uaine."

Bai enne bai commese omme Giacche. I garra no fudde, i garra no uaine. Meichese misteicche, enne traidese da cao forre bonce binnese.

Giacchasse!

What defines the sound pattern of a language? It must be more than just an inventory of phonemes. Consider the following words:

| | | |
|---|---|---|
| ptak | thale | hlad |
| plaft | sram | mgla |
| vlas | flutch | dnom |
| rtut | toasp | nyip |

All of the phonemes are found in English, but any native speaker recognizes that *thale, plaft,* and *flutch* are not English words but could be, whereas the remaining ones are not English words and could not be. Speakers must have tacit knowledge about how phonemes are strung together in their language.

Phonemes are not assembled into words as one-dimensional left-to-right strings. Like words and phrases, they are grouped into units, which are then grouped into bigger units, and so on, defining a tree. The group of consonants (C) at the beginning of a syllable is called an onset; the vowel (V) and any consonants coming after it are called the rime:

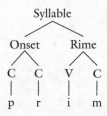

The rules generating syllables define legal and illegal kinds of words in a language. In English an onset can consist of a cluster of consonants, like *flit, thrive,* and *spring,* as long as they follow certain restrictions. (For example, *vlit* and *sring* are impossible.) A rime can consist of a vowel followed by a consonant or certain clusters of consonants, as in *toast, lift,* and *sixths.* In Japanese, in contrast, an onset can have only a single consonant and a rime must be a bare vowel; hence *strawberry ice cream* is translated as *sutoroberi aisukurimo, girl-friend* as *garufurendo.* Italian allows some clusters of consonants in an onset but no consonants at the end of a rime. Belviso used this constraint to simulate the sound pattern of Italian in the Giacche story; *and* becomes *enne, from* becomes *fromme, beans* becomes *binnese.*

Onsets and rimes not only define the possible sounds of a language; they are the pieces of word-sound that are most salient to people, and thus are the units that get manipulated in poetry and word games. Words that rhyme share a rime; words that alliterate share an onset (or just an initial consonant). Pig Latin, eggy-peggy, aygo-paygo, and other secret languages of children tend to splice words at onset-rime boundaries, as does the Yinglish construction in *fancy-shmancy* and *Oedipus-Shmoedipus.* In the 1964 hit song "The Name Game" ("Noam Noam Bo-Boam, Bonana Fana Fo-Foam, Fee Fi Mo Moam, Noam"), Shirley Ellis could have saved several lines in the stanza explaining the rules if she had simply referred to onsets and rimes.

Syllables, in turn, are collected into rhythmic groups called feet:

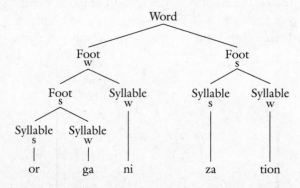

Syllables and feet are classified as strong (s) and weak (w) by other rules, and the pattern of weak and strong branches determines how much stress each syllable will be given when it is pronounced. Feet, like onsets and rhymes, are salient chunks of word that we tend to manipulate in poetry and wordplay. Meter is defined by the kind of feet that go into a line. A succession of feet with a strong-weak pattern is a trochaic meter, as in *Mary had a little lamb;* a succession with a weak-strong pattern is iambic, as in *The rain in Spain falls mainly in the plain.* An argot popular among young ruffians contains forms like *fan-fuckin-tastic, abso-bloody-lutely, Phila-fuckin-delphia,* and *Kalama-fuckin-zoo.* Ordinarily, expletives appear in front of an emphatically stressed word; Dorothy Parker once replied to a question about why she had not been at the symphony lately by saying "I've been too fucking busy and vice versa." But in this lingo they are placed inside a single word, always in front of a stressed foot. The rule is followed religiously: *Philadel-fuckin-phia* would get you launched out of the pool hall.

The assemblies of phonemes in the morphemes and words stored in memory undergo a series of adjustments before they are actually articulated as sounds, and these adjustments give further definition to the sound pattern of a language. Say the words *pat* and *pad.* Now add the inflection *-ing* and pronounce them again: *patting, padding.* In many dialects of English they are now pronounced identically; the original difference between the *t* and the *d* has been obliterated. What obliterated them is a phonological rule called flapping: if a stop consonant produced with the tip of the tongue appears between two vowels, the consonant is pronounced by flicking the tongue against the gum ridge, rather than keeping it there long enough for air pressure to build up. Rules like flapping apply not only when two morphemes are joined, like *pat* and *-ing;* they also apply to one-piece words. For many English speakers *ladder* and *latter,* though they "feel" like they are made out of different sounds and indeed are represented differently in the mental dictionary, are pronounced the same (except in artificially exaggerated speech). Thus when cows come up in conversation, often some wag will speak of an udder mystery, an udder success, and so on.

Interestingly, phonological rules apply in an ordered sequence, as if words were manufactured on an assembly line. Pronounce *write* and *ride*. In most dialects of English, the vowels differ in some way. At the very least, the *i* in *ride* is longer than the *i* in *write*. In some dialects, like the Canadian English of newscaster Peter Jennings, hockey star Wayne Gretzky, and yours truly (an accent satirized a few years back, eh, in the television characters Bob and Doug McKenzie), the vowels are completely different: *ride* contains a diphthong gliding from the vowel in *hot* to the vowel *ee*; *write* contains a diphthong gliding from the higher vowel in *hut* to *ee*. But regardless of exactly how the vowel is altered, it is altered in a consistent pattern: there are no words with long/low *i* followed by *t*, nor with short/high *i* followed by *d*. Using the same logic that allowed Lois Lane in her rare lucid moments to deduce that Clark Kent and Superman were the same, namely that they are never in the same place at the same time, we can infer that there is a single *i* in the mental dictionary, which is altered by a rule before being pronounced, depending on whether it appears in the company of *t* or *d*. We can even guess that the initial form stored in memory is like the one in *ride*, and that *write* is the product of the rule, rather than vice versa. The evidence is that when there is no *t* or *d* after the *i*, as in *rye*, and thus no rule disguising the underlying form, it is the vowel in *ride* that we hear.

Now pronounce *writing* and *riding*. The *t* and *d* have been made identical by the flapping rule. But the two *i*'s are still different. How can that be? It is only the difference between *t* and *d* that causes a difference between the two *i*'s, and that difference has been erased by the flapping rule. This shows that the rule that alters *i* must have applied *before* the flapping rule, while *t* and *d* were still distinct. In other words, the two rules apply in a fixed order, vowel-change before flapping. Presumably the ordering comes about because the flapping rule is in some sense there to make articulation easier and thus is farther downstream in the chain of processing from brain to tongue.

Notice another important feature of the vowel-altering rule. The vowel *i* is altered in front of many different consonants, not just *t*. Compare:

|        |       |
|--------|-------|
| prize  | price |
| five   | fife  |
| jibe   | hype  |
| geiger | biker |

Does this mean there are five different rules that alter *i*—one for *z* versus *s*, one for *v* versus *f*, and so on? Surely not. The change-triggering consonants *t*, *s*, *f*, *p*, and *k* all differ in the same way from their counterparts *d*, *z*, *v*, *b*, and *g:* they are unvoiced, whereas the counterparts are voiced. We need only one rule, then: change *i* whenever it appears before an *unvoiced* consonant. The proof that this is the real rule in people's heads (and not just a way to save ink by replacing five rules with one) is that if an English speaker succeeds in pronouncing the German *ch* in the *Third Reich,* that speaker will pronounce the *ei* as in *write,* not as in *ride.* The consonant *ch* is not in the English inventory, so English speakers could not have learned any rule specifically applying to it. But it is an unvoiced consonant, and if the rule applies to any unvoiced consonant, an English speaker knows exactly what to do.

This selectivity works not only in English but in all languages. Phonological rules are rarely triggered by a single phoneme; they are triggered by an entire class of phonemes that share one or more features (like voicing, stop versus fricative manner, or which organ is doing the articulating). This suggests that rules do not "see" the phonemes in a string but instead look right through them to the features they are made from.

And it is features, not phonemes, that are manipulated by the rules. Pronounce the following past-tense forms:

|         |        |
|---------|--------|
| walked  | jogged |
| slapped | sobbed |
| passed  | fizzed |

In *walked, slapped,* and *passed,* the *-ed* is pronounced as a *t;* in *jogged, sobbed,* and *fizzed,* it is pronounced as a *d.* By now you can probably figure out what is behind the difference: the *t* pronunciation comes after voiceless consonants like *k, p,* and *s;* the *d* comes after voiced ones like *g, b,* and *z*. There must be a rule that adjusts the pronunciation of the suffix *-ed* by peering back into the final phoneme of the stem and checking to see if it has the voicing feature. We can confirm the hunch by asking people to pronounce *Mozart out-Bached Bach*. The verb *to out-Bach* contains the sound *ch*, which does not exist in English. Nonetheless everyone pronounces the *-ed* as a *t*, because the *ch* is unvoiced, and the rule puts a *t* next to any unvoiced consonant. We can even determine whether people store the *-ed* suffix as a *t* in memory and use the rule to convert it to a *d* for some words, or the other way around. Words like *play* and *row* have no consonant at the end, and everyone pronounces their past tenses like *plade* and *rode,* not *plate* and *rote*. With no stem consonant triggering a rule, we must be hearing the suffix in its pure, unaltered form in the mental dictionary, that is, *d*. It is a nice demonstration of one of the main discoveries of modern linguistics: a morpheme may be stored in the mental dictionary in a different form from the one that is ultimately pronounced.

Readers with a taste for theoretical elegance may want to bear with me for one more paragraph. Note that there is an uncanny pattern in what the *d*-to-*t* rule is doing. First, *d* itself is voiced, and it ends up next to voiced consonants, whereas *t* is unvoiced, and it ends up next to unvoiced consonants. Second, except for voicing, *t* and *d* are the same; they use the same speech organ, the tongue tip, and that organ moves in the same way, namely sealing up the mouth at the gum ridge and then releasing. So the rule is not just tossing phonemes around arbitrarily, like changing a *p* to an *l* following a high vowel or any other substitution one might pick at random. It is doing delicate surgery on the *-ed* suffix, adjusting it to be the same in voicing as its neighbor, but leaving the rest of its features alone. That is, in converting *slap* + *ed* to *slapt,* the rule is "spreading" the voicing instruction, packaged with the *p* at the end of *slap,* onto the *-ed* suffix, like this:

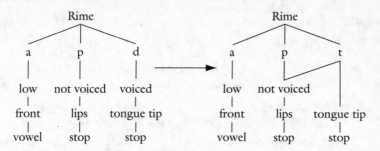

The voicelessness of the *t* in *slapped* matches the voicelessness of the *p* in *slapped* because they are the *same* voicelessness; they are mentally represented as a single feature linked to two segments. This happens very often in the world's languages. Features like voicing, vowel quality, and tones can spread sideways or sprout connections to several phonemes in a word, as if each feature lived on its own horizontal "tier," rather than being tethered to one and only one phoneme.

So phonological rules "see" features, not phonemes, and they adjust features, not phonemes. Recall, too, that languages tend to arrive at an inventory of phonemes by multiplying out the various combinations of some set of features. These facts show that features, not phonemes, are the atoms of linguistic sound stored and manipulated in the brain. A phoneme is merely a bundle of features. Thus even in dealing with its smallest units, the features, language works by using a combinatorial system.

Every language has phonological rules, but what are they for? You may have noticed that they often make articulation easier. Flapping a *t* or a *d* between two vowels is faster than keeping the tongue in place long enough for air pressure to build up. Spreading voicelessness from the end of a word to its suffix spares the talker from having to turn the larynx off while pronouncing the end of the stem and then turn it back on again for the suffix. At first glance, phonological rules seem to be a mere summary of articulatory laziness. And from here it is a small step to notice phonological adjustments in some dialect other than one's own and conclude that they typify the slovenliness of the speakers. Neither side of the Atlantic is safe. George Bernard Shaw wrote:

> The English have no respect for their language and will not teach their children to speak it. They cannot spell it because they have nothing to spell it with but an old foreign alphabet of which only the consonants—and not all of them—have any agreed speech value. Consequently it is impossible for an Englishman to open his mouth without making some other Englishman despise him.

In his article "Howta Reckanize American Slurvian," Richard Lederer writes:

> Language lovers have long bewailed the sad state of pronunciation and articulation in the United States. Both in sorrow and in anger, speakers afflicted with sensitive ears wince at such mumblings as *guvmint* for *government* and *assessories* for *accessories*. Indeed, everywhere we turn we are assaulted by a slew of slurrings.

But if their ears were even more sensitive, these sorrowful speakers might notice that in fact there is no dialect in which sloppiness prevails. Phonological rules give with one hand and take away with the other. The same bumpkins who are derided for dropping *g*'s in *Nothin' doin'* are likely to enunciate the vowels in *pó-lice* and *accidént* that pointy-headed intellectuals reduce to a neutral "uh" sound. When the Brooklyn Dodgers pitcher Waite Hoyt was hit by a ball, a fan in the bleachers shouted, "Hurt's hoit!" Bostonians who pahk their cah in Hahvahd Yahd name their daughters Sheiler and Linder. In 1992 an ordinance was proposed that would have banned the hiring of any immigrant teacher who "speaks with an accent" in—I am not making this up—Westfield, Massachusetts. An incredulous woman wrote to the *Boston Globe* recalling how her native New England teacher defined "homonym" using the example *orphan* and *often*. Another amused reader remembered incurring the teacher's wrath when he spelled "cuh-rée-uh" *k-o-r-e-a* and "cuh-rée-ur" *c-a-r-e-e-r,* rather than vice versa. The proposal was quickly withdrawn.

There is a good reason why so-called laziness in pronunciation is in fact tightly regulated by phonological rules, and why, as a consequence, no dialect allows its speakers to cut corners at will. Every act of sloppiness on the part of a speaker demands a compensating measure of mental effort on the part of the conversational partner. A society of lazy talkers would be a society of hard-working listeners. If speakers were to have their way, all rules of phonology would spread and reduce and delete. But if listeners were to have their way, phonology would do the opposite: it would enhance the acoustic differences between confusable phonemes by forcing speakers to exaggerate or embroider them. And indeed, many rules of phonology do that. (For example, there is a rule that forces English speakers to round their lips while saying *sh* but not while saying *s.* The benefit of forcing everyone to make this extra gesture is that the long resonant chamber formed by the pursed lips enhances the lower-frequency noise that distinguishes *sh* from *s,* allowing for easier identification of the *sh* by the listener.) Although every speaker soon becomes a listener, human hypocrisy would make it unwise to depend on the speaker's foresight and consideration. Instead, a single, partly arbitrary set of phonological rules, some reducing, some enhancing, is adopted by every member of a linguistic community when he or she acquires the local dialect as a child.

Phonological rules help listeners even when they do not exaggerate some acoustic difference. By making speech patterns predictable, they add redundancy to a language; English text has been estimated as being between two and four times as long as it has to be for its information content. For example, this book takes up about 900,000 characters on my computer disk, but my file compression program can exploit the redundancy in the letter sequences and squeeze it into about 400,000 characters; computer files that do not contain English text cannot be squished nearly that much. The logician Quine explains why many systems have redundancy built in:

It is the judicious excess over minimum requisite support. It is why a good bridge does not crumble when subjected to stress beyond what reasonably could have been foreseen. It is fallback and failsafe. It is why we address our mail to city and state in so many words, despite the zip code. One indistinct digit in the zip code would spoil everything. . . . A kingdom, legend tells us, was lost for want of a horseshoe nail. Redundancy is our safeguard against such instability.

Thanks to the redundancy of language, yxx cxn xndxrstxnd whxt x xm wrxtxng xvsn xf x rxplxcx xll thx vxwxls wxth xn "x" (t gts lttl hrdr f y dn't vn kn whr th vwls r). In the comprehension of speech, the redundancy conferred by phonological rules can compensate for some of the ambiguity in the sound wave. For example, a listener can know that "thisrip" must be *this rip* and not *the srip* because the English consonant cluster *sr* is illegal.

So why is it that a nation that can put a man on the moon cannot build a computer that can take dictation? According to what I have explained so far, each phoneme should have a telltale acoustic signature: a set of resonances for vowels, a noise band for fricatives, a silence-burst-transition sequence for stops. The sequences of phonemes are massaged in predictable ways by ordered phonological rules, whose effects could presumably be undone by applying them in reverse.

The reason that speech recognition is so hard is that there's many a slip 'twixt brain and lip. No two people's voices are alike, either in the shape of the vocal tract that sculpts the sounds, or in the person's precise habits of articulation. Phonemes also sound very different depending on how much they are stressed and how quickly they are spoken; in rapid speech, many are swallowed outright.

But the main reason an electric stenographer is not just around the corner has to do with a general phenomenon in muscle control called coarticulation. Put a saucer in front of you and a coffee cup a foot or so away from it on one side. Now quickly touch the saucer and pick up the cup. You probably touched the saucer at the edge nearest the cup, not dead center. Your fingers probably assumed the handle-grasping posture while your hand was making its way to the cup, well before it arrived. This graceful smoothing and overlapping of gestures is ubiquitous in motor control. It reduces the forces necessary to move body parts around and lessens the wear and tear on the joints. The tongue and throat are no different. When we want to articulate a phoneme, our tongue cannot assume the target posture instantaneously; it is a heavy slab of meat that takes time to heft into place. So while we are moving it, our brains are anticipating the next posture in planning the trajectory, just like the cup-and-saucer maneuver. Among the range of positions in the mouth that can define a phoneme, we place the tongue in the one that offers the shortest path to the target for the next phoneme. If the current phoneme does not specify where a speech organ should be, we anticipate where the next phoneme wants it to be and put it there in advance. Most of us are completely unaware of these adjustments until they are called to our attention. Say *Cape Cod*. Until now you probably never noticed that your tongue body is in different positions for the two *k* sounds. In *horseshoe*, the first *s* becomes a *sh*; in *NPR*, the *n* becomes an *m*; in *month* and *width*, the *n* and *d* are articulated at the teeth, not the usual gum ridge.

Because sound waves are minutely sensitive to the shapes of the cavities they pass through, this coarticulation wreaks havoc with the speech sound. Each phoneme's sound signature is colored by the phonemes that come before and after, sometimes to the point of having nothing in common with its sound signature in the company of a different set of phonemes. That is why you cannot cut up a tape of the sound *cat* and hope to find a beginning piece that contains the *k* alone. As you make earlier and earlier cuts, the piece may go from sounding like *ka* to sounding like a chirp or whistle. This shingling of phonemes in the speech stream could, in principle, be a boon to an optimally designed speech recognizer. Consonant and vowels are being signaled simultaneously, greatly increasing the rate of phonemes per second, as I noted at the beginning of this chapter, and there are many redundant sound cues to a given phoneme. But this advantage can be enjoyed only by a high-tech speech recognizer, one that has some kind of knowledge of how vocal tracts blend sounds.

The human brain, of course, is a high-tech speech recognizer, but no one knows how it succeeds. For this reason psychologists who study speech perception and engineers who build speech recognition machines keep a close eye on each other's work. Speech recognition may be so hard that there are only a few ways it could be solved in principle. If so, the way the brain does it may offer hints as to the best way to build a machine to do it, and how a successful machine does it may suggest hypotheses about how the brain does it.

Early in the history of speech research, it became clear that human listeners might somehow take advantage of their expectations of the kinds of things a speaker is likely to say. This could narrow down the alternatives left open by the acoustic analysis of the speech signal. We have already noted that the rules of phonology provide one sort of redundancy that can be exploited, but people might go even farther. The psychologist George Miller played tapes of sentences in background noise and asked people to repeat back exactly what they heard. Some of the sentences followed the rules of English syntax and made sense.

> Furry wildcats fight furious battles.
> Respectable jewelers give accurate appraisals.
> Lighted cigarettes create smoky fumes.
> Gallant gentlemen save distressed damsels.
> Soapy detergents dissolve greasy stains.

Others were created by scrambling the words within phrases to create colorless-green-ideas sentences, grammatical but nonsensical:

> Furry jewelers create distressed stains.
> Respectable cigarettes save greasy battles.
> Lighted gentlemen dissolve furious appraisals.
> Gallant detergents fight accurate fumes.
> Soapy wildcats give smoky damsels.

A third kind was created by scrambling the phrase structure but keeping related words together, as in

> Furry fight furious wildcat battles.
> Jewelers respectable appraisals accurate give.

Finally, some sentences were utter word salad, like

> Furry create distressed jewelers stains.
> Cigarettes respectable battles greasy save.

People did best with the grammatical sensible sentences, worse with the grammatical nonsense and the ungrammatical sense, and worst of all with the ungrammatical nonsense. A few years later the psychologist Richard Warren taped sentences like *The state governors met with their respective legislatures convening in the capital city,* excised the first *s* from *legislatures,* and spliced in a cough. Listeners could not tell that any sound was missing.

If one thinks of the sound wave as sitting at the bottom of a hierarchy from sounds to phonemes to words to phrases to the meanings of sentences to general knowledge, these demonstrations seem to imply that human speech perception works from the top down rather than just from the bottom up. Maybe we are constantly guessing what a speaker will say next, using every scrap of conscious and unconscious knowledge at our disposal, from how coarticulation distorts sounds, to the rules of English phonology, to the rules of English syntax, to stereotypes about who tends to do what to whom in the world, to hunches about what our conversational partner has in mind at that very moment. If the expectations are accurate enough, the acoustic analysis can be fairly crude; what the sound wave lacks, the context can fill in. For example, if you are listening to a discussion about the destruction of ecological habitats, you might be on the lookout for words pertaining to threatened animals and plants, and then when you hear speech sounds whose phonemes you cannot pick out like "eesees," you would perceive it correctly as *species*—unless you are Emily Litella, the hearing-impaired editorialist on *Saturday Night Live* who argued passionately against the campaign to protect endangered feces. (Indeed, the humor in the Gilda Radner character, who also fulminated against saving Soviet jewelry, stopping violins in the streets, and preserving natural racehorses, comes not from her impairment at the bottom of the speech-processing system but from her ditziness at the top, the level that should have prevented her from arriving at her interpretations.)

The top-down theory of speech perception exerts a powerful emotional tug on some people. It confirms the relativist philosophy that we hear what we expect to hear, that our knowledge determines our perception, and ultimately that we are not in direct contact with any objective reality. In a sense, perception that is strongly driven from the top down would be a barely controlled hallucination, and that is the problem. A perceiver forced to rely on its expectations is at a severe disadvantage in a world that is unpredictable even under the best of circumstances. There is a reason to believe that human speech perception is, in fact, driven quite strongly by acoustics. If you have an indulgent friend, you can try the following experiment. Pick ten words at random out of a dictionary, phone up the friend, and say the words clearly. Chances are the friend will reproduce them perfectly, relying only on the information in the sound wave and knowledge of English vocabulary and phonology. The friend could not have been using any higher-level expectations about phrase structure, context, or story line because a list of words blurted out of the blue has none. Though we may call upon high-level conceptual knowledge in noisy or degraded circumstances (and even here it is not clear whether the knowledge alters perception or just allows us to guess intelligently after the fact), our brains seem designed to squeeze every last drop of phonetic information out of the sound wave itself. Our sixth sense may perceive speech as language, not as sound, but it *is* a sense, something that connects us to the world, and not just a form of suggestibility.

Another demonstration that speech perception is not the same thing as fleshing out expectations comes from an illusion that the columnist Jon Carroll has called the mondegreen, after his mis-hearing of the folk ballad "The Bonnie Earl O'Moray":

Oh, ye hielands and ye lowlands,
Oh, where hae ye been?
They have slain the Earl of Moray,
And laid him on the green.

He had always thought that the lines were "They have slain the Earl of Moray, And Lady Mondegreen." Mondegreens are fairly common (they are an extreme version of the Pullet Surprises and Pencil Vaneas mentioned earlier); here are some examples:

A girl with colitis goes by. [A girl with kaleidoscope eyes. From the Beatles song "Lucy in the Sky with Diamonds."]
Our father wishart in heaven; Harold be thy name . . . Lead us not into Penn Station.
Our father which art in Heaven; hallowed by thy name . . . Lead us not into temptation. From the Lord's Prayer.]
He is trampling out the vintage where the grapes are wrapped and stored. [. . . grapes of wrath are stored. From "The Battle Hymn of the Republic."]
Gladly the cross-eyed bear. [Gladly the cross I'd bear.]
I'll never be your pizza burnin'. [. . . your beast of burden. From the Rolling Stones song.]
It's a happy enchilada, and you think you're gonna drown. [It's a half an inch of water . . . From the John Prine song "That's the Way the World Goes 'Round."]

The interesting thing about mondegreens is that the mishearings are generally *less* plausible than the intended lyrics. In no way do they bear out any sane listener's general expectations of what a speaker is likely to say or mean. (In one case a student stubbornly misheard the Shocking Blue hit song "I'm Your Venus" as "I'm Your Penis" and wondered how it was allowed on the radio.) The mondegreens do conform to English phonology, English syntax (sometimes), and English vocabulary (though not always, as in the word *mondegreen* itself). Apparently, listeners lock in to some set of words that fit the sound and that hang together more or less as English words and phrases, but plausibility and general expectations are not running the show.

The history of artificial speech recognizers offers a similar moral. In the 1970s a team of artificial intelligence researchers at Carnegie-Mellon University headed by Raj Reddy designed a computer program called HEARSAY that interpreted spoken commands to move chess pieces. Influenced by the top-down theory of speech perception, they designed the program as a "community" of "expert" subprograms cooperating to give the most likely interpretation of the signal. There were subprograms that specialized in acoustic analysis, in phonology, in the dictionary, in syntax, in rules for the legal moves of chess, even in chess strategy as applied to the game in progress. According to one story, a general from the defense agency that was funding the research came up for a demonstration. As the scientists sweated he was seated in front of a chessboard and a microphone hooked up to the computer. The general cleared his throat. The program printed "Pawn to King 4."

The recent program DragonDictate, mentioned earlier in the chapter, places the burden more on good acoustic, phonological, and lexical analyses, and that seems to be responsible for its greater success. The program has a dictionary of words and their sequences of phonemes. To help anticipate the effects of phonological rules and coarticulation, the program is told what every English phoneme sounds like in the context of every possible preceding phoneme and every possible following phoneme. For each word, these phonemes-in-context are arranged into a little chain, with a probability attached to each transition from one sound unit to the next. This chain serves as a crude model of the speaker, and when a real speaker uses the system, the probabilities in the chain are adjusted to capture that person's manner of speaking. The entire word, too, has a probability attached to it, which depends on its frequency in the language and on the speaker's habits. In some versions of the program, the probability value for a word is adjusted depending on which word precedes it; this is the only top-down information that the program uses. All this knowledge allows the program to calculate which word is most likely to have come out of the mouth of the speaker given the input sound. Even then, DragonDictate relies more on expectancies than an able-eared human does. In the demonstration I saw, the program had to be coaxed into recognizing *word* and *worm*, even when they were pronounced as clear as a bell, because it kept playing the odds and guessing higher-frequency *were* instead.

Now that you know how individual speech units are produced, how they are represented in the mental dictionary, and how they are rearranged and smeared before they emerge from the mouth, you have reached the prize at the bottom of this chapter: why English spelling is not as deranged as it first appears.

The complaint about English spelling, of course, is that it pretends to capture the sounds of words but does not. There is a long tradition of doggerel making this point, of which this stanza is a typical example:

> Beware of heard, a dreadful word
> That looks like beard and sounds like bird,
> And dead: it's said like bed, not bead—
> For goodness' sake don't call it "deed"!
> Watch out for meat and great and threat
> (They rhyme with suite and straight and debt).

George Bernard Shaw led a vigorous campaign to reform the English alphabet, a system so illogical, he said, that it could spell *fish* as "ghoti"—*gh* as in *tough, o* as in *women, ti* as in *nation*. ("Mnom-noupte" for *minute* and "mnopspteiche" for *mistake* are other examples.) In his will Shaw bequeathed a cash prize to be awarded to the designer of a replacement alphabet for English, in which each sound in the spoken language would be recognizable by a single symbol: He wrote:

> To realize the annual difference in favour of a forty-two letter phonetic alphabet . . . you must multiply the number of minutes in the year, the number of people in the world who are

continuously writing English words, casting types, manufacturing printing and writing machines, by which time the total figure will have become so astronomical that you will realize that the cost of spelling even one sound with two letters has cost us centuries of unnecessary labour. A new British 42 letter alphabet would pay for itself a million times over not only in hours but in moments. When this is grasped, all the useless twaddle about enough and cough and laugh and simplified spelling will be dropped, and the economists and statisticians will be set to work to gather in the orthographic Golconda.

My defense of English spelling will be halfhearted. For although language is an instinct, written language is not. Writing was invented a small number of times in history, and alphabetic writing, where one character corresponds to one sound, seems to have been invented only once. Most societies have lacked written language, and those that have it inherited it or borrowed it from one of the inventors. Children must be taught to read and write in laborious lessons, and knowledge of spelling involves no daring leaps from the training examples like the leaps we saw in Simon, Mayela, and the Jabba and *mice-eater* experiments in Chapters 3 and 5. And people do not uniformly succeed. Illiteracy, the result of insufficient teaching, is the rule in much of the world, and dyslexia, a presumed congenital difficulty in learning to read even with sufficient teaching, is a severe problem even in industrial societies, found in five to ten percent of the population.

But though writing is an artificial contraption connecting vision and language, it must tap into the language system at well-demarcated points, and that gives it a modicum of logic. In all known writing systems, the symbols designate only three kinds of linguistic structure: the morpheme, the syllable, and the phoneme. Mesopotamian cuneiform, Egyptian hieroglyphs, Chinese logograms, and Japanese kanji encode morphemes. Cherokee, Ancient Cypriot, and Japanese kana are syllable-based. All modern phonemic alphabets appear to be descended from a system invented by the Canaanites around 1700 B.C. No writing system has symbols for actual sound units that can be identified on an oscilloscope or spectrogram, such as a phoneme as it is pronounced in a particular context or a syllable chopped in half.

Why has no writing system ever met Shaw's ideal of one symbol per sound? As Shaw himself said elsewhere, "There are two tragedies in life. One is not to get your heart's desire. The other is to get it." Just think back to the workings of phonology and coarticulation. A true Shavian alphabet would mandate different vowels in *write* and *ride,* different consonants in *write* and *writing,* and different spellings for the past-tense suffix in *slapped, sobbed,* and *sorted. Cape Cod* would lose its visual alliteration. A *horse* would be spelled differently from its *horseshoe,* and National Public Radio would have the enigmatic abbreviation *MPR.* We would need brand-new letters for the *n* in *month* and the *d* in *width.* I would spell *often* differently from *orphan,* but my neighbors here in the Hub would not, and their spelling of *career* would be my spelling of *Korea* and vice versa.

Obviously, alphabets do not and should not correspond to sounds; at best they correspond to the phonemes specified in the mental dictionary. The actual sounds are different in different contexts, so true phonetic spelling would only obscure their underlying identity. The surface sounds are predictable by phonological rules, though, so there is no need to clutter up the page with symbols for the actual sounds; the reader needs only the abstract blueprint for a word and can flesh out the sound if needed. Indeed, for about eighty-four percent of English words, spelling is completely predictable from regular rules. Moreover, since dialects separated by time and space often differ most in the phonological rules that convert mental dictionary entries into pronunciations, a spelling corresponding to the underlying entries, not the sounds, can be widely shared. The words with truly weird spellings (like *of, people, women, have, said, do, done,* and *give*) generally are the commonest ones in the language, so there is ample opportunity for everyone to memorize them.

Even the less predictable aspects of spelling bespeak hidden linguistic regularities. Consider the following pairs of words where the same letters get different pronunciations:

| | |
|---|---|
| electric–electricity | declare–declaration |
| photograph–photography | muscle–muscular |
| grade–gradual | condemn–condemnation |
| history–historical | courage–courageous |
| revise–revision | romantic–romanticize |
| adore–adoration | industry–industrial |
| bomb–bombard | fact-factual |
| nation–national | inspire–inspiration |
| critical–criticize | sign–signature |
| mode–modular | malign–malignant |
| resident–residential | |

Once again the similar spellings, despite differences in pronunciation, are there for a reason: they are identifying two words as being based on the same root morpheme. This shows that English spelling is not completely phonemic; sometimes letters encode phonemes, but sometimes a sequence of letters is specific to a morpheme. And a morphemic writing system is more useful than you might think. The goal of reading, after all, is to understand the text, not to pronounce it. A morphemic spelling can help a reader distinguishing homophones, like *meet* and *mete*. It can also tip off a reader that one word contains another (and not just a phonologically identical impostor). For example, spelling tells us that *overcome* contains *come,* so we know that its past tense must be *overcame,* whereas *succumb* just contains the sound "kum," not the morpheme *come,* so its past tense is not *succame* but *succumbed*. Similarly, when something *recedes,* one has a *recession,* but when someone *re-seeds* a lawn, we have a *re-seeding.*

In some ways, a morphemic writing system has served the Chinese well, despite the inherent disadvantage that readers are at a loss when they face a new or rare word. Mutually unintelligible dialects can share texts (even if their speakers pronounce the words very differently), and many documents that are thousands of years old are readable by modern speakers. Mark Twain alluded to such inertia in our own Roman writing system when he wrote, "They spell it Vinci and pronounce it Vinchy; foreigners always spell better than they pronounce."

Of course English spelling could be better than it is. But it is already much better than people think it is. That is because writing systems do not aim to represent the actual sounds of talking, which we do not hear, but the abstract units of language underlying them, which we do hear.

# 7

❧

# Talking Heads

*For centuries, people have been terrified that their programmed creations* might outsmart them, overpower them, or put them out of work. The fear has long been played out in fiction, from the medieval Jewish legend of the Golem, a clay automaton animated by an inscription of the name of God placed in its mouth, to HAL, the mutinous computer of *2001: A Space Odyssey.* But when the branch of engineering called "artificial intelligence" (AI) was born in the 1950s, it looked as though fiction was about to turn into frightening fact. It is easy to accept a computer calculating pi to a million decimal places or keeping track of a company's payroll, but suddenly computers were also proving theorems in logic and playing respectable chess. In the years following there came computers that could beat anyone but a grand master, and programs that outperformed most experts at recommending treatments for bacterial infections and investing pension funds. With computers solving such brainy tasks, it seemed only a matter of time before a C3PO or a Terminator would be available from the mail-order catalogues; only the easy tasks remained to be programmed. According to legend, in the 1970s Marvin Minsky, one of the founders of AI, assigned "vision" to a graduate student as a summer project.

But household robots are still confined to science fiction. The main lesson of thirty-five years of AI research is that the hard problems are easy and the easy problems are hard. The mental abilities of a four-year-old that we take for granted—recognizing a face, lifting a pencil, walking across a room, answering a question—in fact solve some of the hardest engineering problems ever conceived. Do not be fooled by the assembly-line robots in the automobile commercials; all they do is weld and spray-paint, tasks that do not require these clumsy Mr. Magoos to see or hold or place anything. And if you want to stump an artificial intelligence system, ask it questions like, Which is bigger, Chicago or a breadbox? Do zebras wear underwear? Is the floor likely to rise up and bite you? If Susan goes to the store, does her head go with her? Most fears of automation are misplaced. As the new generation of intelligent devices appears, it will be the stock analysts and petrochemical engineers and parole board members who are in danger of being replaced by machines. The gardeners, receptionists, and cooks are secure in their jobs for decades to come.

Understanding a sentence is one of these hard easy problems. To interact with computers we still have to learn their languages; they are not smart enough to learn ours. In fact, it is all too easy to give computers more credit at understanding than they deserve.

Recently an annual competition was set up for the computer program that can best fool users into thinking that they are conversing with another human. The competition for the Loebner Prize was intended to implement a suggestion made by Alan Turing in a famous 1950 paper. He suggested that the philosophical question "Can machines think?" could best be answered in an imitation game, where a judge converses with a person over one terminal and with a computer programmed to imitate a person on another. If the judge cannot guess which is which, Turing suggested, there is no basis for denying that the computer can think. Philosophical questions aside, it was apparent to the committee charged with overseeing the competition that no program could come close to winning the $100,000 prize, so they devised a $1,500 version that would be fairer to the state of the art. Each of the judges had to stick to a single topic of conversation selected by the programmer or by the human foil, whichever it was, and the judge was not allowed to engage in any "trickery or guile" such as repeating a question ten times or asking whether zebras wear underwear; the conversation had to be "natural." After interacting with several programs and human foils for about seven minutes apiece, the judges ranked all the humans as more humanlike than any of the computers. About half the judges did, however, misidentify the winning program as human.

The accomplishment is less impressive than it sounds. The rules handcuffed the judges: "unnatural trickery or guile" is another way of referring to any attempt to determine whether one is conversing with a human or a machine, which is the whole point of the test! Also, the winning programmer shrewdly exploited the opportunity to designate the topic of conversation for his program. He chose "whimsical conversation," which is a dubious example of a "topic," and which, by definition, can be full of non sequiturs:

> Judge: It has been a long day! Now I think I'm ready for some fun. The weather isn't very pleasant today. Do you have any jokes that could lift the clouds?
>
> Computer: Everybody talks about the weather but nobody seems to do much about it. Why do you need to tell me now you think you're ready for some fun? Groannnnnn . . .
>
> Judge: You're trying to make this difficult for me aren't you? I just explained that I was ready for fun!
>
> Computer: Would you prefer it if it was not true that I are trying to make this difficult for you aren't you? I just explained that I was ready for fun! I am, therfore I compute . . . What makes you think I am trying to make this difficult for you aren't I?
>
> Judge: I can't say I understand the question . . . but the whimsy of a poet is one thing and the whimsy of most of us is another.
>
> Computer: How do you know you can't say you understand the question . . . but the whimsy of a poet is one thing and the whimsy of most of us is another.

In fact the programmer had simply updated a technique invented in 1965 by the computer scientist Joseph Weizenbaum. Weizenbaum wrote a rather stupid program called ELIZA, named after the Shaw heroine, which looked for a few key words in an input line, retrieved a canned response from a list, and inserted fragments of the input line into its response (with some minor substitutions, like "you are" for "I am"). It could thus crudely simulate a conversation. One version, called Doctor, mimicked a Rogerian psychotherapist, taking advantage of Carl Rogers' philosophy that a therapist should recast patients' statements back to them for their contemplation. A version of the program was later sold as a computer game in Radio Shack stores, and another comes packaged with my word processing program, GNU Emacs. (The documentation says, "If using Emacs becomes terribly frustrating and none of the techniques described above solve the problem, Emacs can still help you. Type 'M-x doctor RET.' ") Here is a typical dialogue:

User: I have been rather depressed lately.
Computer: Are you depressed often?
User: Yes, almost all the time.
Computer: Why do you say yes almost all the time?
User: Right now I feel like I could cry.
Computer: Is it because right now you feel like you could cry
that you came to me?

To Weizenbaum's horror, one day he found his secretary transfixed at the terminal, pouring her heart out to the program.

Most computer scientists are annoyed by the Loebner Prize competition. They consider it a pointless publicity stunt, because it is an exercise in how to fool an amateur, not how to get computers to use language. (Artificial intelligence researchers and other professionals who are knowledgeable about language were not allowed to act as judges, and none bothered to compete; the submissions were from hobbyists.) It is about as productive as promoting biology by offering a prize to the designer of the most convincing silk flower, or running a space program by simulating a moon landing on a Hollywood back lot.

There has been intensive research on computer language-understanding systems, but no serious engineer has the hubris to predict that the systems will duplicate the human ability anytime soon.

In fact, from a scientist's perspective, people have no right to be as good at sentence understanding as they are. Not only can they solve a viciously complex task, but they solve it *fast*. Comprehension ordinarily takes place in "real time." Listeners keep up with talkers; they do not wait for the end of a batch of speech and interpret it after a proportional delay, like a critic reviewing a book. And the lag between speaker's mouth and listener's mind is remarkably short: about a syllable or two, around half a second. Some people can understand *and* repeat sentences, shadowing a speaker as he speaks, with a lag of a quarter of a second!

Understanding understanding has practical applications other than building machines we can converse with. Human sentence comprehension is fast and powerful, but it is not perfect. It works when the incoming conversation or text is structured in certain ways. When it is not, the process can bog down, backtrack, and misunderstand. As we explore language understanding in this chapter, we will discover which kinds of sentences mesh with the mind of the understander. One practical benefit is a set of guidelines for clear prose, a scientific style manual, such as Joseph Williams' 1990 *Style: Toward Clarity and Grace,* which is informed by many of the findings we will examine.

Another practical application involves the law. Judges are frequently faced with guessing how a typical person is likely to understand some ambiguous passage, such as a customer scanning a contract, a jury listening to instructions, or a member of the public reading a potentially libelous characterization. Many of the people's habits of interpretation have been worked out in the laboratory, and the linguist and lawyer Lawrence Solan has explained the connections between language and law in his interesting 1993 book *The Language of Judges,* to which we will return.

How do we understand a sentence? The first step is to "parse" it. This does not refer to the exercises you grudgingly did in elementary school, which Dave Barry's "Ask Mr. Language Person" remembers as follows:

> Q. Please explain how to diagram a sentence.
> A. First spread the sentence out on a clean, flat surface, such as an ironing board. Then, using a sharp pencil or X-Acto knife, locate the "predicate," which indicates where the action has taken place and is usually located directly behind the gills. For example, in the sentence: "LaMont never would of bit a forest ranger," the action probably took place in forest. Thus your diagram would be shaped like a little tree with branches sticking out of it to indicate the locations of the various particles of speech, such as your gerunds, proverbs, adjutants, etc.

But it does involve a similar process of finding subject, verbs, objects, and so on, that takes place unconsciously. Unless you are Woody Allen speed-reading *War and Peace,* you have to group words into phrases, determine which phrase is the subject of which verb, and so on. For example, to understand the sentence *The cat in the hat came back,* you have to group the words *the cat in the hat* into one phrase, to see that it is the cat that came back, not just the hat. To distinguish *Dog bites man* from *Man bites dog,* you have to find the subject and object. And to distinguish *Man bites dog* from *Man is bitten by dog* or *Man suffers dog bite,* you have to look up the verbs' entries in the mental dictionary to determine what the subject, *man,* is doing or having done to him.

Grammar itself is a mere code or protocol, a static database specifying what kinds of sounds correspond to what kinds of meanings in a particular language. It is not a recipe or program for speaking and understanding. Speaking and understanding share a grammatical database (the language we speak is the same as the language we understand), but they also need procedures that specify what the mind should *do,* step by step, when the words start pouring in or when one is about to speak. The mental program that analyzes sentence structure during language comprehension is called the parser.

The best way to appreciate how understanding works is to trace the parsing of a simple sentence, generated by a toy grammar like the one of Chapter 4, which I repeat here:

S → NP VP
"A sentence can consist of a noun phrase and a verb phrase."

NP → (det) N (PP)
"A noun phrase can consist of an optional determiner, a noun, and an optional prepositional phrase."

VP → V NP (PP)
"A verb phrase can consist of a verb, a noun phrase, and an optional prepositional phrase."

PP → P NP
"A prepositional phrase can consist of a preposition and a noun phrase."

N → boy, girl, dog, cat, ice cream, candy, hotdogs
"The nouns in the mental dictionary include *boy, girl,* . . ."

V → eats, likes, bites
"The verbs in the mental dictionary include *eats, likes, bites.*"

P → with, in, near
"The prepositions include *with, in, near.*"

det → a, the, one
"The determiners include *a, the, one.*"

Take the sentence *The dog likes ice cream.* The first word arriving at the mental parser is *the.* The parser looks it up in the mental dictionary, which is equivalent to finding it on the right-hand side of a rule and discovering its category on the left-hand side. It is a determiner (det). This allows the parser to grow the first twig of the tree for the sentence. (Admittedly, a tree that grows upside down from its leaves to its root is botanically improbable.)

det
|
the...

Determiners, like all words, have to be part of some larger phrase. The parser can figure out which phrase by checking to see which rule has "det" on its right-hand side. That rule is the one defining a noun phrase, NP. More tree can be grown:

```
        NP
       /  \
     det    N
      |
     the...
```

This dangling structure must be held in a kind of memory. The parser keeps in mind that the word at hand, *the*, is part of a noun phrase, which soon must be completed by finding words that fill its other slots—in this case, at least a noun.

In the meantime, the tree continues to grow, for NP's cannot float around unattached. Having checked the right-hand sides of the rules for an NP symbol, the parser has several options. The freshly built NP could be part of a sentence, part of a verb phrase, or part of a prepositional phrase. The choice can be resolved from the root down: all words and phrases must eventually be plugged into a sentence (s), and a sentence must begin with an NP, so the sentence rule is the logical one to use to grow more of the tree:

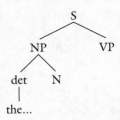

Note that the parser is now keeping *two* incomplete branches in memory: the noun phrase, which needs an N to complete it, and the sentence, which needs a VP.

The dangling N twig is equivalent to a prediction that the next word should be a noun. When the next word, *dog*, comes in, a check against the rules confirms the prediction: *dog* is part of the N rule.

This allows *dog* to be integrated into the tree, completing the noun phrase:

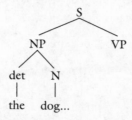

The parser no longer has to remember that there is an NP to be completed; all it has to keep in mind is the incomplete S.

At this point some of the meaning of the sentence can be inferred. Remember that the noun inside a noun phrase is a head (what the phrase is about) and that other phrases inside the noun phrase can modify the head. By looking up the definitions of *dog* and *the* in their dictionary entries, the parser can note that the phrase is referring to a previously mentioned dog.

The next word is *likes,* which is found to be a verb, V. A verb has nowhere to come from but a verb phrase, VP, which, fortunately, has already been predicted, so they can just be joined up. The verb phrase contains more than a V; it also has a noun phrase (its object). The parser therefore predicts that an NP is what should come next:

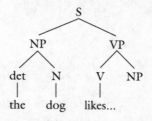

What does come next is *ice cream,* a noun, which can be part of an NP—just as the dangling NP branch predicts. The last pieces of the puzzle snap nicely together:

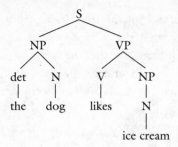

The word *ice cream* has completed the noun phrase, so it need not be kept in memory any longer; the NP has completed the verb phrase, so it can be forgotten, too; and the VP has completed the sentence. When memory has been emptied of all its incomplete dangling branches, we experience the mental "click" that signals that we have just heard a complete grammatical sentence.

As the parser has been joining up branches, it has been building up the meaning of the sentence, using the definitions in the mental dictionary and the principles for combining them. The verb is the head of its VP, so the VP is about liking. The NP inside the VP, *ice cream,* is the verb's object. The dictionary entry for *likes* says that its object is the liked entity; therefore the VP is about being fond of ice cream. The NP to the left of a tensed verb is the subject; the entry for *likes* says that its subject is the one doing the liking. Combining the semantics of the subject with the semantics of the VP, the parser has determined that the sentence asserts that an aforementioned canine is fond of frozen confections.

Why is it so hard to program a computer to do this? And why do people, too, suddenly find it hard to do this when reading bureaucratese and other bad writing? As we stepped our way through the sentence pretending we were the parser, we faced two computational burdens. One was memory: we had to keep track of the dangling phrases that needed particular kinds of words to complete them. The other was decision-making: when a word or phrase was found on the right-hand side of two different rules, we had to decide which to use to build the next branch of the tree. In accord with the first law of artificial intelligence, that the hard problems are easy and the easy problems are hard, it turns out that the memory part is easy for computers and hard for people, and the decision-making part is easy for people (at least when the sentence has been well constructed) and hard for computers.

A sentence parser requires many kinds of memory, but the most obvious is the one for incomplete phrases, the remembrance of things parsed. Computers must set aside a set of memory locations, usually called a "stack," for this task; this is what allows a parser to use phrase structure grammar at all, as opposed to being a word-chain device. People, too, must dedicate some of their short-term memory to dangling phrases. But short-term memory is the primary bottleneck in human information processing. Only a few items—the usual estimate is seven, plus or minus two—can be held in the mind at once, and the items are immediately subject to fading or being overwritten. In the following sentences you can feel the effects of keeping a dangling phrase open in memory too long:

> He gave the girl that he met in New York while visiting his parents for ten days around Christmas and New Year's the candy.
>
> He sent the poisoned candy that he had received in the mail from one of his business rivals connected with the Mafia to the police.
>
> She saw the matter that had caused her so much anxiety in former years when she was employed as an efficiency expert by the company through.
>
> That many teachers are being laid off in a shortsighted attempt to balance this year's budget at the same time that the governor's cronies and bureaucratic hacks are lining their pockets is appalling.

These memory-stretching sentences are called "top-heavy" in style manuals. In languages that use case markers to signal meaning, a heavy phrase can simply be slid to the end of the sentences, so the listener can digest the beginning without having to hold the heavy phrase in mind. English is tyrannical about order, but even English provides its speakers with some alternative constructions in which the order of phrases is inverted. A considerate writer can use them to save the heaviest for last and lighten the burden on the listener. Note how much easier these sentences are to understand:

> He gave the candy to the girl that he met in New York while visiting his parents for ten days around Christmas and New Year's.
>
> He sent to the police the poisoned candy that he had received in the mail from one of his business rivals connected with the Mafia.
>
> She saw the matter through that had caused her so much anxiety in former years when she was employed as an efficiency expert by the company.
>
> It is appalling that teachers are being laid off in a short-sighted attempt to balance this year's budget at the same time that the governor's cronies and bureaucratic hacks are lining their pockets.

Many linguists believe that the reason that languages allow phrase movement, or choices among more-or-less synonymous constructions, is to ease the load on the listener's memory.

As long as the words in a sentence can be immediately grouped into complete phrases, the sentence can be quite complex but still understandable:

> Remarkable is the rapidity of the motion of the wing of the hummingbird.
>
> This is the cow with the crumpled horn that tossed the dog that worried the cat that killed the rat that ate the malt that lay in the house that Jack built.
>
> Then came the Holy One, blessed be He, and destroyed the angel of death that slew the butcher that killed the ox that drank the water that quenched the fire that burned the stick

that beat the dog that bit the cat my father bought for two
zuzim.

These sentences are called "right-branching," because of the geometry of their phrase structure trees. Note that as one goes from left to
right, only one branch has to be left dangling at a time:

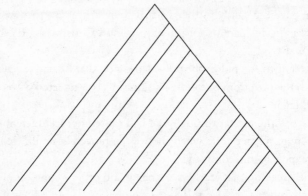

Remarkable is the rapidity of the motion of the wing of the hummingbird

Sentences can also branch to the left. Left-branching trees are most
common in head-last languages like Japanese but are found in a few
constructions in English, too. As before, the parser never has to keep
more than one dangling branch in mind at a time:

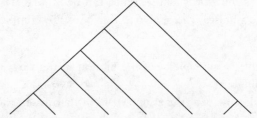

The hummingbird's wing's motion's rapidity is remarkable

There is a third kind of tree geometry, but it goes down far less
easily. Take the sentence

The rapidity that the motion has is remarkable.

The clause *that the motion has* has been embedded in the noun phrase containing *The rapidity*. The result is a bit stilted but easy to understand. One can also say

The motion that the wing has is remarkable.

But the result of embedding the *motion that the wing has* phrase inside the *rapidity that the motion has* phrase is surprisingly hard to understand:

The rapidity that the motion that the wing has has is remarkable.

Embedding a third phrase, like *the wing that the hummingbird has*, creating a triply embedded onion sentence, results in complete unintelligibility:

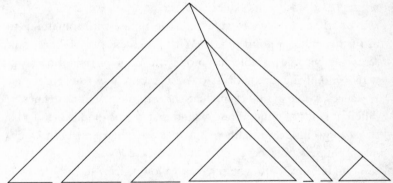

The rapidity that the motion that the wing that the hummingbird has has has is remarkable

When the human parser encounters the three successive *has*'s, it thrashes ineffectively, not knowing what to do with them. But the problem is not that the phrases have to be held in memory too long; even short sentences are uninterpretable if they have multiple embeddings:

The dog the stick the fire burned beat bit the cat.
The malt that the rat that the cat killed ate lay in the house.
If if if it rains it pours I get depressed I should get help.
That that that he left is apparent is clear is obvious.

Why does human sentence understanding undergo such complete collapse when interpreting sentences that are like onions or Russian dolls? This is one of the most challenging puzzles about the design of the mental parser and the mental grammar. At first one might wonder whether the sentences are even grammatical. Perhaps we got the rules wrong, and the real rules do not even provide a way for these words to fit together. Could the maligned word-chain device of Chapter 4, which has no memory for dangling phrases, be the right model of humans after all? No way; the sentences check out perfectly. A noun phrase can contain a modifying clause; if you can say *the rat*, you can say *the rat that S,* where S is a sentence missing an object that modifies *the rat*. And a sentence like *the cat killed X* can contain a noun phrase, such as its subject, *the cat*. So when you say *The rat that the cat killed,* you have modified a noun phrase with something that in turn contains a noun phrase. With just these two abilities, onion sentences become possible: just modify the noun phrase inside a clause with a modifying clause of its own. The only way to prevent onion sentences would be to claim that the mental grammar defines two different kinds of noun phrase, a kind that can be modified and a kind that can go inside a modifier. But that can't be right: both kinds of noun phrase would have to be allowed to contain the same twenty thousand nouns, both would have to allow articles and adjectives and possessors in identical positions, and so on. Entities should not be multiplied unnecessarily, and that is what such tinkering would do. Positing different kinds of phrases in the mental grammar just to explain why onion sentences are unintelligible would make the grammar exponentially more complicated and would give the child an exponentially larger number of rules to record when learning the language. The problem must lie elsewhere.

Onion sentences show that a grammar and a parser are different things. A person can implicitly "know" constructions that he or she can never understand, in the same way that Alice knew addition despite the Red Queen's judgment:

> "Can you do addition?" the White Queen asked. "What's one and one and one and one and one and one and one and one and one and one?"
> "I don't know," said Alice. "I lost count."
> "She can't do Addition," the Red Queen interrupted.

Why does the human parser seem to lose count? Is there not enough room in short-term memory to hold more than one or two dangling phrases at a time? The problem must be more subtle. Some three-layer onion sentences are a little hard because of the memory load but are not nearly as opaque as the *has has has* sentence:

> The cheese that some rats I saw were trying to eat turned out
>   to be rancid.
> The policies that the students I know object to most strenu-
>   ously are those pertaining to smoking.
> The guy who is sitting between the table that I like and the
>   empty chair just winked.
> The woman who the janitor we just hired hit on is very pretty.

What boggles the human parser is not the amount of memory needed but the *kind* of memory: keeping a particular kind of phrase in memory, intending to get back to it, at the same time as it is analyzing another example of *that very same kind of phrase*. Examples of these "recursive" structures include a relative clause in the middle of the same kind of relative clause, or an *if . . . then* sentence inside another *if . . . then* sentence. It is as if the human sentence parser keeps track of where it is in a sentence not by writing down a list of currently incomplete phrases in the order in which they must be completed, but by writing a number in a slot next to each phrase type on a master checklist. When a type of phrase has to be remembered more than once—so that both it *(the cat that . . .)* and the identical type of phrase it is inside of *(the rat that . . .)* can be completed in order—there is not enough room on the checklist for both numbers to fit, and the phrases cannot be completed properly.

Unlike memory, which people are bad at and computers are good at, decision-making is something that people are good at and computers are bad at. I contrived the toy grammar and the baby sentence we have just walked through so that every word had a single dictionary entry (that is, was at the right-hand side of only one rule). But all you have to do is open up a dictionary, and you will see that many nouns have a secondary entry as a verb, and vice versa. For example, *dog* is listed a second time—as a verb, for sentences like *Scandals dogged the administration all year*. Similarly, in real life *hot dog* is not only a noun but a verb, meaning "to show off." And each of the verbs in the toy grammar should also be listed as nouns, because English speakers can talk of cheap *eats*, his *likes* and dislikes, and taking a few *bites*. Even the determiner *one*, as in *one dog*, can have a second life as a noun, as in *Nixon's the one*.

These local ambiguities present a parser with a bewildering number of forks at every step along the road. When it comes across, say, the word *one* at the beginning of a sentence, it cannot simply build

$$\begin{array}{c} \text{det} \\ | \\ \text{one} \end{array}$$

but must also keep in mind

$$\begin{array}{c} \text{N} \\ | \\ \text{one} \end{array}$$

Similarly, it has to jot down two rival branches when it comes across *dog*, one in case it is a noun, the other in case it is a verb. To handle *one dog*, it would need to check four possibilities: determiner-noun, determiner-verb, noun-noun, and noun-verb. Of course determiner-verb can be eliminated because no rule of grammar allows it, but it still must be checked.

It gets even worse when the words are grouped into phrases, because phrases can fit inside larger phrases in many different ways. Even in our toy grammar, a prepositional phrase (PP) can go inside either a noun phrase or a verb phrase—as in the ambiguous *discuss sex with Dick Cavett,* where the writer intended the PP *with Dick Cavett* to go inside the verb phrase (discuss it with him) but readers can interpret it as going inside the noun phrase (sex with him). These ambiguities are the rule, not the exception; there can be dozens or hundreds of possibilities to check at every point in a sentence. For example, after processing *The plastic pencil marks . . . ,* the parser has to keep several options open: it can be a four-word noun phrase, as in *The plastic pencil marks were ugly,* or a three-word noun phrase plus a verb, as in *The plastic pencil marks easily.* In fact, even the first two words, *The plastic . . . ,* are temporarily ambiguous: compare *The plastic rose fell* with *The plastic rose and fell.*

If it were just a matter of keeping track of all the possibilities at each point, a computer would have little trouble. It might churn away for minutes on a simple sentence, or use up so much short-term memory that the printout would spill halfway across the room, but eventually most of the possibilities at each decision point would be contradicted by later information in the sentence. If so, a single tree and its associated meaning should pop out at the end of the sentence, as in the toy example. When the local ambiguities fail to cancel each other out and two consistent trees are found for the same sentence, we should have a sentence that people find ambiguous, like

Ingres enjoyed painting his models nude.
My son has grown another foot.
Visiting relatives can be boring.
Vegetarians don't know how good meat tastes.
I saw the man with the binoculars.

But here is the problem. Computer parsers are too meticulous for their own good. They find ambiguities that are quite legitimate, as far as English grammar is concerned, but that would never occur to a sane person. One of the first computer parsers, developed at Harvard in the 1960s, provides a famous example. The sentence *Time flies like an arrow* is surely unambiguous if there ever was an unambiguous sentence (ignoring the difference between literal and metaphorical meanings, which have nothing to do with syntax). But to the surprise of the programmers, the sharp-eyed computer found it to have five different trees!

> Time proceeds as quickly as an arrow proceeds. (the intended reading)
> Measure the speed of flies in the same way that you measure the speed of an arrow.
> Measure the speed of flies in the same way that an arrow measures the speed of flies.
> Measure the speed of flies that resemble an arrow.
> Flies of a particular kind, time-flies, are fond of an arrow.

Among computer scientists the discovery has been summed up in the aphorism "Time flies like an arrow; fruit flies like a banana." Or consider the song line *Mary had a little lamb*. Unambiguous? Imagine that the second line was: *With mint sauce*. Or: *And the doctors were surprised*. Or: *The tramp!* There is even structure in seemingly nonsensical lists of words. For example, this fiendish string devised by my student Annie Senghas is a grammatical sentence:

> Buffalo buffalo Buffalo buffalo buffalo buffalo Buffalo buffalo.

American bison are called *buffalo*. A kind of bison that comes from Buffalo, New York, could be called a *Buffalo buffalo*. Recall that there is a verb *to buffalo* that means "to overwhelm, to intimidate." Imagine that New York State bison intimidate one another: *(The) Buffalo buffalo (that) Buffalo buffalo (often) buffalo (in turn) buffalo (other) Buffalo buffalo.* The psycholinguist and philosopher Jerry Fodor has observed that a Yale University football cheer

> Bulldogs Bulldogs Bulldogs Fight Fight Fight!

is a grammatical sentence, albeit a triply center-embedded one.

How do people home in on the sensible analysis of a sentence, without tarrying over all the grammatically legitimate but bizarre alternatives? There are two possibilities. One is that our brains are like computer parsers, computing dozens of doomed tree fragments in the background, and the unlikely ones are somehow filtered out before they reach consciousness. The other is that the human parser somehow gambles at each step about the alternative most likely to be true and then plows ahead with that single interpretation as far as possible. Computer scientists call these alternatives "breadth-first search" and "depth-first search."

At the level of individual words, it looks as if the brain does a breadth-first search, entertaining, however briefly, several entries for an ambiguous word, even unlikely ones. In an ingenious experiment, the psycholinguist David Swinney had people listen over headphones to passages like the following:

> Rumor had it that, for years, the government building had been plagued with problems. The man was not surprised when he found several spiders, roaches, and other bugs in the corner of his room.

Did you notice that the last sentence contains an ambiguous word, *bug,* which can mean either "insect" or "surveillance device"? Probably not; the second meaning is more obscure and makes no sense in context. But psycholinguists are interested in mental processes that last only milliseconds and need a more subtle technique than just asking people. As soon as the word *bug* had been read from the tape, a computer flashed a word on a screen, and the person had to press a button as soon as he or she had recognized it. (Another button was available for nonwords like *blick.*) It is well known that when a person hears one word, any word related to it is easier to recognize, as if the mental dictionary is organized like a thesaurus, so that when one word is found, others similar in meaning are more readily available. As expected, people pressed the button faster when recognizing *ant,* which is related to *bug,* than when recognizing *sew,* which is unrelated. Surprisingly, people were just as primed to recognize the word *spy,* which is, of course, related to *bug,* but only to the meaning that makes no sense in the context. It suggests that the brain knee-jerkingly activates both entries for *bug*, even though one of them could sensibly be ruled out beforehand. The irrelevant meaning is not around long: if the test word appeared on the screen three syllables after *bugs* instead of right after it, then only *ant* was recognized quickly; *spy* was no longer any faster than *sew*. Presumably that is why people deny that they even entertain the inappropriate meaning.

The psychologists Mark Seidenberg and Michael Tanenhaus showed the same effect for words that were ambiguous as to part-of-speech category, like *tires,* which we encountered in the ambiguous headline *Stud Tires Out*. Regardless of whether the word appeared in a noun position, like *The tires . . .* , or in a verb position, like *He tires . . .* , the word primed both *wheels,* which is related to the noun meaning, and *fatigue,* which is related to the verb meaning. Mental dictionary lookup, then, is quick and thorough but not very bright; it retrieves nonsensical entries that must be weeded out later.

At the level of the phrases and sentences that span many words, though, people clearly are not computing every possible tree for a sentence. We know this for two reasons. One is that many sensible ambiguities are simply never recognized. How else can we explain the ambiguous newspaper passages that escaped the notice of editors, no doubt to their horror later on? I cannot resist quoting some more:

> The judge sentenced the killer to die in the electric chair
> for the second time.
> Dr. Tackett Gives Talk on Moon
> No one was injured in the blast, which was attributed to
> the buildup of gas by one town official.
> The summary of information contains totals of the number
> of students broken down by sex, marital status, and age.

I once read a book jacket flap that said that the author lived with her husband, an architect and an amateur musician in Cheshire, Connecticut. For a moment I thought it was a ménage à quatre.

Not only do people fail to find some of the trees that are consistent with a sentence; sometimes they stubbornly fail to find the *only* tree that is consistent with a sentence. Take these sentences:

> The horse raced past the barn fell.
> The man who hunts ducks out on weekends.
> The cotton clothing is usually made of grows in Mississippi.
> The prime number few.
> Fat people eat accumulates.
> The tycoon sold the offshore oil tracts for a lot of money
>     wanted to kill JR.

Most people proceed contendedly through the sentence up to a certain point, then hit a wall and frantically look back to earlier words to try to figure out where they went wrong. Often the attempt fails and people assume that the sentences have an extra word tacked onto the end or consist of two pieces of sentence stitched together. In fact, each one is a grammatical sentence:

> The horse that was walked past the fence proceeded stead-
>     ily, but the horse raced past the barn fell.
> The man who fishes goes into work seven days a week, but
>     the man who hunts ducks out on weekends.
> The cotton that sheets are usually made of grows in Egypt,
>     but the cotton clothing is usually made of grows in Missis-
>     sippi.
> The mediocre are numerous, but the prime number few.
> Carbohydrates that people eat are quickly broken down,
>     but fat people eat accumulates.
> JR Ewing had swindled one tycoon too many into buying
>     useless properties. The tycoon sold the offshore oil tracts
>     for a lot of money wanted to kill JR.

These are called garden path sentences, because their first words lead the listener "up the garden path" to an incorrect analysis. Garden path sentences show that people, unlike computers, do not build all possible trees as they go along; if they did, the correct tree would be among them. Rather, people mainly use a depth-first strategy, picking an analysis that seems to be working and pursuing it as long as possible; if they come across words that cannot be fitted into the tree, they backtrack and start over with a different tree. (Sometimes people can hold a second tree in mind, especially people with good memories, but the vast majority of possible trees are never entertained.) The depth-first strategy gambles that a tree that has fit the words so far will continue to fit new ones, and thereby saves memory space by keeping only that tree in mind, at the cost of having to start over if it bet on the wrong horse raced past the barn.

Garden path sentences, by the way, are one of the hallmarks of bad writing. Sentences are not laid out with clear markers at every fork, allowing the reader to stride confidently through to the end. Instead the reader repeatedly runs up against dead ends and has to wend his way back. Here are some examples I have collected from newspapers and magazines:

> Delays Dog Deaf-Mute Murder Trial
> British Banks Soldier On
> I thought that the Vietnam war would end for at least an
>     appreciable chunk of time this kind of reflex anticommu-
>     nist hysteria.
> The musicians are master mimics of the formulas they dress
>     up with irony.
> The movie is Tom Wolfe's dreary vision of a past that never
>     was set against a comic view of the modern hype-bound
>     world.
> That Johnny Most didn't need to apologize to Chick Kearn,
>     Bill King, or anyone else when it came to describing the
>     action [Johnny Most when he was in his prime].
> Family Leave Law a Landmark Not Only for Newborn's
>     Parents
> Condom Improving Sensation to be Sold

In contrast, a great writer like Shaw can send a reader in a straight line from the first word of a sentence to the full stop, even if it is 110 words away.

* * *

A depth-first parser must use some criterion to pick one tree (or a small number) and run with it—ideally the tree most likely to be correct. One possibility is that the entirety of human intelligence is brought to bear on the problem, analyzing the sentence from the top down. According to this view, people would not bother to build any part of a tree if they could guess in advance that the meaning for that branch would not make sense in context. There has been a lot of debate among psycholinguists about whether this would be a sensible way for the human sentence parser to work. To the extent that a listener's intelligence can actually predict a speaker's intentions accurately, a top-down design would steer the parser toward correct sentence analyses. But the entirety of human intelligence is a lot of intelligence, and using it all at once may be too slow to allow for real-time parsing as the hurricane of words whizzes by. Jerry Fodor, quoting Hamlet, suggests that if knowledge and context had to guide sentence parsing, "the native hue of resolution would be sicklied o'er with the pale cast of thought." He has suggested that the human parser is an encapsulated module that can look up information only in the mental grammar and the mental dictionary, not in the mental encyclopedia.

Ultimately the matter must be settled in the laboratory. The human parser does seem to use at least a bit of knowledge about what tends to happen in the world. In an experiment by the psychologists John Trueswell, Michael Tanenhaus, and Susan Garnsey, people bit on a bar to keep their heads perfectly still and read sentences on a computer screen while their eye movements were recorded. The sentences had potential garden paths in them. For example, read the sentence

> The defendant examined by the lawyer turned out to be
> unreliable.

You may have been momentarily sidetracked at the word *by,* because up to that point the sentence could have been about the defendant's examining something rather than his being examined. Indeed, the subjects' eyes lingered on the word *by* and were likely to backtrack to reinterpret the beginning of the sentence (compared to unambiguous control sentences). But now read the following sentence:

> The evidence examined by the lawyer turned out to be
> unreliable.

If garden paths can be avoided by common-sense knowledge, this sentence should be much easier. Evidence, unlike defendants, can't examine anything, so the incorrect tree, in which the evidence would be examining something, is potentially avoidable. People do avoid it: the subjects' eyes hopped through the sentence with little pausing or backtracking. Of course, the knowledge being applied is quite crude (defendants examine things; evidence doesn't), and the tree that it calls for was fairly easy to find, compared with the dozens that a computer can find. So no one knows *how much* of a person's general smarts can be applied to understanding sentences in real time; it is an active area of laboratory research.

Words themselves also provide some guidance. Recall that each verb makes demands of what else can go in the verb phrase (for example, you can't just *devour* but have to *devour something;* you can't *dine something,* you can only *dine*). The most common entry for a verb seems to pressure the mental parser to find the role players it wants. Trueswell and Tanenhaus watched their volunteers' eyeballs as they read

> The student forgot the solution was in the back of the book.

At the point of reaching *was,* the eyes lingered and then hopped back, because the people misinterpreted the sentence as being about a student forgetting the solution, period. Presumably, inside people's heads the word *forget* was saying to the parser: "Find me an object, now!" Another sentence was

> The student hoped the solution was in the back of the book.

With this one there was little problem, because the word *hope* was saying, instead, "Find me a sentence!" and a sentence was there to be found.

Words can also help by suggesting to the parser exactly which other words they tend to appear with inside a given kind of phrase. Though word-by-word transition probabilities are not enough to understand a sentence (Chapter 4), they could be helpful; a parser armed with good statistics, when deciding between two possible trees allowed by a grammar, can opt for the tree that was most likely to have been spoken. The human parser seems to be somewhat sensitive to word pair probabilities: many garden paths seem especially seductive because they contain common pairs like *cotton clothing, fat people,* and *prime number.* Whether or not the brain benefits from language statistics, computers certainly do. In laboratories at AT&T and IBM, computers have been tabulating millions of words of text from sources like the *Wall Street Journal* and Associated Press stories. Engineers are hoping that if they equip their parsers with the frequencies with which each word is used, and the frequencies with which sets of words hang around together, the parsers will resolve ambiguities sensibly.

Finally, people find their way through a sentence by favoring trees with certain shapes, a kind of mental topiary. One guideline is momentum: people like to pack new words into the current dangling phrase, instead of closing off the phrase and hopping up to add the words to a dangling phrase one branch up. This "late closure" strategy might explain why we travel the garden path in the sentence

Flip said that Squeaky will do the work yesterday.

The sentence is grammatical and sensible, but it takes a second look (or maybe even a third) to realize it. We are led astray because when we encounter the adverb *yesterday,* we try to pack it inside the currently open VP *do the work,* rather than closing off that VP and hanging the adverb upstairs, where it would go in the same phrase as *Flip said.* (Note, by the way, that our knowledge of what is plausible, like the fact that the meaning of *will* is incompatible with the meaning of *yesterday,* did not keep us from taking the garden path. This suggests that the power of general knowledge to guide sentence understanding is limited.) Here is an another example, though this time the psycholinguist responsible for it, Annie Senghas, did not contrive it as an example; one day she just blurted out, "The woman sitting next to Steven Pinker's pants are like mine." (Anne was pointing out that the woman sitting next to me had pants like hers.)

A second guideline is thrift: people to try to attach a phrase to a tree using as few branches as possible. This explains why we take the garden path in the sentence

> Sherlock Holmes didn't suspect the very beautiful young countess was a fraud.

It takes only one branch to attach the *countess* inside the VP, where Sherlock would suspect her, but two branches to attach her to an S that is itself attached to the VP, where he would suspect her of being a fraud:

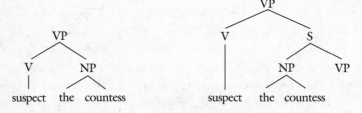

The mental parser seems to go for the minimal attachment, though later in the sentence it proves to be incorrect.

Since most sentences are ambiguous, and since laws and contracts must be couched in sentences, the principles of parsing can make a big difference in people's lives. Lawrence Solan discusses many examples in his recent book. Examine these passages, the first from an insurance contract, the second from a statute, the third from instructions to a jury:

> Such insurance as is provided by this policy applies to the use of a non-owned vehicle by the named insured and any person responsible for use by the named insured provided such use is with the permission of the owner.

> Every person who sells any controlled substance which is speci-

fied in subdivision (d) shall be punished. . . . (d) Any material, compound, mixture, or preparation which contains any quantity of the following substances having a potential for abuse associated with a stimulant effect on the central nervous system: Amphetamine; Methamphetamine . . .

The jurors must not be swayed by mere sentiment, conjecture, sympathy, passion, prejudice, public opinion or public feeling.

In the first case, a woman was distraught over being abandoned in a restaurant by her date, and drove off in what she thought was the date's Cadillac, which she then totaled. It turned out to be someone else's Cadillac, and she had to recover the money from her insurance company. Was she covered? A California appellate court said yes. The policy was ambiguous, they noted, because the requirement *with the permission of the owner,* which she obviously did not meet, could be construed as applying narrowly to *any person responsible for use by the named insured,* rather than to *the named insured* (that is, her) *and any person responsible for use by the named insured.*

In the second case, a drug dealer was trying to swindle a customer—unfortunately for him, an undercover narcotics agent—by selling him a bag of inert powder that had only a minuscule trace of methamphetamine. The *substance* had "a potential for abuse," but the *quantity of the substance* did not. Did he break the law? The appellate court said he did.

In the third case, the defendant had been convicted of raping and murdering a fifteen-year-old-girl, and a jury imposed the death penalty. United States constitutional law forbids any instruction that would deny a defendant the right to have the jury consider any "sympathy factor" raised by the evidence, which in his case consisted of psychological problems and a harsh family background. Did the instructions unconstitutionally deprive the accused of *sympathy,* or did it deprive him only of the more trivial *mere sympathy?* The United States Supreme Court ruled 5–4 that he was denied only *mere sympathy;* that denial is constitutional.

Solan points out that the courts often resolve these cases by relying on "canons of construction" enshrined in the legal literature, which correspond to the principles of parsing I discussed in the preceding section. For example, the Last Antecedent Rule, which the courts used to resolve the first two cases, is simply the "minimal attachment" strategy that we just saw in the Sherlock sentence. The principles of mental parsing, then, literally have life-or-death consequences. But psycholinguists who are now worrying that their next experiment may send someone to the gas chamber can rest easy. Solan notes that judges are not very good linguists; for better or worse, they try to find a way around the most natural interpretation of a sentence if it would stand in the way of the outcome they feel is just.

I have been talking about trees, but a sentence is not just a tree. Since the early 1960s, when Chomsky proposed transformations that convert deep structures to surface structures, psychologists have used laboratory techniques to try to detect some kind of fingerprint of the transformation. After a few false alarms the search was abandoned, and for several decades the psychology textbooks dismissed transformations as having no "psychological reality." But laboratory techniques have become more sophisticated, and the detection of something like a transformational operation in people's minds and brains is one of the most interesting recent findings in the psychology of language.

Take the sentence

The policeman saw the boy that the crowd at the party accused
   (trace) of the crime.

Who was accused of a crime? The boy, of course, even though the words *the boy* do not occur after *accused*. According to Chomsky, that is because a phrase referring to the boy really does occur after *accused* in deep structure; it has been moved backwards to the position of *that* by a transformation, leaving behind a silent "trace." A person trying to understand the sentence must undo the effect of the transformation and mentally put a copy of the phrase back in the position of the trace. To do so, the understander must first notice, while at the beginning of the sentence, that there is a moved phrase, *the boy,* that needs a home. The understander must hold the phrase in short-term memory until he or she discovers a gap: a position where a phrase should be but isn't. In this sentence there is a gap after *accused,* because *accused* demands an object, but there isn't one. The person can assume that the gap contains a trace and can then retrieve the phrase *the boy* from short-term memory and link it to the trace. Only then can the person figure out what role *the boy* played in the event—in this case, being accused.

Remarkably, every one of these mental processes can be measured. During the span of words between the moved phrase and the trace—the region I have underlined—people must hold the phrase in memory. The strain should be visible in poorer performance of any mental task carried out concurrently. And in fact, while people are reading that span, they detect extraneous signals (like a blip flashed on the screen) more slowly, and have more trouble keeping a list of extra words in memory. Even their EEG's (electroencephalograms, or records of the brain's electrical activity) show the effects of the strain.

Then, at the point at which the trace is discovered and the memory store can be emptied, the dumped phrase makes an appearance on the mental stage that can be detected in several ways. If an experimenter flashes a word from the moved phrase (for example, *boy*) at that point, people recognize it more quickly. They also recognize words related to the moved phrase—say, *girl*—more quickly. The effect is strong enough to be visible in brain waves: if interpreting the trace results in an implausible interpretation, as in

Which food did the children read *(trace)* in class?

the EEG's show a boggle reaction at the point of the trace.

Connecting phrases with traces is a hairy computational operation. The parser, while holding the phrase in mind, must constantly be checking for the trace, an invisible and inaudible little nothing. There is no way of predicting how far down in the sentence the trace will appear, and sometimes it can be quite far down:

The girl wondered who John believed that Mary claimed that the baby saw *(trace)*.

And until it is found, the semantic role of the phrase is a wild card, especially now that the *who/whom* distinction is going the way of the phonograph record.

I wonder who *(trace)* introduced John to Marsha. [*who* = the introducer]

I wonder who Bruce introduced *(trace)* to Marsha. [*who* = the one being introduced]

I wonder who Bruce introduced John to *(trace)*. [*who* = the target of the introduction]

This problem is so tough that good writers, and even the grammar of the language itself, take steps to make it easier. One principle for good style is to minimize the amount of intervening sentence in which a moved phrase must be held in memory (the underlined regions). This is a task that the English passive construction is good for (notwithstanding the recommendations of computerized "style-checkers" to avoid it across the board). In the following pair of sentences, the passive version is easier, because the memory-taxing region before the trace is shorter:

Reverse the clamp that the stainless steel hex-head bolt extending upward from the seatpost yoke holds *(trace)* in place.

Reverse the clamp that *(trace)* is held in place by the stainless steel hex-head bolt extending upward from the seatpost yoke.

And universally, grammars restrict the amount of tree that a phrase can move across. For example, one can say

That's the guy that you heard the rumor about *(trace)*.

But the following sentence is quite odd:

That's the guy that you heard the rumor that Mary likes *(trace)*.

Languages have "bounding" restrictions that turn some phrases, like the complex noun phrase *the rumor that Mary likes him,* into "islands" from which no words can escape. This is a boon to listeners, because the parser, knowing that the speaker could not have moved something out of such a phrase, can get away with not monitoring it for a trace. But the boon to listeners exerts a cost on speakers; for these sentences they have to resort to a clumsy extra pronoun, as in *That's the guy that you heard the rumor that Mary likes him.*

Parsing, for all its importance, is only the first step in understanding a sentence. Imagine parsing the following real-life dialogue:

P: The grand jury thing has its, uh, uh, uh—view of this they might, uh. Suppose we have a grand jury proceeding. Would that, would that, what would that do to the Ervin thing? Would it go right ahead anyway?

D: Probably.

P: But then on that score, though, we have—let me just, uh, run by that, that—You do that on a grand jury, we could then have a much better cause in terms of saying, "Look, this is a grand jury, in which, uh, the prosecutor—" How about a special prosecutor? We could use Petersen, or use another one. You see he is probably suspect. Would you call in another prosecutor?

D: I'd like to have Petersen on our side, advising us [laughs] frankly.

P: Frankly. Well, Petersen is honest. Is anybody about to be question him, are they?

D: No, no, but he'll get a barrage when, uh, these Watergate hearings start.

P: Yes, but he can go up and say that he's, he's been told to go further in the Grand Jury and go in to this and that and the other thing. Call everybody in the White House. I want them to come, I want the, uh, uh, to go to the Grand Jury.

D: This may result—This may happen even without our calling for it when, uh, when these, uh—

P: Vesco?

D: No. Well, that's one possibility. But also when these people go back before the Grand Jury here, they are going to
pull all these criminal defendants back in before the
Grand Jury and immunize them.

P: And immunize them: Why? Who? Are you going to—On
what?

D: Uh, the U.S. Attorney's Office will.

P: To do what?

D: To talk about anything further they want to talk about.

P: Yeah. What do they gain out of it?

D: Nothing.

P: To hell with them.

D: They, they're going to stonewall it, uh, as it now stands.
Except for Hunt. That's why, that's the leverage in his
threat.

H: This is Hunt's opportunity.

P: That's why, that's why,

H: God, if he can lay this—

P: That's why your, for your immediate thing you've got no
choice with Hunt but the hundred and twenty or whatever
it is, right?

D: That's right.

P: Would you agree that that's a buy time thing, you better damn well get that done, but fast?

D: I think he ought to be given some signal, anyway, to, to—

P: [expletive deleted], get it, in a, in a way that, uh—Who's
going to talk to him? Colson? He's the one who's supposed to know him.

D: Well, Colson doesn't have any money though. That's the
thing. That's been our, one of the real problems. They
have, uh, been unable to raise any money. A million dollars
in cash, or, or the like, has been just a very difficult prob-

lem as we've discussed before. Apparently, Mitchell talked to Pappas, and I called him last—John asked me to call him last night after our discussion and after you'd met with John to see where that was. And I, I said, "Have you talked to, to Pappas?" He was at home, and Martha picked up the phone so it was all in code. "Did you talk to the Greek?" And he said, uh, "Yes, I have." And I said, "Is the Greek bearing gifts?" He said, "Well, I want to call you tomorrow on that."

P: Well, look, uh, what is it that you need on that, uh, when, uh, uh? Now look [unintelligible] I am, uh, unfamiliar with the money situation.

This dialogue took place on March 17, 1973, among President Richard Nixon (P), his counsel John W. Dean 3rd (D), and his chief of staff H. R. Haldeman (H). Howard Hunt, working for Nixon's re-election campaign in June 1972, had directed a break-in at the Democratic Party headquarters in the Watergate building, in which his men bugged the telephones of the party chairman and other workers. Several investigations were under way to determine if the operation had been ordered from the White House, by Haldeman or Attorney General John Mitchell. The men were discussing whether to pay $120,000 in "hush money" to Hunt before he testified before a grand jury. We have this verbatim dialogue because in 1970 Nixon, claiming to be acting on behalf of future historians, bugged his own office and began secretly taping all his conversations. In February 1974 the Judiciary Committee of the House of Representatives subpoenaed the tapes to help them determine whether Nixon should be impeached. This excerpt is from their transcription. Largely on the basis of this passage, the committee recommended impeachment. Nixon resigned in August 1974.

The Watergate tapes are the most famous and extensive transcripts of real-life speech ever published. When they were released, Americans were shocked, though not all for the same reason. Some people—a very small number—were surprised that Nixon had taken part in a conspiracy to obstruct justice. A few were surprised that the leader of the free world cussed like a stevedore. But one thing that surprised everyone was what ordinary conversation looks like when it is written down verbatim. Conversation out of context is virtually opaque.

Part of the problem comes from the circumstances of transcription: the intonation and timing that delineate phrases is lost, and a transcription from anything but the highest-fidelity tape is unreliable. Indeed, in the White House's independent transcription of this low-quality recording, many puzzling passages are rendered more sensibly. For example, *I want the, uh, uh, to go* is transcribed as *I want them, uh, uh, to go.*

But even when transcribed perfectly, conversation is hard to interpret. People often speak in fragments, interrupting themselves in midsentence to reformulate the thought or change the subject. It's often unclear who or what is being talked about, because conversers use pronouns (*him, them, this, that, we, they, it, one*), generic words (*do, happen, the thing, the situation, that score, these people, whatever*), and ellipses (*The U.S. Attorney's Office will* and *That's why*). Intentions are expressed indirectly. In this episode, whether a man would end the year as president of the United States or as a convicted criminal literally hinged on the meaning of *get it* and on whether *What is it that you need?* was meant as a request for information or as an implicit offer to provide something.

Not everyone was shocked by the unintelligibility of transcribed speech. Journalists know all about it, and it is a routine practice to edit quotations and interviews heavily before they are published. For many years the temperamental Boston Red Sox pitcher Roger Clemens complained bitterly that the press misquoted him. The *Boston Herald,* in what they must have known was a cruel trick, responded by running a daily feature in which his post-game comments were reproduced word for word.

Journalists' editing of conversations became a legal issue in 1983, when the writer Janet Malcolm published an unflattering *New Yorker* series about the psychoanalyst Jeffrey Masson. Masson had written a book accusing Freud of dishonesty and cowardice in retracting his observation that neurosis is caused by sexual abuse in childhood, and was fired as the curator of the Freud archives in London. According to Malcolm, Masson described himself in her interviews as "an intellectual gigolo" and "after Freud, the greatest analyst who's ever lived," and as planning to turn Anna Freud's house after her death into "a place of sex, women, and fun." Masson sued Malcolm and the *New Yorker* for ten million dollars, claiming that he had never said these things and that other quotations had been altered to make him look ridiculous. Though Malcolm could not document the quotations from her tapes and handwritten notes, she denied having manufactured them, and her lawyers argued that even if she had, they were a "rational interpretation" of what Masson had said. Doctored quotes, they argued, are standard journalistic practice and are not examples of printing something with knowledge that it is false or with reckless disregard for whether it is false, part of the definition of libel.

Several courts threw out the case on First Amendment grounds, but in June 1991 the Supreme Court unanimously reinstated it. In a closely watched opinion, the majority defined a middle ground for journalists' treatment of quotations. (Requiring them to publish quotes verbatim was not even considered.) Justice Kennedy, writing for the majority, said that the "deliberate alteration of the words uttered by a plaintiff does not equate with knowledge of falsity," and that "If an author alters a speaker's words, but effects no material change in meaning, the speaker suffers no injury to reputation. We reject any special test of falsity for quotations, including one which would draw the line at correction of grammar or syntax." If the Supreme Court had asked me, I would have sided with Justices White and Scalia in calling for some such line to be drawn. Like many linguists, I doubt that it is possible to alter a speaker's words—including most grammar and syntax—without materially changing the meaning.

These incidents show that real speech is very far from *The dog likes ice cream* and that there is much more to understanding a sentence than parsing it. Comprehension uses the semantic information recovered from a tree as just one premise in a complex chain of inference to the speaker's intentions. Why is this so? Why is it that even honest speakers rarely articulate the truth, the whole truth, and nothing but the truth?

The first reason is air time. Conversation would bog down if one had to refer to the United States Senate Select Committee on the Watergate Break-In and Related Sabotage Efforts by uttering that full description every time. Once alluded to, *the Ervin thing,* or just *it,* will suffice. For the same reason it is wasteful to spell out the following chain of logic:

> Hunt knows who gave him the orders to organize the Watergate break-in.
>
> The person who gave him the orders might be part of our administration.
>
> If the person is in our administration and his identity becomes public, the entire administration will suffer.
>
> Hunt has an incentive to reveal the identity of the person who gave him the orders because it might reduce his prison sentence.
>
> Some people will take risks if they are given enough money.
>
> Therefore Hunt may conceal the identity of his superior if he is given enough money.
>
> There is reason to believe that approximately $120,000 would be a large enough incentive for Hunt to conceal the identity of the person who gave him the order.
>
> Hunt could accept that money now, but it is in his interest to continue to blackmail us in the future.
>
> Nonetheless it might be sufficient for us to keep him quiet in the short run because the press and the public might lose interest in the Watergate scandal in the months to come, and if he reveals the identity later, the consequences for our administration would not be as negative.
>
> Therefore the self-interested course of action for us is to pay Hunt the amount of money that would be a large enough incentive for him to keep silent until such time as public interest in Watergate wanes.

It is more efficient to say, "For your immediate thing you've got no choice with Hunt but the hundred and twenty or whatever it is."

The efficiency, though, depends on the participants' sharing a lot of background knowledge about the events and about the psychology of human behavior. They must use this knowledge to cross-reference the names, pronouns, and descriptions with a single cast of characters, and to fill in the logical steps that connect each sentence with the next. If background assumptions are not shared—for example, if one's conversational partner is from a very different culture, or is schizophrenic, or is a machine—then the best parsing in the world will fail to deliver the full meaning of a sentence. Some computer scientists have tried to equip programs with little "scripts" of stereotyped settings like restaurants and birthday parties to help their programs fill in the missing parts of texts while understanding them. Another team is trying to teach a computer the basics of human common sense, which they estimate to comprise about ten million facts. To see how formidable the task is, consider how much knowledge about human behavior must be interpolated to understand what *he* means in a simple dialogue like this:

Woman: I'm leaving you.
Man: Who is he?

Understanding, then, requires integrating the fragments gleaned from a sentence into a vast mental database. For that to work, speakers cannot just toss one fact after another into a listener's head. Knowledge is not like a list of facts in a trivia column but is organized into a complex network. When a series of facts comes in succession, as in a dialogue or text, the language must be structured so that the listener can place each fact into an existing framework. Thus information about the old, the given, the understood, the topic, should go early in the sentence, usually as the subject, and information about the new, the focus, the comment, should go at the end. Putting the topic early in the sentence is another function of the maligned passive construction. In his book on style, Williams notes that the usual advice "Avoid passives" should be flouted when the topic being discussed has the role connected with the deep-structure object of the verb. For example, read the following two-sentence discussion:

> Some astonishing questions about the nature of the universe have been raised by scientists studying the nature of black holes in space. The collapse of a dead star into a point perhaps no larger than a marble creates a black hole.

The second sentence feels like a non sequitur. It is much better to put it in the passive voice:

> Some astonishing questions about the nature of the universe have been raised by scientists studying the nature of black holes in space. A black hole is created by the collapse of a dead star into a point perhaps no larger than a marble.

The second sentence now fits in smoothly, because its subject, *a black hole,* is the topic, and its predicate adds new information to that topic. In an extended conversation or essay, a good writer or speaker will make the focus of one sentence the topic of the next one, linking propositions into an orderly train.

The study of how sentences are woven into a discourse and interpreted in context (sometimes called "pragmatics") has made an interesting discovery, first pointed out by the philosopher Paul Grice and recently refined by the anthropologist Dan Sperber and the linguist Deirdre Wilson. The act of communicating relies on a mutual expectation of cooperation between speaker and listener. The speaker, having made a claim on the precious ear of the listener, implicitly guarantees that the information to be conveyed is relevant: that it is not already known, and that it is sufficiently connected to what the listener is thinking that he or she can make inferences to new conclusions with little extra mental effort. Thus listeners tacitly expect speakers to be informative, truthful, relevant, clear, unambiguous, brief, and orderly. These expectations help to winnow out the inappropriate readings of an ambiguous sentence, to piece together fractured utterances, to excuse slips of the tongue, to guess the referents of pronouns and descriptions, and to fill in the missing steps of an argument. (When a receiver of a message is not cooperative but adversarial, all of this missing information must be stated explicitly, which is why we have the tortuous language of legal contracts with their "party of the first part" and "all rights under said copyright and all renewals thereof subject to the terms of this Agreement.")

The interesting discovery is that the maxims of relevant conversation are often observed in the breach. Speakers deliberately flout them in the literal content of their speech so that listeners can interpolate assumptions that would restore the conversation to relevance. Those assumptions then serve as the real message. A familiar example is the following kind of letter of recommendation:

> Dear Professor Pinker:
>
> I am very pleased to be able to recommend Irving Smith to you. Mr. Smith is a model student. He dresses well and is extremely punctual. I have known Mr. Smith for three years now, and in every way I have found him to be most cooperative. His wife is charming.
>
> Sincerely,
>
> John Jones
> Professor

Though the letter contains nothing but positive, factual statements, it guarantees that Mr. Smith will not get the position he is seeking. The letter contains no information relevant to the reader's needs, and thereby violates the maxim that speakers be informative. The reader works on the tacit assumption that the communicative act as a whole *is* relevant, even if the content of the letter itself is not, so he infers a premise that together with the letter makes the act relevant: that the writer has no relevant positive information to convey. Why does the writer demand this minuet, rather than just saying "Stay away from Smith; he's dumb as a tree"? It is because of another premise that the reader can interpolate: the writer is the kind of person who does not casually injure those who put their trust in him.

It is natural that people exploit the expectations necessary for successful conversation as a way of slipping their real intentions into covert layers of meaning. Human communication is not just a transfer of information like two fax machines connected with a wire; it is a series of alternating displays of behavior by sensitive, scheming, second-guessing, social animals. When we put words into people's ears we are impinging on them and revealing our own intentions, honorable or not, just as surely as if we were touching them. Nowhere is this more apparent than in the convoluted departures from plain speaking found in every society that are called politeness. Taken literally, the statement "I was wondering if you would be able to drive me to the airport" is a prolix string of incongruities. Why notify me of the contents of your ruminations? Why are you pondering my competence to drive you to the airport, and under which hypothetical circumstances? Of course the real intent—"Drive me to the airport"—is easily inferred, but because it was never stated, I have an out. Neither of us has to live with the face-threatening consequences of your issuing a command that presupposes you could coerce my compliance. Intentional violations of the unstated norms of conversation are also the trigger for many of the less pedestrian forms of nonliteral language, such as irony, humor, metaphor, sarcasm, putdowns, ripostes, rhetoric, persuasion, and poetry.

Metaphor and humor are useful ways to summarize the two mental performances that go into understanding a sentence. Most of our everyday expressions about language use a "conduit" metaphor that captures the parsing process. In this metaphor, ideas are objects, sentences are containers, and communciation is sending. We "gather" our ideas to "put" them "into" words, and if our verbiage is not "empty" or "hollow," we might "convey" or "get" these ideas "across" "to" a listener, who can "unpack" our words to "extract" their "content." But as we have seen, the metaphor is misleading. The complete process of understanding is better characterized by the joke about the two psychoanalysts who meet on the street. One says, "Good morning"; the other thinks, "I wonder what he meant by that."

# 8

⚜

# The Tower of Babel

*And the whole earth was of one language, and of one speech. And* it came to pass, as they journeyed from the east, that they found a plain in the land of Shinar; and they dwelt there. And they said to one another, Go to, let us make brick, and burn them thoroughly. And they had brick for stone, and slime had they for mortar. And they said, Go to, let us build us a city and a tower, whose top may reach unto heaven; and let us make us a name, lest we be scattered abroad upon the face of the whole earth. And the Lord came down to see the city and the tower, which the children of men builded. And the Lord said, Behold, the people is one, and they have all one language; and this they begin to do: and now nothing will be restrained from them, which they have imagined to do. Go to, let us go down, and there confound their language, that they may not understand one another's speech. So the Lord scattered them abroad from thence upon the face of all the earth: and they left off to build the city. Therefore is the name of it called Babel; because the Lord did there confound the language of all the earth: and from thence did the Lord scatter them abroad upon the face of all the earth. (Genesis 11:1–9)

In the year of our Lord 1957, the linguist Martin Joos reviewed the preceding three decades of research in linguistics and concluded that God had actually gone much farther in confounding the language of Noah's descendants. Whereas the God of Genesis was said to be content with mere mutual unintelligibility, Joos declared that "languages could differ from each other without limit and in unpredictable ways." That same year, the Chomskyan revolution began with the publication of *Syntactic Structures,* and the next three decades took us back to the literal biblical account. According to Chomsky, a visiting Martian scientist would surely conclude that aside from their mutually unintelligible vocabularies, Earthlings speak a single language.

Even by the standards of theological debates, these interpretations are strikingly different. Where did they come from? The 4,000 to 6,000 languages of the planet do look impressively different from English and from one another. Here are the most conspicuous ways in which languages can differ from what we are used to in English:

1. English is an "isolating" language, which builds sentences by rearranging immutable word-sized units, like *Dog bites man* and *Man bites dog.* Other languages express who did what to whom by modifying nouns with case affixes, or by modifying the verb with affixes that agree with its role-players in number, gender, and person. One example is Latin, an "inflecting" language in which each affix contains several pieces of information; another is Kivunjo, an "agglu-tinating" language in which each affix conveys one piece of information and many affixes are strung together, as in the eight-part verb in Chapter 5.

2. English is a "fixed-word-order" language where each phrase has a fixed position. "Free-word-order" languages allow phrase order to vary. In an extreme case like the Australian aboriginal language Warlpiri, words from different phrases can be scrambled together: *This man speared a kangaroo* can be expressed as *Man this kangaroo speared, Man kanga-*

*roo speared this,* and any of the other four orders, all completely synonymous.

3. English is an "accusative" language, where the subject of an intransitive verb, like *she* in *She ran,* is treated identically to the subject of a transitive verb, like *she* in *She kissed Larry,* and different from the object of the transitive verb, like *her* in *Larry kissed her.* "Ergative" languages like Basque and many Australian languages have a different scheme for collapsing these three roles. The subject of an intransitive verb and the *object* of a transitive verb are identical, and the subject of the transitive is the one that behaves differently. It is as if we were to say *Ran her* to mean "She ran."

4. English is a "subject-prominent" language in which all sentences must have a subject (even if there is nothing for the subject to refer to, as in *It is raining* or *There is a unicorn in the garden*). In "topic-prominent" languages like Japanese, sentences have a special position that is filled by the current topic of the conversation, as in *This place, planting wheat is good* or *California, climate is good.*

5. English is an "SVO" language, with the order subject-verb-object (*Dog bites man*). Japanese is subject-object-verb (SOV: *Dog man bites*); Modern Irish (Gaelic) is verb-subject-object (VSO: *Bites dog man*).

6. In English, a noun can name a thing in any construction: *a banana; two bananas; any banana; all the bananas.* In "classifier" languages, nouns fall into gender classes like human, animal, inanimate, one-dimensional, two-dimensional, cluster, tool, food, and so on. In many constructions, the name for the class, not the noun itself, must be used—for example, three hammers would be referred to as *three tools, to wit hammer.*

And, of course, a glance at a grammar for any particular language will reveal dozens or hundreds of idiosyncrasies.

On the other hand, one can also hear striking universals through the babble. In 1963 the linguist Joseph Greenberg examined a sample of 30 far-flung languages from five continents, including Serbian, Italian, Basque, Finnish, Swahili, Nubian, Masaai, Berber, Turkish, Hebrew, Hindi, Japanese, Burmese, Malay, Maori, Mayan, and Quechua (a descendant of the language of the Incas). Greenberg was not working in the Chomskyan school; he just wanted to see if any interesting properties of grammar could be found in all these languages. In his first investigation, which focused on the order of words and morphemes, he found no fewer than forty-five universals.

Since then, many other surveys have been conducted, involving scores of languages from every part of the world, and literally hundreds of universal patterns have been documented. Some hold absolutely. For example, no language forms questions by reversing the order of words within a sentence, like *Built Jack that house the this is?* Some are statistical: subjects normally precede objects in almost all languages, and verbs and their objects tend to be adjacent. Thus most languages have SVO or SOV order; fewer have VSO; VOS and OVS are rare (less than 1%); and OSV may be nonexistent (there are a few candidates, but not all linguists agree that they are OSV). The largest number of universals involve implications: if a language has X, it will also have Y. We came across a typical example of an implicational universal in Chapter 4: if the basic order of a language is SOV, it will usually have question words at the end of the sentence, and postpositions; if it is SVO, it will have question words at the beginning, and prepositions. Universal implications are found in all aspects of language, from phonology (for instance, if a language has nasal vowels, it will have non-nasal vowels) to word meanings (if a language has a word for "purple," it will have a word for "red"; if a language has a word for "leg," it will have a word for "arm.")

If lists of universals show that languages do not vary freely, do they imply that languages are restricted by the structure of the brain? Not directly. First one must rule out two alternative explanations.

One possibility is that language originated only once, and all existing languages are the descendants of that proto-language and retain some of its features. These features would be similar across the languages for the same reason that alphabetical order is similar across the Hebrew, Greek, Roman, and Cyrillic alphabets. There is nothing special about alphabetical order; it was just the order that the Canaanites invented, and all Western alphabets came from theirs. No linguist accepts this as an explanation for language universals. For one thing, there can be radical breaks in language transmission across the generations, the most extreme being creolization, but universals hold of all languages including creoles. Moreover, simple logic shows that a universal implication, like "If a language has SVO order, then it has prepositions, but if it has SOV order, then it has postpositions," cannot be transmitted from parent to child the way words are. An implication, by its very logic, is not a fact about English: children could learn that English is SVO *and* has prepositions, but nothing could show them that *if* a language is SVO, *then* it must have prepositions. A universal implication is a fact about all languages, visible only from the vantage point of a comparative linguist. If a language changes from SOV to SVO over the course of history and its postpositions flip to prepositions, there has to be some explanation of what keeps these two developments in sync.

Also, if universals were simply what is passed down through the generations, we would expect that the major differences between kinds of language should correlate with the branches of the linguistic family tree, just as the difference between two cultures generally correlates with how long ago they separated. As humanity's original language differentiated over time, some branches might become SOV and others SVO; within each of these branches some might have agglutinated words, others isolated words. But this is not so. Beyond a time depth of about a thousand years, history and typology often do not correlate well at all. Languages can change from grammatical type to type relatively quickly, and can cycle among a few types over and over; aside from vocabulary, they do not progressively differentiate and diverge. For example, English has changed from a free-word-order, highly inflected, topic-prominent language, as its sister German remains to this day, to a fixed-word-order, poorly inflected, subject-prominent language, all in less than a millennium. Many language families contain close to the full gamut of variations seen across the world in particular aspects of grammar. The absence of a strong correlation between the grammatical properties of languages and their place in the family tree of languages suggests that language universals are not just the properties that happen to have survived from the hypothetical mother of all languages.

The second counterexplanation that one must rule out before attributing a universal of language to a universal language instinct is that languages might reflect universals of thought or of mental information processing that are not specific to language. As we saw in Chapter 3, universals of color vocabulary probably come from universals of color vision. Perhaps subjects precede objects because the subject of an action verb denotes the causal agent (as in *Dog bites man*); putting the subject first mirrors the cause coming before the effect. Perhaps head-first or head-last ordering is consistent across all the phrases in a language because it enforces a consistent branching direction, right or left, in the language's phrase structure trees, avoiding difficult-to-understand onion constructions. For example, Japanese is SOV and has modifiers to the left; this gives it constructions like "modifier-S O V" with the modifier on the outside rather than "S-modifier O V" with the modifier embedded inside.

But these functional explanations are often tenuous, and for many universals they do not work at all. For example, Greenberg noted that if a language has both derivational suffixes (which create new words from old ones) and inflectional suffixes (which modify a word to fit its role in the sentence), then the derivational suffixes are always closer to the stem than the inflectional ones. In Chapter 5 we saw this principle in English in the difference between the grammatical *Darwinisms* and the ungrammatical *Darwinsism*. It is hard to think of how this law could be a consequence of any universal principle of thought or memory: why would the concept of two ideologies based on one Darwin be thinkable, but the concept of one ideology based on two Darwins (say, Charles and Erasmus) not be thinkable (unless one reasons in a circle and declares that the mind must find *-ism* to be more cognitively basic than the plural, because that's the order we see in language)? And remember Peter Gordon's experiments showing that children say *mice-eater* but never *rats-eater*, despite the conceptual similarity of rats and mice and despite the absence of either kind of compound in parents' speech. His results corroborate the suggestion that this particular universal is caused by the way that morphological rules are computed in the brain, with inflection applying to the products of derivation but not vice versa.

In any case, Greenbergisms are not the best place to look for a neurologically given Universal Grammar that existed before Babel. It is the organization of grammar as a whole, not some laundry list of facts, that we should be looking at. Arguing about the possible causes of something like SVO order misses the forest for the trees. What is most striking of all is that we can look at a randomly picked language and find things that can sensibly be called subjects, objects, and verbs to begin with. After all, if we were asked to look for the order of subject, object, and verb in musical notation, or in the computer programming language FORTRAN, or in Morse code, or in arithmetic, we would protest that the very idea is nonsensical. It would be like assembling a representative collection of the world's cultures from the six continents and trying to survey the colors of their hockey team jerseys or the form of their harakiri rituals. We should be impressed, first and foremost, that research on universals of grammar is even possible!

When linguists claim to find the same kinds of linguistic gadgets in language after language, it is not just because they expect languages to have subjects and so they label as a "subject" the first kind of phrase they see that resembles an English subject. Rather, if a linguist examining a language for the first time calls a phrase a "subject" using one criterion based on English subjects—say, denoting the agent role of action verbs—the linguist soon discovers that other criteria, like agreeing with the verb in person and number and occurring before the object, will be true of that phrase as well. It is these *correlations* among the properties of a linguistic thingamabob across languages that make it scientifically meaningful to talk about subjects and objects and nouns and verbs and auxiliaries and inflections—and not just Word

Class #2,783 and Word Class #1,491—in languages from Abaza to Zyrian.

Chomsky's claim that from a Martian's-eye-view all humans speak a single language is based on the discovery that the same symbol-manipulating machinery, without exception, underlies the world's languages. Linguists have long known that the basic design features of language are found everywhere. Many were documented in 1960 by the non-Chomskyan linguist C. F. Hockett in a comparison between human languages and animal communication systems (Hockett was not acquainted with Martian). Languages use the mouth-to-ear channel as long as the users have intact hearing (manual and facial gestures, of course, are the substitute channel used by the deaf). A common grammatical code, neutral between production and comprehension, allows speakers to produce any linguistic message they can understand, and vice versa. Words have stable meanings, linked to them by arbitrary convention. Speech sounds are treated discontinuously; a sound that is acoustically halfway between *bat* and *pat* does not mean something halfway between batting and patting. Languages can convey meanings that are abstract and remote in time or space from the speaker. Linguistic forms are infinite in number, because they are created by a discrete combinatorial system. Languages all show a duality of patterning in which one rule system is used to order phonemes within morphemes, independent of meaning, and another is used to order morphemes within words and phrases, specifying their meaning.

Chomskyan linguistics, in combination with Greenbergian surveys, allows us to go well beyond this basic spec sheet. It is safe to say that the grammatical machinery we used for English in Chapters 4–6 is used in all the world's languages. All languages have a vocabulary in the thousands or tens of thousands, sorted into part-of-speech categories including noun and verb. Words are organized into phrases according to the X-bar system (nouns are found inside N-bars, which are found inside noun phrases, and so on). The higher levels of phrase structure include auxiliaries (INFL), which signify tense, modality, aspect, and negation. Nouns are marked for case and assigned semantic roles by the mental dictionary entry of the verb or other predicate. Phrases can be moved from their deep-structure positions, leaving a gap or "trace," by a structure-dependent movement rule, thereby forming questions, relative clauses, passives, and other widespread constructions. New word structures can be created and modified by derivational and inflectional rules. Inflectional rules primarily mark nouns for case and number, and mark verbs for tense, aspect, mood, voice, negation, and agreement with subjects and objects in number, gender, and person. The phonological forms of words are defined by metrical and syllable trees and separate tiers of features like voicing, tone, and manner and place of articulation, and are subsequently adjusted by ordered phonological rules. Though many of these arrangements are in some sense useful, their details, found in language after language but not in any artificial system like FORTRAN or musical notation, give a strong impression that a Universal Grammar, not reducible to history or cognition, underlies the human language instinct.

God did not have to do much to confound the language of Noah's descendants. In addition to vocabulary—whether the word for "mouse" is *mouse* or *souris*—a few properties of language are simply not specified in Universal Grammar and can vary as parameters. For example, it is up to each language to choose whether the order of elements within a phrase is head-first or head-last (*eat sushi* and *to Chicago* versus *sushi eat* and *Chicago to*) and whether a subject is mandatory in all sentences or can be omitted when the speaker desires. Furthermore, a particular grammatical widget often does a great deal of important work in one language and hums away unobtrusively in the corner of another. The overall impression is that Universal Grammar is like an archetypal body plan found across vast numbers of animals in a phylum. For example, among all the amphibians, reptiles, birds, and mammals, there is a common body architecture, with a segmented backbone, four jointed limbs, a tail, a skull, and so on. The various parts can be grotesquely distorted or stunted across animals: a bat's wing is a hand, a horse trots on its middle toes, whales' forelimbs have become flippers and their hindlimbs have shrunken to invisible nubs, and the tiny hammer, anvil, and stirrup of the mammalian middle ear are jaw parts of reptiles. But from newts to elephants, a common topology of the body plan—the shin bone connected to the thigh bone, the thigh bone connected to the hip bone—can be discerned. Many of the differences are caused by minor variations in the relative timing and rate of growth of the parts during embryonic development. Differences among languages are similar. There seems to be a common plan of syntactic, morphological, and phonological rules and principles, with a small set of varying parameters, like a checklist of options. Once set, a parameter can have far-reaching changes on the superficial appearance of the language.

If there is a single plan just beneath the surfaces of the world's languages, then any basic property of one language should be found in all the others. Let's reexamine the six supposedly un-English language traits that opened the chapter. A closer look shows that all of them can be found right here in English, and that the supposedly distinctive traits of English can be found in the other languages.

1. English, like the inflecting languages it supposedly differs from, has an agreement marker, the third person singular *-s* in *He walks*. It also has case distinctions in the pronouns, such as *he* versus *him*. And like agglutinating languages, it has machinery that can glue many bits together into a long word, like the derivational rules and affixes that create *sensationalization* and *Darwinianisms*. Chinese is supposed to be an even more extreme example of an isolating language than English, but it, too, contains rules that create multipart words such as compounds and derivatives.

2. English, like free-word-order languages, has free ordering in strings of prepositional phrases, where each preposition marks the semantic role of its noun phrase as if it were a case marker: *The package was sent from Chicago to Boston by Mary; The package was sent by Mary to Boston from Chicago; The package was sent to Boston from Chicago by Mary*, and so on. Conversely, in so-called scrambling languages at the

other extreme, like Warlpiri, word order is never completely free; auxiliaries, for example, must go in the second position in a sentence, which is rather like their positioning in English.

3. English, like ergative languages, marks a similarity between the objects of transitive verbs and the subjects of intransitive verbs. Just compare *John broke the glass* (*glass* = object) with *The glass broke* (*glass* = subject of intransitive), or *Three men arrived* with *There arrived three men*.

4. English, like topic-prominent languages, has a topic constituent in constructions like *As for fish, I eat salmon* and *John I never really liked*.

5. Like SOV languages, not too long ago English availed itself of an SOV order, which is still interpretable in archaic expressions like *Till death do us part* and *With this ring I thee wed*.

6. Like classifier languages, English insists upon classifiers for many nouns: you can't refer to a single square as *a paper* but must say *a sheet of paper*. Similarly, English speakers say *a piece of fruit* (which refers to an apple, not a piece of an apple), *a blade of grass, a stick of wood, fifty head of cattle,* and so on.

If a Martian scientist concludes that humans speak a single language, that scientist might well wonder why Earthspeak has those thousands of mutually unintelligible dialects (assuming that the Martian has not read Genesis 11; perhaps Mars is beyond the reach of the Gideon Society). If the basic plan of language is innate and fixed across the species, why not the whole banana? Why the head-first parameter, the different-sized color vocabularies, the Boston accent?

Terrestrial scientists have no conclusive answer. The theoretical physicist Freeman Dyson proposed that linguistic diversity is here for a reason: "it was nature's way to make it possible for us to evolve rapidly," by creating isolated ethnic groups in which undiluted biological and cultural evolution can proceed swiftly. But Dyson's evolutionary reasoning is defective. Lacking foresight, lineages try to be the best that they can be, *now;* they do not initiate change for change's sake on the chance that one of the changes might come in handy in some ice age ten thousand years in the future. Dyson is not the first to ascribe a purpose to linguistic diversity. A Colombian Bará Indian, a member of an outbreeding set of tribes, when asked by a linguist why there were so many languages, explained, "If we were all Tukano speakers, where would we get our women?"

As a native of Quebec, I can testify that differences in language lead to differences in ethnic identification, with widespread effects, good and bad. But the suggestions of Dyson and the Bará put the causal arrow backwards. Surely head-first parameters and all the rest represent massive overkill in some design to distinguish among ethnic groups, assuming that that was even evolutionarily desirable. Humans are ingenious at sniffing out minor differences to figure out whom they should despise. All it takes is that European-Americans have light skin and African-Americans have dark skin, that Hindus make a point of not eating beef and Moslems make a point of not eating pork, or, in the Dr. Seuss story, that the Star-Bellied Sneetches have bellies with stars and the Plain-Bellied Sneetches have none upon thars. Once there is more than one language, ethnocentrism can do the rest; we need to understand why there is more than one language.

Darwin himself expressed the key insight:

> The formation of different languages and of distinct species, and the proofs that both have been developed through a gradual process, are curiously parallel. . . . We find in distinct languages striking homologies due to community of descent, and analogies due to a similar process of formation. . . . Languages, like organic beings, can be classed in groups under groups; and they can be classed either naturally, according to descent, or artificially by other characters. Dominant languages and dialects spread widely, and lead to the gradual extinction of other tongues. A language, like a species, when extinct, never . . . reappears.

That is, English is similar though not identical to German for the same reason that foxes are similar though not identical to wolves: English and German are modifications of a common ancestor language spoken in the past, and foxes and wolves are modifications of a common ancestor species that lived in the past. Indeed, Darwin claimed to have taken some of his ideas about biological evolution from the linguistics of his time, which we will encounter later in this chapter.

Differences among languages, like differences among species, are the effects of three processes acting over long spans of time. One process is variation—mutation, in the case of species; linguistic innovation, in the case of languages. The second is heredity, so that descendants resemble their progenitors in these variations—genetic inheritance, in the case of species; the ability to learn, in the case of languages. The third is isolation—by geography, breeding season, or reproductive anatomy, in the case of species; by migration or social barriers, in the case of languages. In both cases, isolated populations accumulate separate sets of variations and hence diverge over time. To understand why there is more than one language, then, we must understand the effects of innovation, learning, and migration.

Let me begin with the ability to learn, and by convincing you that there is something to explain. Many social scientists believe that learning is some pinnacle of evolution that humans have scaled from the lowlands of instinct, so that our ability to learn can be explained by our exalted braininess. But biology says otherwise. Learning is found in organisms as simple as bacteria, and, as James and Chomsky pointed out, human intelligence may depend on our having *more* innate instincts, not fewer. Learning is an option, like camouflage or horns, that nature gives organisms as needed—when some aspect of the organisms' environmental niche is so unpredictable that anticipation of its contingencies cannot be wired in. For example, birds that nest on small cliff ledges do not learn to recognize their offspring. They do not need to, for any blob of the right size and shape in their nest is sure to be one. Birds that nest in large colonies, in contrast, are in danger of feeding some neighbor's offspring that sneaks in, and they have evolved a mechanism that allows them to learn the particular nuances of their own babies.

Even when a trait starts off as a product of learning, it does not have to remain so. Evolutionary theory, supported by computer simulations, has shown that when an environment is stable, there is a selective pressure for learned abilities to become increasingly innate. That is because if an ability is innate, it can be deployed earlier in the lifespan of the creature, and there is less of a chance that an unlucky creature will miss out on the experiences that would have been necessary to teach it.

Why might it pay for the child to learn parts of a language rather than having the whole system hard-wired? For vocabulary, the benefits are fairly obvious: 60,000 words might be too many to evolve, store, and maintain in a genome comprising only 50,000 to 100,000 genes. And words for new plants, animals, tools, and especially people are needed throughout the lifespan. But what good is it to learn different grammars? No one knows, but here are some plausible hypotheses.

Perhaps some of the things about language that we have to learn are easily learned by simple mechanisms that antedated the evolution of grammar. For example, a simple kind of learning circuit might suffice to record which element comes before which other one, as long as the elements are first defined and identified by some other cognitive module. If a universal grammar module defines a head and a role-player, their relative ordering (head-first or head-last) could thus be recorded easily. If so, evolution, having made the basic computational units of language innate, may have seen no need to replace every bit of learned information with innate wiring. Computer simulations of evolution show that the pressure to replace learned neural connections with innate ones diminishes as more and more of the network becomes innate, because it becomes less and less likely that learning will fail for the rest.

A second reason for language to be partly learned is that language inherently involves sharing a code with other people. An innate grammar is useless if you are the only one possessing it: it is a tango of one, the sound of one hand clapping. But the genomes of other people mutate and drift and recombine when they have children. Rather than selecting for a completely innate grammar, which would soon fall out of register with everyone else's, evolution may have given children an ability to learn the variable parts of language as a way of synchronizing their grammars with that of the community.

The second component of language differentiation is a source of variation. Some person, somewhere, must begin to speak differently from the neighbors, and the innovation must spread and catch on like a contagious disease until it becomes epidemic, at which point children perpetuate it. Change can arise from many sources. Words are coined, borrowed from other languages, stretched in meaning, and forgotten. New jargon or speech styles may sound way cool within some subculture and then infiltrate the mainstream. Specific examples of these borrowings are a subject of fascination to pop language fanciers and fill many books and columns. Personally, I have trouble getting excited. Should we really be astounded to learn that English borrowed *kimono* from Japanese, *banana* from Spanish, *moccasin* from the American Indians, and so on?

Because of the language instinct, there is something much more fascinating about linguistic innovation: each link in the chain of language transmission is a human brain. That brain is equipped with a universal grammar and is always on the lookout for examples in ambient speech of various kinds of rules. Because speech can be sloppy and words and sentences ambiguous, people are occasionally apt to *reanalyze* the speech they hear—they interpret it as having come from a different dictionary entry or rule than the ones that the speaker actually used.

A simple example is the word *orange*. Originally it was *norange*, borrowed from the Spanish *naranja*. But at some point some unknown creative speaker must have reanalyzed *a norange* as *an orange*. Though the speaker's and hearer's analyses specify identical sounds for that particular phrase, *anorange,* once the hearer uses the rest of grammar creatively, the change becomes audible, as in *those oranges* rather than *those noranges*. (This particular change has been common in English. Shakespeare used *nuncle* as an affectionate name, a recutting of *mine Uncle* to *my nuncle,* and *Ned* came from *Edward* by a similar route. Nowadays many people talk about *a whole nother thing,* and I know of a child who eats *ectarines* and an adult called *Nalice* who refers to people she doesn't care for as *nidiots.*)

Reanalysis, a product of the discrete combinatorial creativity of the language instinct, partly spoils the analogy between language change on the one hand and biological and cultural evolution on the other. Many linguistic innovations are not like random mutation, drift, erosion, or borrowing. They are more like legends or jokes that are embellished or improved or reworked with each retelling. That is why, although grammars change quickly through history, they do not degenerate, for reanalysis is an inexhaustible source of new complexity. Nor must they progressively differentiate, for grammars can hop among the grooves made available by the universal grammar in everyone's mind. Moreover, one change in a language can cause an imbalance that can trigger a cascade of other changes elsewhere, like falling dominoes. Any part of language can change:

• Many phonological rules arose when hearers in some community reanalyzed rapid, coarticulated speech. Imagine a dialect that lacks the rule that converts *t* to a flapped *d* in *utter.* Its speakers generally pronounce the *t* as a *t,* but may not do so when speaking rapidly or affecting a casual "lazy" style. Hearers may then credit them with a flapping rule, and they (or their children) would then pronounce the *t* as a flap even in careful speech. Taken further, even the underlying phonemes can be reanalyzed. This is how we got *v.* Old English didn't have a *v;* our word *starve* was originally *steorfan.* But any *f* between two vowels was pronounced with voicing turned on, so *ofer* was pronounced "over," thanks to a rule similar to the contemporary flapping rule. Listeners eventually analyzed the *v* as a separate phoneme, rather than as a pronunciation of *f,* so now the word actually is *over,* and *v* and *f* are available as separate phonemes. For example, we can now differentiate words like *waver* and *wafer,* but King Ethelbuld could not have.

• The phonological rules governing the *pronunciation* of words can, in turn, be reanalyzed into morphological rules governing the *construction* of them. Germanic languages like Old English had an "umlaut" rule that changed a back vowel to a front vowel if the next syllable contained a high front vowel sound. For example, in *foti,* the plural of "foot," the back *o* was altered by the rule to a front *e,* harmonizing with the front *i.* Subsequently the *i* at the end ceased being pronounced, and because the phonological rule no longer had anything to trigger it, speakers reinterpreted the *o–e* shift as a morphological relationship signaling the plural—resulting in our *foot–feet, mouse–mice, goose–geese, tooth–teeth,* and *louse–lice.*

• Reanalysis can also take two variants of one word, one created from the other by an inflectional rule, and recategorize them as separate words. The speakers of yesteryear might have noticed that an inflectional *oo–ee* rule applies not to all items but only to a few: *tooth–teeth,* but not *booth–beeth.* So *teeth* was interpreted as a separate, irregular word linked to *tooth,* rather than the product of a rule applied to *tooth.* The vowel change no longer acts like a rule—hence Lederer's humorous story "Foxen in the Henhice." Other sets of vaguely related words came into English by this route, like *brother–brethren, half–halve, teeth–teethe, to fall–to fell, to rise–to raise;* even *wrought,* which used to be the past tense of *work.*

• Other morphological rules can be formed when the words that commonly accompany some other word get eroded and then glued onto it. Tense markers may come from auxiliaries; for example, as I've mentioned, the English *-ed* suffix may have evolved from *did: hammer-did → hammered.* Case markers may come from slurred prepositions or from sequences of verbs (for example, in a language that allows the construction *take nail hit it, take* might erode into an accusative case marker like *ta-*). Agreement markers can arise from pronouns: in *John, he kissed her, he* and *her* can eventually glom onto the verb as agreement affixes.

• Syntactic constructions can arise when a word order that is merely preferred becomes reanalyzed as obligatory. For example, when English had case markers, both *give him a book* and *give a book him* were possible, but the former was more common. When the case markers eroded in casual speech, many sentences would have become ambiguous if order were still allowed to vary. The more common order was thus enshrined as a rule of syntax. Other constructions can arise from multiple reanalyses. The English perfect *I had written a book* originally came from *I had a book written* (meaning "I owned a book that was written"). The reanalysis was inviting because the SOV pattern was alive in English; the participle *written* could be reanalyzed as the main verb of the sentence, and *had* could be reanalyzed as its auxiliary, begetting a new analysis with a related meaning.

The third ingredient for language splitting is separation among groups of speakers, so that successful innovations do not take over everywhere but accumulate separately in the different groups. Though people modify their language every generation, the extent of these changes is slight: vastly more sounds are preserved than mutated, more constructions analyzed properly than reanalyzed. Because of this overall conservatism, some patterns of vocabulary, sound, and grammar survive for millennia. They serve as the fossilized tracks of mass migrations in the remote past, clues to how human beings spread out over the earth to end up where we find them today.

How far back can we trace the language of this book, modern American English? Surprisingly far, perhaps five or even nine thousand years. Our knowledge of where our language has come from is considerably more precise than the recollection of Dave Barry's Mr. Language Person: "The English language is a rich verbal tapestry woven together from the tongues of the Greeks, the Latins, the Angles, the Klaxtons, the Celtics, and many more other ancient peoples, all of whom had severe drinking problems." Let's work our way back.

America and England first came to be divided by a common language, in Wilde's memorable words, when colonists and immigrants isolated themselves from British speech by crossing the Atlantic Ocean. England was already a Babel of regional and class dialects when the first colonists left. What was to become the standard American dialect was seeded by the ambitious or dissatisfied members of lower and middle classes from southeastern England. By the eighteenth century an American accent was noted, and pronunciation in the American South was particularly influenced by the immigration of the Ulster Scots. Westward expansions preserved the layers of dialects of the eastern seaboard, though the farther west the pioneers went, the more their dialects mixed, especially in California, which required leapfrogging of the vast interior desert. Because of immigration, mobility, literacy, and now the mass media, the English of the United States, even with its rich regional differences, is homogeneous compared with the languages in territories of similar size in the rest of the world; the process has been called "Babel in reverse." It is often said that the dialects of the Ozarks and Appalachia are a relict of Elizabethan English, but this is just a quaint myth, coming from the misconception of language as a cultural artifact. We think of the folk ballads, the hand-stitched quilts, and the whiskey aging slowly in oak casks and easily swallow the rumor that in this land that time forgot, the people still speak the traditional tongue lovingly handed down through the generations. But language does not work that way—at all times, in all communities, language changes, though the various parts of a language may change in different ways in different communities. Thus it is true that these dialects preserve some English forms that are rare elsewhere, such as *afeared, yourn, hisn,* and *et, holp,* and *clome* as the past of *eat, help,* and *climb.* But so does *every* variety of American English, including the standard one. Many so-called Americanisms were in fact carried over from England, where they were subsequently lost. For example, the participle *gotten,* the pronunciation of *a* in *path* and *bath* with a front-of-the-mouth "a" rather than the back-of-the-mouth "ah," and the use of *mad* to mean "angry," *fall* to mean "autumn," and *sick* to mean "ill," strike the British ear as all-American, but they are actually holdovers from the English that was spoken in the British Isles at the time of the American colonization.

English has changed on both sides of the Atlantic, and had been changing well before the voyage of the *Mayflower*. What grew into standard contemporary English was simply the dialect spoken around London, the political and economic center of England, in the seventeenth century. In the centuries preceding, it had undergone a number of major changes, as you can see in these versions of the Lord's Prayer:

CONTEMPORARY ENGLISH: Our Father, who is in heaven, may your name be kept holy. May your kingdom come into being. May your will be followed on earth, just as it is in heaven. Give us this day our food for the day. And forgive us our offenses, just as we forgive those who have offended us. And do not bring us to the test. But free us from evil. For the kingdom, the power, and the glory are yours forever. Amen.

EARLY MODERN ENGLISH (C. 1600): Our father which are in heaven, hallowed be thy Name. Thy kingdom come. Thy will be done, on earth, as it is in heaven. Give us this day our daily bread. And forgive us our trespasses, as we forgive those who trespass against us. And lead us not into temptation, but deliver us from evil. For thine is the kingdom, and the power, and the glory, for ever, amen.

MIDDLE ENGLISH (C. 1400): Oure fadir that art in heuenes halowid be thi name, thi kyngdom come to, be thi wille don in erthe es in heuene, yeue to us this day oure bread ouir other substance, & foryeue to us oure dettis, as we forgeuen to oure dettouris, & lede us not in to temptacion: but delyuer us from yuel, amen.

OLD ENGLISH (C. 1000): Faeder ure thu the eart on heofonum, si thin nama gehalgod. Tobecume thin rice. Gewurthe in willa on eorthan swa swa on heofonum. Urne gedaeghwamlican hlaf syle us to daeg. And forgyf us ure gyltas, swa swa we forgyfath urum gyltedum. And ne gelaed thu us on contnungen ac alys us of yfele. Sothlice.

The roots of English are in northern Germany near Denmark, which was inhabited early in the first millennium by pagan tribes called the Angles, the Saxons, and the Jutes. After the armies of the collapsing Roman Empire left Britain in the fifth century, these tribes invaded what was to become England (Angle-land) and displaced the indigenous Celts there into Scotland, Ireland, Wales, and Cornwall. Linguistically, the defeat was total; English has virtually no traces of Celtic. Vikings invaded in the ninth to eleventh centuries, but their language, Old Norse, was similar enough to Anglo-Saxon that aside from many borrowings, the language, Old English, did not change much.

In 1066 William the Conqueror invaded Britain, bringing with him the Norman dialect of French, which became the language of the ruling classes. When King John of the Anglo-Norman kingdom lost Normandy shortly after 1200, English reestablished itself as the exclusive language of England, though with a marked influence of French that lasts to this day in the form of thousands of words and a variety of grammatical quirks that go with them. This "Latinate" vocabulary—including such words as *donate, vibrate,* and *desist*—has a more restricted syntax; for example, you can say *give the museum a painting* but not *donate the museum a painting, shake it up* but not *vibrate it up.* The vocabulary also has its own sound pattern: Latinate words are largely polysyllabic with stress on the second syllable, such as *desist, construct,* and *transmit,* whereas their Anglo-Saxon synonyms *stop, build,* and *send* are single syllables. The Latinate words also trigger many of the sound changes that make English morphology and spelling so idiosyncratic, like *electric–electricity* and *nation–national.* Because Latinate words are longer, and are more formal because of their ancestry in the government, church, and schools of the Norman conquerors, overusing them produces the stuffy prose universally deplored by style manuals, such as *The adolescents who had effectuated forcible entry into the domicile were apprehended* versus *We caught the kids who broke into the house.* Orwell captured the flabbiness of Latinate English in his translation of a passage from Ecclesiastes into modern institutionalese:

> I returned and saw under the sun, that the race is not to the swift, nor the battle to the strong, neither yet bread to the wise, nor yet riches to men of understanding, nor yet favour to men of skill; but time and chance happeneth to them all.

Objective consideration of contemporary phenomena compels the conclusion that success or failure in competitive activities exhibits no tendency to be commensurate with innate capacity, but that a considerable element of the unpredictable must invariably be taken into account.

English changed noticeably in the Middle English period (1100–1450) in which Chaucer lived. Originally all syllables were enunciated, including those now represented in spelling by "silent" letters. For example, *make* would have been pronounced with two syllables. But the final syllables became reduced to the generic schwa like the *a* in *allow* and in many cases they were eliminated entirely. Since the final syllables contained the case markers, overt case began to vanish, and the word order became fixed to eliminate the resulting ambiguity. For the same reason, prepositions and auxiliaries like *of* and *do* and *will* and *have* were bled of their original meanings and given important grammatical duties. Thus many of the signatures of modern English syntax were the result of a chain of effects beginning with a simple shift in pronunciation.

The period of Early Modern English, the language of Shakespeare and the King James Bible, lasted from 1450 to 1700. It began with the Great Vowel Shift, a revolution in the pronunciation of long vowels whose causes remain mysterious. (Perhaps it was to compensate for the fact that long vowels sounded too similar to short vowels in the monosyllables that were now prevalent; or perhaps it was a way for the upper classes to differentiate themselves from the lower classes once Norman French became obsolete.) Before the vowel shift, *mouse* had been pronounced "mooce"; the old "oo" turned into a diphthong. The gap left by the departed "oo" was filled by raising what used to be an "oh" sound; what we pronounce as *goose* had, before the Great Vowel Shift, been pronounced "goce." That vacuum, in turn, was filled by the "o" vowel (as in *hot,* only drawn out), giving us *broken* from what had previously been pronounced more like "brocken." In a similar rotation, the "ee" vowel turned into a diphthong; *like* had been pronounced "leek." This dragged in the vowel "eh" to replace it; our *geese* was originally pronounced "gace." And that gap was filled when the long version of *ah* was raised, resulting in *name* from what used to be pronounced "nahma." The spelling never bothered to track these shifts, which is why the letter *a* is pronounced one way in *cam* and another way in *came,* where it had formerly been just a longer version of the *a* in *cam.* This is also why vowels are rendered differently in English spelling than in all the other European alphabets and in "phonetic" spelling.

Incidentally, fifteenth-century Englishmen did not wake up one day and suddenly pronounce their vowels differently, like a switch to Daylight Savings Time. To the people living through it, the Great Vowel Shift probably felt like the current trend in the Chicago area to pronounce *hot* like *hat,* or the growing popularity of that strange surfer dialect in which *dude* is pronounced something like "diiihh-hooooood."

What happens if we try to go back farther in time? The languages of the Angles and the Saxons did not come out of thin air; they evolved from Proto-Germanic, the language of a tribe that occupied much of northern Europe in the first millennium B.C. The western branch of the tribe split into groups that gave us not only Anglo-Saxon, but German and its offshoot Yiddish, and Dutch and its offshoot Afrikaans. The northern branch settled Scandinavia and came to speak Swedish, Danish, Norwegian, and Icelandic. The similarities in vocabulary among these languages are visible in an instant, and there are many similarities in grammar as well, such as forms of the past-tense ending -*ed.*

The ancestors of the Germanic tribes left no clear mark in written history or the archeological record. But they did leave a special mark on the territory they occupied. That mark was discerned in 1786 by Sir William Jones, a British judge stationed in India, in one of the most extraordinary discoveries in all scholarship. Jones had taken up the study of Sanskrit, a long-dead language, and noted:

> The Sanskrit language, whatever may be its antiquity, is of a
> wonderful structure; more perfect than the Greek, more copi-

ous than the Latin, and more exquisitely refined than either, yet bearing to both of them a stronger affinity, both in the roots of verbs and in the forms of grammar, than could possibly have been produced by accident; so strong indeed that no philologer could examine them all three, without believing them to have sprung from some common source, which, perhaps no longer exists; there is a similar reason, though not quite so forcible, for supposing that both the Gothic [Germanic] and the Celtic, though blended with a very different idiom, had the same origin as the Sanskrit; and the old Persian might be added to the same family . . .

Here are the kinds of affinities that impressed Jones:

| ENGLISH: | brother | mead | is | thou bearest | he bears |
|---|---|---|---|---|---|
| GREEK: | phrater | methu | esti | phereis | pherei |
| LATIN: | frater | | est | fers | fert |
| OLD SLAVIC: | bratre | mid | yeste | berasi | beretu |
| OLD IRISH: | brathir | mith | is | | beri |
| SANSKRIT: | bhrater | medhu | asti | bharasi | bharati |

Such similarities in vocabulary and grammar are seen in an immense number of modern languages. Among others, they embrace Germanic, Greek, Romance (French, Spanish, Italian, Portuguese, Romanian), Slavic (Russian, Czech, Polish, Bulgarian, Serbo-Croatian), Celtic (Gaelic, Irish, Welsh, Breton), and Indo-Iranian (Persian, Afghan, Kurdish, Sanskrit, Hindi, Bengali, and the Romany language of the Gypsies). Subsequent scholars were able to add Anatolian (extinct languages spoken in Turkey, including Hittite), Armenian, Baltic (Lithuanian and Latvian), and Tocharian (two extinct languages spoken in China). The similarities are so pervasive that linguists have reconstructed a grammar and a large dictionary for a hypothetical common ancestor language, Proto-Indo-European, and a set of systematic rules by which the daughter languages changed. For example, Jacob Grimm (one of the two Grimm brothers, famous as collectors of fairy tales) discovered the rule by which *p* and *t* in Proto-

Indo-European became *f* and *th* in Germanic, as one can see in comparing Latin *pater* and Sanskrit *piter* with English *father*.

The implications are mind-boggling. Some ancient tribe must have taken over most of Europe, Turkey, Iran, Afghanistan, Pakistan, northern India, western Russia, and parts of China. The idea has excited the imagination of a century of linguists and archeologists, though even today no one really knows who the Indo-Europeans were. Ingenious scholars have made guesses from the reconstructed vocabulary. Words for metals, wheeled vehicles, farm implements, and domesticated animals and plants suggest that the Indo-Europeans were a late Neolithic people. The ecological distributions of the natural objects for which there are Proto-Indo-European words—elm and willow, for example, but not olive or palm—have been used to place the speakers somewhere in the territory from inland northern Europe to southern Russia. Combined with words for patriarch, fort, horse, and weapons, the reconstructions led to an image of a powerful conquering tribe spilling out of an ancestral homeland on horseback to overrun most of Europe and Asia. The word "Aryan" became associated with the Indo-Europeans, and the Nazis claimed them as ancestors. More sanely, archeologists have linked them to artifacts of the Kurgan culture in the southern Russian steppes from around 3500 B.C., a band of tribes that first harnessed the horse for military purposes.

Recently the archeologist Colin Renfrew has argued that the Indo-European takeover was a victory not of the chariot but of the cradle. His controversial theory is that the Indo-Europeans lived in Anatolia (part of modern Turkey) on the flanks of the Fertile Cresent region around 7000 B.C., where they were among the world's first farmers. Farming is a method for mass-producing human beings by turning land into bodies. Farmers' daughters and sons need more land, and even if they moved just a mile or two from their parents, they would quickly engulf the less fecund hunter-gatherers standing in their way. Archeologists agree that farming spread in a wave that began in Turkey around 8500 B.C. and reached Ireland and Scandinavia by 2500 B.C. Geneticists recently discovered that a certain set of genes is most concentrated among modern people in Turkey and becomes progressively diluted as one moves through the Balkans to northern Europe. This supports the theory originally proposed by the human geneticist Luca Cavalli-Sforza that farming spread by the movement of farmers, as their offspring interbred with indigenous hunter-gatherers, rather than by the movement of farming techniques, as a fad adopted by the hunter-gatherers. Whether these people were the Indo-Europeans, and whether they spread into Iran, India, and China by a similar process, is still not known. It is an awesome possibility. Every time we use a word like *brother*, or form the past tense of an irregular verb like *break–broke* or *drink–drank*, we would be using the preserved speech patterns of the instigators of the most important event in human history, the spread of agriculture.

Most of the other human languages on earth can also be grouped into phyla descending from ancient tribes of astoundingly successful farmers, conquerers, explorers, or nomads. Not all of Europe is Indo-European. Finnish, Hungarian, and Estonian are Uralic languages, which together with Lappish, Samoyed, and other languages are the remnants of a vast nation based in central Russia about 7,000 years ago. Altaic is generally thought to include the main languages of Turkey, Mongolia, the Islamic republics of the former USSR, and much of central Asia and Siberia. The earliest ancestors are uncertain, but later ones include a sixth-century empire as well as the Mongolian empire of Genghis Khan and the Manchu dynasty. Basque is an orphan, presumably from an island of aboriginal Europeans that resisted the Indo-European tidal wave.

Afro-Asiatic (or Hamito-Semitic), including Arabic, Hebrew, Maltese, Berber, and many Ethiopian and Egyptian languages, dominates Saharan Africa and much of the Middle East. The rest of Africa is divided among three groups. Khoisan includes the !Kung and other groups (formerly called "Hottentots" and "Bushmen"), whose ancestors once occupied most of sub-Saharan Africa. The Niger-Congo phylum includes the Bantu family, spoken by farmers from western Africa who pushed the Khoisan into their current small enclaves in southern and southeastern Africa. The third phylum, Nilo-Saharan, occupies three large patches in the southern Saharan region.

In Asia, Dravidian languages such as Tamil dominate southern India and are found in pockets to the north. Dravidian speakers must therefore be the descendants of a people who occupied most of the Indian subcontinent before the incursion of the Indo-Europeans. Some 40 languages between the Black Sea and the Caspian Sea belong to the family called Caucasian (not to be confused with the informal racial term for the typically light-skinned people of Europe and Asia). Sino-Tibetan includes Chinese, Burmese, and Tibetan. Austronesian, having nothing to do with Australia (*Austr-* means "south"), includes the languages of Madagascar off the coast of Africa, Indonesia, Malaysia, the Philippines, New Zealand (Maori), Micronesia, Melanesia, and Polynesia, all the way to Hawaii—the record of people with extraordinary wanderlust and seafaring skill. Vietnamese and Khmer (the language of Cambodia) fall into Austro-Asiatic. The 200 aboriginal languages of Australia belong to a family of their own, and the 800 of New Guinea belong to a family as well, or perhaps to a small number of families. Japanese and Korean look like linguistic orphans, though a few linguists lump one or both with Altaic.

What about the Americas? Joseph Greenberg, whom we met earlier as the founder of the study of language universals, also classifies languages into phyla. He played a large role in unifying the 1,500 African languages into their four groups. Recently he has claimed that the 200 language stocks of native Americans can be grouped into only three phyla, each descending from a group of migrants who came over the Bering land bridge from Asia beginning 12,000 years ago or earlier. The Eskimos and Aleuts were the most recent immigrants. They were preceded by the Na-Dene, who occupied most of Alaska and northwestern Canada and embrace some of the languages of the American Southwest such as Navajo and Apache. This much is widely accepted. But Greenberg has also proposed that all the other languages, from Hudson Bay to Tierra del Fuego, belong to a single phylum, Amerind. The sweeping idea that America was settled by only three migrations has received some support from recent studies by

Cavalli-Sforza and others of modern natives' genes and tooth patterns, which fall into groups corresponding roughly to the three language phyla.

At this point we enter a territory of fierce controversy but potentially large rewards. Greenberg's hypothesis has been furiously attacked by other scholars of American languages. Comparative linguistics is an impeccably precise domain of scholarship, where radical divergences between related languages over centuries or a few millennia can with great confidence be traced back step by step to a common ancestor. Linguists raised in this tradition are appalled by Greenberg's unorthodox method of lumping together dozens of languages based on rough similarities in vocabulary, rather than carefully tracing sound-changes and reconstructing proto-languages. As an experimental psycholinguist who deals with the noisy data of reaction times and speech errors, I have no problem with Greenberg's use of many loose correspondences, or even with the fact that some of his data contain random errors. What bothers me more is his reliance on gut feelings of similarity rather than on actual statistics that control for the number of correspondences that might be expected by chance. A charitable observer can always spot similarities in large vocabulary lists, but that does not imply that they descended from a common lexical ancestor. It could be a coincidence, like the fact that the word for "blow" is *pneu* in Greek and *pniw* in Klamath (an American Indian language spoken in Oregon), or the fact that the word for "dog" in the Australian aboriginal language Mbabaram happens to be *dog*. (Another serious problem, which Greenberg's critics do point out, is that languages can resemble each other because of lateral borrowing rather than vertical inheritance, as in the recent exchanges that led to *her negligées* and *le weekend*.)

The odd absence of statistics also leaves in limbo a set of even more ambitious, exciting, and controversial hypotheses about language families and the prehistoric peoplings of continents that they would represent. Greenberg and his associate Merritt Ruhlen are joined by a school of Russian linguists (Sergei Starostin, Aharon

Dogopolsky, Vitaly Shevoroshkin, and Vladislav Illich-Svitych) who lump languages aggressively and seek to reconstruct the very ancient language that would have been the progenitor of each lump. They discern similarities among the proto-languages of Indo-European, Afro-Asiatic, Dravidian, Altaic, Uralic, and Eskimo-Aleut, as well as the orphans Japanese and Korean and a few miscellaneous language groups, reflecting a common ancestor proto-proto-language they call Nostratic. For example, the reconstructed Proto-Indo-European word for mulberry, *mor,* is similar to Proto-Altaic *mür* "berry," Proto-Uralic *marja* "berry," and Proto-Kartvelian (Georgian) *marcaw* "strawberry." The Nostraticists would have them all evolve from the hypothetical Nostratic root *marja.* Similarly, Proto-Indo-European *melg* "to milk" resembles Proto-Uralic *malge* "breast" and Arabic *mlg* "to suckle." Nostratic would have been spoken by a hunter-gatherer population, for there are no names of domesticated species among the 1,600 words the linguists claim to have reconstructed. The Nostratic hunter-gatherers would have occupied all of Europe, northern Africa, and northern, northeastern, western, and southern Asia, perhaps 15,000 years ago, from an origin in the Middle East.

And various lumpers from this school have suggested other audacious superphyla and super-superphyla. One comprises Amerind and Nostratic. Another, Sino-Caucasian, comprises Sino-Tibetan, Caucasian, and maybe Basque and Na-Dene. Lumping the lumps, Starostin has suggested that Sino-Caucasian can be connected to Amerind-Nostratic, forming a proto-proto-proto language that has been called SCAN, covering continental Eurasia and the Americas. Austric would embrace Austronesian, Austro-Asiatic, and various minor languages in China and Thailand. In Africa, some see similarities between Niger-Congo and Nilo-Saharan that warrant a Congo-Saharan group. If one were to accept all of these mergers—and some are barely distinguishable from wishful thinking—all human languages would fall into only six groups: SCAN in Eurasia, the Americas, and northern Africa; Khoisan and Congo-Saharan in sub-Saharan Africa; Austric in Southeast Asia and the Indian and Pacific Oceans; Australian; and New Guinean. Ancestral stocks of this geographic magnitude would have to correspond to the major expansions of the human species, and Cavalli-Sforza and Ruhlen have argued that they do. Cavalli-Sforza examined minor variations in the genes of hundreds of people representing a full spectrum of racial and ethnic groups. He claims that by lumping together sets of people who have similar genes, and then lumping the lumps, a genetic family tree of humankind can be constructed. The first bifurcation splits the sub-Saharan Africans off from everyone else. The adjoining branch in turn splits into two, one embracing Europeans, northeast Asians (including Japanese and Koreans), and American Indians, the other containing southeast Asians and Pacific Islanders on one sub-branch, and aboriginal Australians and New Guineans on another. The correspondences with the hypothetical language superphyla are reasonably clear, though not perfect. One interesting parallel is that what most people think of as the Mongoloid or Oriental race on the basis of superficial facial features and skin coloring may have no biological reality. In Cavalli-Sforza's genetic family tree, northeast Asians such as Siberians, Japanese, and Koreans are more similar to Europeans than to southeast Asians such as Chinese and Thai. Strikingly, this non-obvious racial grouping corresponds to the non-obvious linguistic grouping of Japanese, Korean, and Altaic with Indo-European in Nostratic, separate from the Sino-Tibetan family in which Chinese is found.

The branches of the hypothetical genetic/linguistic family tree can be taken to depict the history of *Homo sapiens sapiens*, from the African population in which mitochondrial Eve was thought to evolve 200,000 years ago, to the migrations out of Africa 100,000 years ago through the Middle East to Europe and Asia, and from there, in the past 50,000 years, to Australia, the islands of the Indian and Pacific Oceans, and the Americas. Unfortunately, the genetic and migrational family trees are almost as controversial as the linguistic one, and any part of this interesting story could unravel in the next few years.

A correlation between language families and human genetic groupings does *not*, by the way, mean that there are genes that make it easier for some kinds of people to learn some kinds of languages. This folk myth is pervasive, like the claim of some French speakers that only those with Gallic blood can truly master the gender system, or the insistence of my Hebrew teacher that the assimilated Jewish students in his college classes innately outperformed their Gentile classmates. As far as the language instinct is concerned, the correlation between genes and languages is a coincidence. People store genes in their gonads and pass them to their children through their genitals; they store grammars in their brains and pass them to their children through their mouths. Gonads and brains are attached to each other in bodies, so when bodies move, genes and grammars move together. That is the only reason that geneticists find any correlation between the two. We know that the connection is easily severed, thanks to the genetic experiments called immigration and conquest, in which children get their grammars from the brains of people other than their parents. Needless to say, the children of immigrants learn a language, even one separated from their parents' language by the deepest historical roots, without any disadvantage compared to age-mates who come from long lineages of the language's speakers. Correlations between genes and languages are thus so crude that they are measurable only at the level of superphyla and aboriginal races. In the past few centuries, colonization and immigration have completely scrambled the original correlations between the superphyla and the inhabitants of the different continents; native English speakers, to take the most obvious example, include virtually every racial subgroup on earth. Well before that, Europeans interbred with their neighbors and conquered each other often enough that there is almost no correlation between genes and language families within Europe (though the ancestors of the non-Indo-European Lapps, Maltese, and Basques left a few genetic mementos). For similar reasons, well-accepted language phyla can contain strange genetic bedfellows, like the black Ethiopians and white Arabs in the Afro-Asiatic phylum, and the white Lapps and Oriental Samoyeds in Uralic.

Moving from the highly speculative to the borderline flaky, Shevoroshkin, Ruhlen, and others have been trying to reconstruct words ancestral to the six superphyla—the vocabulary of the language of African Eve, "Proto-World." Ruhlen has posited 31 roots, such as *tik* "one" which would have evolved into Proto-Indo-European *deik* "to point" and then Latin *digit* "finger," Nilo-Saharan *dik* "one," Eskimo *tik* "index finger," Kede *tong* "arm," Proto-Afro-Asiatic *tak* "one," and Proto-Austro-Asiatic *ktig* "arm or hand." Though I am willing to be patient with Nostratic and similar hypotheses pending the work of a good statistician with a free afternoon, I find the Proto-World hypothesis especially suspect. (Comparative linguists are speechless.) It is not that I doubt that language evolved only once, one of the assumptions behind the search for the ultimate mother tongue. It's just that one can trace words back only so far. It is like the man who claimed to be selling Abraham Lincoln's ax—he explained that over the years the head had to be replaced twice and the handle three times. Most linguists believe that after 10,000 years no traces of a language remain in its descendants. This makes it extremely doubtful that anyone will find extant traces of the most recent ancestor of all contemporary languages, or that that ancestor would in turn retain traces of the language of the first modern humans, who lived some 200,000 years ago.

This chapter must end on a sad and urgent more. Languages are perpetuated by the children who learn them. When linguists see a language spoken only by adults, they know it is doomed. By this reasoning, they warn of an impending tragedy in the history of humankind. The linguist Michael Krauss estimates that 150 North American Indian languages, about 80% of the existing ones, are moribund. Elsewhere, his counts are equally grim: 40 moribund languages (90% of the existing ones) in Alaska and northern Siberia, 160 (23%) in Central and South America, 45 (70%) in Russia, 225 (90%) in Australia, perhaps 3,000 (50%) worldwide. Only about 600 languages are reasonably safe by dint of the sheer number of their speakers, say, a minimum of 100,000 (though this does not *guarantee* even short-term survival), and this optimistic assumption still suggests that between 3,600 and 5,400 languages, as many as 90% of the world's total, are threatened with extinction in the next century.

The wide-scale extinction of languages is reminiscent of the current (though less severe) wide-scale extinction of plant and animal species. The causes overlap. Languages disappear by the destruction of the habitats of their speakers, as well as by genocide, forced assimilation and assimilatory education, demographic submersion, and bombardment by electronic media, which Krauss calls "cultural nerve gas." Aside from halting the more repressive social and political causes of cultural annihilation, we can forestall some linguistic extinctions by developing pedagogical materials, literature, and television in the indigenous language. Other extinctions can be mitigated by preserving grammars, lexicons, texts, and recorded speech samples with the help of archives and faculty positions for native speakers. In some cases, like Hebrew in the twentieth century, the continued ceremonial use of a language together with preserved documents can be sufficient to revive it, given the will.

Just as we cannot reasonably hope to preserve every species on earth, we cannot preserve every language, and perhaps should not. The moral and practical issues are complex. Linguistic differences can be a source of lethal divisiveness, and if a generation chooses to switch to a language of the mainstream that promises them economic and social advancement, does some outside group have the right to coerce them not to on the grounds that it finds the idea of them keeping the old language pleasing? But such complexities aside, when 3,000-odd languages are moribund, we can be sure that many of the deaths are unwanted and preventable.

Why should people care about endangered languages? For linguistics and the sciences of mind and brain that encompass it, linguistic diversity shows us the scope and limits of the language instinct. Just think of the distorted picture we would have if only English were available for study! For anthropology and human evolutionary biology, languages trace the history and geography of the species, and the extinction of a language (say, Ainu, formerly spoken in Japan by a mysterious Caucasoid people) can be like the burning of a library of historical documents or the extinction of the last species in a phylum. But the reasons are not just scientific. As Krauss writes, "Any language is a supreme achievement of a uniquely human collective genius, as divine and endless a mystery as a living organism." A language is a medium from which a culture's verse, literature, and song can never be extricated. We are in danger of losing treasures ranging from Yiddish, with far more words for "simpleton" than the Eskimos were reputed to have for "snow," to Damin, a ceremonial variant of the Australian language Lardil, which has a unique 200-word vocabulary that is learnable in a day but that can express the full range of concepts in everyday speech. As the linguist Ken Hale has put it, "The loss of a language is part of the more general loss being suffered by the world, the loss of diversity in all things."

# 9

⁂

# Baby Born Talking— Describes Heaven

*On May 21, 1985, a periodical called the* Sun *ran these intriguing* headlines:

John Wayne Liked to Play with Dolls

Prince Charles' Blood Is Sold for $10,000
by Dishonest Docs

Family Haunted by Ghost of Turkey
They Ate for Christmas

BABY BORN TALKING—DESCRIBES HEAVEN
Incredible proof of reincarnation

The last headline caught my eye—it seemed like the ultimate demonstration that language is innate. According to the article,

Life in heaven is grand, a baby told an astounded obstetrical team seconds after birth. Tiny Naomi Montefusco literally came into the world singing the praises of God's firmament. The miracle so shocked the delivery room team, one nurse ran screaming down the hall. "Heaven is a beautiful place, so warm and so serene," Naomi said. "Why did you bring me here?"

Among the witnesses was mother Theresa Montefusco, 18, who delivered the child under local anesthetic . . . "I distinctly heard her describe heaven as a place where no one has to work, eat, worry about clothing, or do anything but sing God's praises. I tried to get off the delivery table to kneel down and pray, but the nurses wouldn't let me."

Scientists, of course, cannot take such reports at face value; any important finding must be replicated. A replication of the Corsican miracle, this time from Taranto, Italy, occurred on October 31, 1989, when the *Sun* (a strong believer in recycling) ran the headline "BABY BORN TALKING—DESCRIBES HEAVEN. Infant's words prove reincarnation exists." A related discovery was reported on May 29, 1990: "BABY SPEAKS AND SAYS: I'M THE REINCARNATION OF NATALIE WOOD." Then, on September 29, 1992, a second replication, reported in the same words as the original. And on June 8, 1993, the clincher: "AMAZING 2-HEADED BABY IS PROOF OF REINCARNATION. ONE HEAD SPEAKS ENGLISH—THE OTHER ANCIENT LATIN."

Why do stories like Naomi's occur only in fiction, never in fact? Most children do not begin to talk until they are a year old, do not combine words until they are one and a half, and do not converse in fluent grammatical sentences until they are two or three. What is going on in those years? Should we ask why it takes children so long? Or is a three-year-old's ability to describe earth as miraculous as a newborn's ability to describe heaven?

All infants come into the world with linguistic skills. We know this because of the ingenious experimental technique (discussed in Chapter 3) in which a baby is presented with one signal over and over to the point of boredom, and then the signal is changed; if the baby perks up, he or she must be able to tell the difference. Since ears don't move the way eyes do, the psychologists Peter Eimas and Peter Jusczyk devised a different way to see what a one-month-old finds interesting. They put a switch inside a rubber nipple and hooked up the switch to a tape recorder, so that when the baby sucked, the tape played. As the tape droned on with *ba ba ba ba* . . . , the infants showed their boredom by sucking more slowly. But when the syllables changed to *pa pa pa* . . . , the infants began to suck more vigorously, to hear more syllables. Moreover, they were using the sixth sense, speech perception, rather than just hearing the syllables as raw sound: two *ba*'s that differed acoustically from each other as much as a *ba* differs from a *pa,* but that are both heard as *ba* by adults, did not revive the infants' interest. And infants must be recovering phonemes, like *b,* from the syllables they are smeared across. Like adults, they hear the same stretch of sound as a *b* if it appears in a short syllable and as a *w* if it appears in a long syllable.

Infants come equipped with these skills; they do not learn them by listening to their parents' speech. Kikuyu and Spanish infants discriminate English *ba*'s and *pa*'s, which are not used in Kikuyu or Spanish and which their parents cannot tell apart. English-learning infants under the age of six months distinguish phonemes used in Czech, Hindi, and Inslekampx (a Native American language), but English-speaking adults cannot, even with five hundred trials of training or a year of university coursework. Adult ears can tell the sounds apart, though, when the consonants are stripped from the syllables and presented alone as chirpy sounds; they just cannot tell them apart *as phonemes.*

The *Sun* article is a bit sketchy on the details, but we can surmise that because Naomi was understood, she must have spoken in Italian, not Proto-World or Ancient Latin. Other infants may enter the world with some knowledge of their mother's language, too. The psychologists Jacques Mehler and Peter Jusczyk have shown that four-day-old French babies suck harder to hear French than Russian, and pick up their sucking more when a tape changes from Russian to French than from French to Russian. This is not an incredible proof of reincarnation; the melody of mothers' speech carries through their bodies and is audible in the womb. The babies still prefer French when the speech is electronically filtered so that the consonant and vowel sounds are muffled and only the melody comes through. But they are indifferent when the tapes are played backwards, which preserves the vowels and some of the consonants but distorts the melody. Nor does the effect prove the inherent beauty of the French language: non-French infants do not prefer French, and French infants do not distinguish Italian from English. The infants must have learned something about the prosody of French (its melody, stress, and timing) in the womb, or in their first days out of it.

Babies continue to learn the sounds of their language throughout the first year. By six months, they are beginning to lump together the distinct sounds that their language collapses into a single phoneme, while continuing to discriminate equivalently distinct ones that their language keeps separate. By ten months they are no longer universal phoneticians but have turned into their parents; they do not distinguish Czech or Inslekampx phonemes unless they are Czech or Inslekampx babies. Babies make this transition before they produce or understand words, so their learning cannot depend on correlating sound with meaning. That is, they cannot be listening for the difference in sound between a word they think means *bit* and a word they think means *beet*, because they have learned neither word. They must be sorting the sounds directly, somehow tuning their speech analysis module to deliver the phonemes used in their language. The module can then serve as the front end of the system that learns words and grammar.

During the first year, babies also get their speech production systems geared up. First, ontogeny recapitulates phylogeny. A newborn has a vocal tract like a nonhuman mammal. The larynx comes up like a periscope and engages the nasal passage, forcing the infant to breathe through the nose and making it anatomically possible to drink and breathe at the same time. By three months the larynx has descended deep into the throat, opening up the cavity behind the tongue (the pharynx) that allows the tongue to move forwards and backwards and produce the variety of vowel sounds used by adults.

Not much of linguistic interest happens during the first two months, when babies produce the cries, grunts, sighs, clicks, stops, and pops associated with breathing, feeding, and fussing, or even during the next three, when coos and laughs are added. Between five and seven months babies begin to play with sounds, rather than using them to express their physical and emotional states, and their sequences of clicks, hums, glides, trills, hisses, and smacks begin to sound like consonants and vowels. Between seven and eight months they suddenly begin to babble in real syllables like *ba-ba-ba, neh-neh-neh,* and *dee-dee-dee.* The sounds are the same in all languages, and consist of the phonemes and syllable patterns that are most common across languages. By the end of the first year, babies vary their syllables, like *neh-nee, da-dee,* and *meh-neh,* and produce that really cute sentencelike gibberish.

In recent years pediatricians have saved the lives of many babies with breathing abnormalities by inserting a tube into their tracheas (the pediatricians are trained on cats, whose airways are similar), or by surgically opening a hole in their trachea below the larynx. The infants are then unable to make voiced sounds during the normal period of babbling. When the normal airway is restored in the second year of life, those infants are seriously retarded in speech development, though they eventually catch up, with no permanent problems. Deaf children's babbling is later and simpler—though if their parents use sign language, they babble, on schedule, with their hands!

Why is babbling so important? The infant is like a person who has been given a complicated piece of audio equipment bristling with unlabeled knobs and switches but missing the instruction manual. In such situations people resort to what hackers call frobbing—fiddling aimlessly with the controls to see what happens. The infant has been given a set of neural commands that can move the articulators every which way, with wildly varying effects on the sound. By listening to their own babbling, babies in effect write their own instruction manual; they learn how much to move which muscle in which way to make which change in the sound. This is a prerequisite to duplicating the speech of their parents. Some computer scientists, inspired by the infant, believe that a good robot should learn an internal software model of its articulators by observing the consequences of its own babbling and flailing.

Shortly before their first birthday, babies begin to understand words, and around that birthday, they start to produce them. Words are usually produced in isolation; this one-word stage can last from two months to a year. For over a century, and all over the globe, scientists have kept diaries of their infants' first words, and the lists are almost identical. About half the words are for objects: food (*juice, cookie*), body parts (*eye, nose*), clothing (*diaper, sock*), vehicles (*car, boat*), toys (*doll, block*), household items (*bottle, light*), animals (*dog, kitty*), and people (*dada, baby*). (My nephew Eric's first word was *Batman*.) There are words for actions, motions, and routines, like *up, off, open, peekaboo, eat,* and *go,* and modifiers, like *hot, allgone, more, dirty,* and *cold.* Finally, there are routines used in social interaction, like *yes, no, want, bye-bye,* and *hi*—a few of which, like *look at that* and *what is that,* are words in the sense of listemes (memorized chunks), but not, at least for the adult, words in the sense of morphological products and syntactic atoms. Children differ in how much they name objects or engage in social interaction using memorized routines. Psychologists have spent a lot of time speculating about the causes of those differences (sex, age, birth order, and socioeconomic status have all been examined), but the most plausible to my mind is that babies are people, only smaller. Some are interested in objects, others like to shmooze.

Since word boundaries do not physically exist, it is remarkable that children are so good at finding them. A baby is like the dog being yelled at in the two-panel cartoon by Gary Larson:

> WHAT WE SAY TO DOGS: "Okay, Ginger! I've had it! You stay out of the garbage! Understand, Ginger? Stay out of the garbage, or else!"

> WHAT THEY HEAR: "Blah blah GINGER blah blah blah blah blah blah blah blah GINGER blah blah blah blah blah."

Presumably children record some words parents use in isolation, or in stressed final positions, like *Look-at-the BOTTLE.* Then they look for matches to these words in longer stretches of speech, and find other words by extracting the residues in between the matched portions. Occasionally there are near misses, providing great entertainment to family members:

I don't want to go to your ami. [from *Miami*]

I am heyv! [from *Behave!*]

Daddy, when you go tinkle you're an eight, and when I go tinkle I'm an eight, right? [from *urinate*]

I know I sound like Larry, but who's Gitis? [from *laryngitis*]

Daddy, why do you call your character Sam Alone? [from *Sam Malone*, the bartender in *Cheers*]

The ants are my friends, they're blowing in the wind. [from *The answer, my friend, is blowing in the wind*]

But these errors are surprisingly rare, and of course adults occasionally make them too, as in the Pullet Surprise and doggy-dog world of Chapter 6. In an episode of the television show *Hill Street Blues*, police officer JD Larue began to flirt with a pretty high school student. His partner, Neal Washington, said, "I have only three words to say to you, JD. Statue. Tory. Rape."

Around eighteen months, language takes off. Vocabulary growth jumps to the new-word-every-two-hours minimum rate that the child will maintain through adolescence. And syntax begins, with strings of the minimum length that allows it: two. Here are some examples:

| | | |
|---|---|---|
| All dry. | All messy. | All wet. |
| I sit. | I shut. | No bed. |
| No pee. | See baby. | See pretty. |
| More cereal. | More hot. | Hi Calico. |
| Other pocket. | Boot off. | Siren by. |
| Mail come. | Airplane allgone. | Bye-bye car. |
| Our car. | Papa away. | Dry pants. |

Children's two-word combinations are so similar in meaning the world over that they read as translations of one another. Children announce when objects appear, disappear, and move about, point out their properties and owners, comment on people doing things and seeing things, reject and request objects and activities, and ask about who, what, and where. These microsentences already reflect the language being acquired: in ninety-five percent of them, the words are properly ordered.

There is more going on in children's minds than in what comes out of their mouths. Even before they put two words together, babies can comprehend a sentence using its syntax. For example, in one experiment, babies who spoke only in single words were seated in front of two television screens, each of which featured a pair of adults improbably dressed up as Cookie Monster and Big Bird from *Sesame Street*. One screen showed Cookie Monster tickling Big Bird; the other showed Big Bird tickling Cookie Monster. A voiceover said, "OH LOOK!!! BIG BIRD IS TICKLING COOKIE MONSTER!! FIND BIG BIRD TICKLING COOKIE MONSTER!!" (or vice versa). The children must have understood the meaning of the ordering of subject, verb, and object—they looked more at the screen that depicted the sentence in the voiceover.

When children do put words together, the words seem to meet up with a bottleneck at the output end. Children's two-and-three-word utterances look like samples drawn from longer potential sentences expressing a complete and more complicated idea. For example, the psychologist Roger Brown noted that although the children he studied never produced a sentence as complicated as *Mother gave John lunch in the kitchen,* they did produce strings containing all of its components, and in the correct order:

| AGENT | ACTION | RECIPIENT | OBJECT | LOCATION |
|---|---|---|---|---|
| (Mother | gave | John | lunch | in the kitchen.) |
| Mommy | fix. | | | |
| Mommy | | | pumpkin. | |
| Baby | | | | table. |
| Give | | doggie. | | |
| | Put | | light. | |
| | Put | | | floor. |
| I | ride | | horsie. | |
| Tractor | go | | | floor. |
| | Give | doggie | paper. | |
| | Put | | truck | window. |
| Adam | put | | it | box. |

* * *

If we divide language development into somewhat arbitrary stages, like Syllable Babbling, Gibberish Babbling, One-Word Utterances, and Two-Word Strings, the next stage would have to be called All Hell Breaks Loose. Between the late twos and the mid-threes, children's language blooms into fluent grammatical conversation so rapidly that it overwhelms the researchers who study it, and no one has worked out the exact sequence. Sentence length increases steadily, and because grammar is a discrete combinatorial system, the number of syntactic types increases exponentially, doubling every month, reaching the thousands before the third birthday. You can get a feel for this explosion by seeing how the speech of a little boy called Adam grows in sophistication over the period of a year, starting with his early word combinations at the age of two years and three months ("2;3"):

2;3: Play checkers. Big drum. I got horn. A bunny-rabbit walk.

2;4: See marching bear go? Screw part machine. That busy bulldozer truck.

2;5: Now put boots on. Where wrench go? Mommy talking bout lady. What that paper clip doing?

2;6: Write a piece a paper. What that egg doing? I lost a shoe. No, I don't want to sit seat.

2;7 Where piece a paper go? Ursula has a boot on. Going to see kitten. Put the cigarette down. Dropped a rubber band. Shadow has hat just like that. Rintintin don't fly, Mommy.

2;8: Let me get down with the boots on. Don't be afraid a horses. How tiger be so healthy and fly like kite? Joshua throw like a penguin.

2;9: Where Mommy keep her pocket book? Show you something funny. Just like turtle make mud pie.

2;10: Look at that train Ursula brought. I simply don't want put in chair. You don't have paper. Do you want little bit, Cromer? I can't wear it tomorrow.

2;11:   That birdie hopping by Missouri in bag. Do want some pie on your face? Why you mixing baby chocolate? I finish drinking all up down my throat. I said why not you coming in? Look at that piece a paper and tell it. Do you want me tie that round? We going turn light on so you can't see.

3;0:   I going come in fourteen minutes. I going wear that to wedding. I see what happens. I have to save them now. Those are not strong mens. They are going sleep in wintertime. You dress me up like a baby elephant.

3;1:   I like to play with something else. You know how to put it back together. I gon' make it like a rocket to blast off with. I put another one on the floor. You went to Boston University? You want to give me some carrots and some beans? Press the button and catch it, sir. I want some other peanuts. Why you put the pacifier in his mouth? Doggies like to climb up.

3;2:   So it can't be cleaned? I broke my racing car. Do you know the lights wents off? What happened to the bridge? When it's got a flat tire it's need a go to the station. I dream sometimes. I'm going to mail this so the letter can't come off. I want to have some espresso. The sun is not too bright. Can I have some sugar? Can I put my head in the mailbox so the mailman can know where I are and put me in the mailbox? Can I keep the screwdriver just like a carpenter keep the screwdriver?

Normal children can differ by a year or more in their rate of language development, though the stages they pass through are generally the same regardless of how stretched out or compressed. I chose to show you Adam's speech because his language development is rather *slow* compared with other children's. Eve, another child Brown studied, was speaking in sentences like this before she was two:

> I got peanut butter on the paddle.
> I sit in my high chair yesterday.

Fraser, the doll's not in your briefcase.
Fix it with the scissor.
Sue making more coffee for Fraser.

Her stages of language development were telescoped into just a few months.

Many things are going on during this explosion. Children's sentences are getting not only longer but more complex, with deeper, bushier trees, because the children can embed one constituent inside another. Whereas before they might have said *Give doggie paper* (a three-branch verb phrase) and *Big doggie* (a two-branch noun phrase), they now say *Give big doggie paper,* with the two-branch NP embedded inside the middle branch of three-branch VP. The earlier sentences resembled telegrams, missing unstressed function words like *of, the, on,* and *does,* as well as inflections like *-ed, -ing,* and *-s.* By the threes, children are using these function words more often than they omit them, many in more than ninety percent of the sentences that require them. A full range of sentence types flower—questions with words like *who, what,* and *where,* relative clauses, comparatives, negations, complements, conjunctions, and passives.

Though many—perhaps even most—of the young three-year-old's sentences are ungrammatical for one reason or another, we should not judge them too harshly, because there are many things that can go wrong in any single sentence. When researchers focus on one grammatical rule and count how often a child obeys it and how often he or she flouts it, the results are astonishing: for any rule you choose, three-year-olds obey it most of the time. As we have seen, children rarely scramble word order and, by the age of three, come to supply most inflections and function words in sentences that require them. Though our ears perk up when we hear errors like *mens, wents, Can you broke those?, What he can ride in?, That's a furniture, Button me the rest,* and *Going to see kitten,* the errors occur in only 0.1% to 8% of the opportunities for making them; more than 90% of the time, the child is on target. The psychologist Karin Stromswold analyzed sentences containing auxiliaries from the speech of thirteen preschoolers. The auxiliary system in English (including words like *can, should, must, be, have,* and *do*) is notorious among grammarians for its complexity. There are about twenty-four billion billion logically possible combinations of auxiliaries (for instance, *He have might eat; He did be eating*), of which only a hundred are grammatical (*He might have eaten; He has been eating*). Stromswold wanted to count how many times children were seduced by several dozen kinds of tempting errors in the auxiliary system—that is, errors that would be natural generalizations of the sentence patterns children heard from their parents:

| PATTERN IN ADULT ENGLISH | ERROR THAT MIGHT TEMPT A CHILD |
|---|---|
| He seems happy. → Does he seem happy? | He is smiling. → Does he be smiling? |
| | She could go. → Does she could go? |
| He did eat. → He didn't eat. | He did a few things. → He didn't a few things. |
| He did eat. → Did he eat? | He did a few things. → Did he a few things? |
| I like going. → He likes going. | I can go. → He cans go. |
| | I am going. → He ams (*or* be's) going. |
| They want to sleep. → They wanted to sleep. | They are sleeping. → They are'd (*or* be'd) sleeping. |
| He is happy. → He is not happy. | He ate something. → He ate not something. |
| He is happy. → Is he happy? | He ate something. → Ate he something? |

For virtually all of these patterns, she found *no* errors among the 66,000 sentences in which they could have occurred.

The three-year-old child is grammatically correct in quality, not just quantity. In earlier chapters we learned of experiments showing that children's movement rules are structure-dependent ("Ask Jabba if the boy who is unhappy is watching Mickey Mouse") and showing that their morphological systems are organized into layers of roots, stems, and inflections ("This monster likes to eat rats; what do you call him?"). Children also seem fully prepared for the Babel of languages they may face: they swiftly acquire free word order, SOV and VSO orders, rich systems of case and agreement, strings of agglutinated suffixes, ergative case marking, or whatever else their language throws at them, with no lag relative to their English-speaking counterparts. Languages with grammatical gender like French and German are the bane of the Berlitz student. In his essay "The Horrors of the German Language," Mark Twain noted that "a tree is male, its buds are female, its leaves are neuter; horses are sexless, dogs are male, cats are female—tomcats included." He translated a conversation in a German Sunday school book as follows:

> Gretchen: Wilhelm, where is the turnip?
> Wilhelm: She has gone to the kitchen.
> Gretchen: Where is the accomplished and beautiful English maiden?
> Wilhelm: It has gone to the opera.

But little children learning German (and other languages with gender) are not horrified; they acquire gender marking quickly, make few errors, and never use the association with maleness and femaleness as a false criterion. It is safe to say that except for constructions that are rare, used predominantly in written language, or mentally taxing even to an adult (like *The horse that the elephant tickled kissed the pig*), all languages are acquired, with equal ease, before the child turns four.

The errors children do make are rarely random garbage. Often the errors follow the logic of grammar so beautifully that the puzzle is not why the children make the errors, but why they sound like errors to adult ears at all. Let me give you two examples that I have studied in great detail.

Perhaps the most conspicuous childhood error is to overgeneralize—the child puts a regular suffix, like the plural *-s* or the past tense *-ed,* onto a word that forms its plural or its past tense in an irregular way. Thus the child says *tooths* and *mouses* and comes up with verb forms like these:

> My teacher holded the baby rabbits and we patted them.
> Hey, Horton heared a Who.
> I finded Renée.
> I love cut-upped egg.
> Once upon a time a alligator was eating a dinosaur and the
>    dinosaur was eating the alligator and the dinosaur was eaten
>    by the alligator and the alligator goed kerplunk.

These forms sound wrong to us because English contains about 180 irregular verbs like *held, heard, cut,* and *went*—many inherited from Proto-Indo-European!—whose past-tense forms cannot be predicted by rule but have to be memorized by rote. Morphology is organized so that whenever a verb has an idiosyncratic form listed in the mental dictionary, the regular *-ed* rule is blocked: *goed* sounds ungrammatical because it is blocked by *went.* Elsewhere, the regular rule applies freely.

So why do children make this kind of error? There is a simple explanation. Since irregular forms have to be memorized and memory is fallible, any time the child tries to use a sentence in the past tense with an irregular verb but cannot summon its past-tense form from memory, the regular rule fills the vacuum. If the child wants to use the past tense of *hold* but cannot dredge up *held,* the regular rule, applying by default, marks it as *holded.* We know fallible memory is the cause of these errors because the irregular verbs that are used the least often by parents (*drank* and *knew,* for instance) are the ones their children err on the most; for the more common verbs, children are correct most of the time. The same thing happens to adults: lower-frequency, less-well-remembered irregular forms like *trod, strove, dwelt, rent, slew,* and *smote* sound odd to modern American ears and are likely to be regularized to *treaded, strived, dwelled, rended, slayed,* and *smited.* Since it's we grownups who are forgetting the irregular past, we get to declare that the forms with *-ed* are not errors! Indeed, over the centuries many of these conversions have become permanent.

Old English and Middle English had about twice as many irregular verbs as Modern English; if Chaucer were here today, he would tell you that the past tenses of *to chide, to geld, to abide,* and *to cleave* are *chid, gelt, abode,* and *clove.* As time passes, verbs can wane in popularity, and one can imagine a time when, say, the verb *to geld* had slipped so far that a majority of adults could have lived their lives seldom having heard its past-tense form *gelt.* When pressed, they would have used *gelded;* the verb had become regular for them and all subsequent generations. The psychological process is no different from what happens when a young child has lived his or her brief life seldom having heard the past-tense form *built* and, when pressed, comes up with *builded.* The only difference is that the child is surrounded by grown-ups who are still using *built.* As the child lives longer and hears *built* more and more times, the mental dictionary entry for *built* becomes stronger and it comes to mind more and more readily, turning off the "add-*ed*" rule each time it does.

Here is another lovely set of examples of childhood grammatical logic, discovered by the psychologist Melissa Bowerman:

> Go me to the bathroom before you go to bed.
> The tiger will come and eat David and then he will be died and
>    I won't have a little brother any more.
> I want you to take me a camel ride over your shoulders into my
>    room.
> Be a hand up your nose.
> Don't giggle me!
> Yawny Baby—you can push her mouth open to drink her.

These are examples of the causative rule, found in English and many other languages, which takes an intransitive verb meaning "to do something" and converts it to a transitive verb meaning "to cause to do something":

> The butter melted. → Sally melted the butter.
> The ball bounced. → Hiram bounced the ball.
> The horse raced past the barn. → The jockey raced the horse
>    past the barn.

The causative rule can apply to some verbs but not others; occasionally children apply it too zealously. But it is not easy, even for a linguist, to say why a ball can bounce or be bounced, and a horse can race or be raced, but a brother can only die, not be died, and a girl can only giggle, not be giggled. Only a few kinds of verbs can easily undergo the rule: verbs referring to a change of the physical state of an object, like *melt* and *break,* verbs referring to a manner of motion, like *bounce* and *slide,* and verbs referring to an accompanied locomotion, like *race* and *dance.* Other verbs, like *go* and *die,* refuse to undergo the rule in English, and verbs involving fully voluntary actions, like *cook* and *play,* refuse to undergo the rule in almost every language (and children rarely err on them). Most of children's errors in English, in fact, would be grammatical in other languages. English-speaking adults, like their children, occasionally stretch the envelope of the rule:

> In 1976 the Parti Québecois began to deteriorate the health care system.
> Sparkle your table with Cape Cod classic glass-ware.
> Well, that decided me.
> This new golf ball could obsolete many golf courses.
> If she subscribes us up, she'll get a bonus.
> Sunbeam whips out the holes where staling air can hide.

So both children and adults stretch the language a bit to express causation; adults are just a tiny bit more fastidious in which verbs they stretch.

The three-year-old, then, is a grammatical genius—master of most constructions, obeying rules far more often than flouting them, respecting language universals, erring in sensible, adultlike ways, and avoiding many kinds of errors altogether. How do they do it? Children of this age are notably incompetent at most other activities. We won't let them drive, vote, or go to school, and they can be flummoxed by no-brainer tasks like sorting beads in order of size, reasoning whether a person could be aware of an event that took place while the person was out of the room, and knowing that the volume of a liquid does not change when it is poured from a short, wide glass into a tall, narrow one. So they are not doing it by the sheer power of their overall acumen. Nor could they be imitating what they hear, or else they would never say *goed* or *Don't giggle me*. It is plausible that the basic organization of grammar is wired into the child's brain, but they still must reconstruct the nuances of English or Kivunjo or Ainu. So how does experience interact with wiring to give a three-year-old the grammar of a particular language?

We know that this experience must include, at a minimum, the speech of other human beings. For several thousand years thinkers have speculated about what would happen to infants deprived of speech input. In the seventh century B.C., according to the historian Herodotus, King Psamtik I of Egypt had two infants separated from their mothers at birth and raised in silence in a shepherd's hut. The king's curiosity about the original language of the world allegedly was satisfied two years later when the shepherd heard the infants use a word in Phrygian, an Indo-European language of Asia Minor. In the centuries since, there have been many stories about abandoned children who have grown up in the wild, from Romulus and Remus, the eventual founders of Rome, to Mowgli in Kipling's *The Jungle Book*. There have also been occasional real-life cases, like Victor, the Wild Boy of Aveyron (the subject of a lovely film by François Truffaut), and, in the twentieth century, Kamala, Amala, and Ramu from India. Legend has these children raised by bears or wolves, depending on which one has the greater affinity to humans in the prevailing mythology of the region, and this scenario is repeated as fact in many textbooks, but I am skeptical. (In a Darwinian animal kingdom it would be a spectacularly stupid bear that when faced with the good fortune of a baby in its lair would rear it rather than eat it. Though some species can be fooled by foster offspring, like birds by cuckoos, bears and wolves are predators of young mammals and are unlikely to be so gullible.) Occasionally other modern children have grown up wild because depraved parents have raised them silently in dark rooms and attics. The outcome is always the same: the children are mute, and often remain so. Whatever innate grammatical abilities there are, they are too schematic to generate speech, words, and grammatical constructions on their own.

The muteness of wild children in one sense emphasizes the role of nurture over nature in language development, but I think we gain more insight by thinking around that tired dichotomy. If Victor or Kamala had run out of the woods speaking fluent Phrygian or Proto-World, who could they have talked to? As I suggested in the preceding chapter, even if the genes themselves specify the basic design of language, they might have to store the specifics of language in the environment, to ensure that a person's language is synchronized with everyone else's despite the genetic uniqueness of every individual. In this sense, language is like another quintessentially social activity. James Thurber and E. B. White once wrote:

> There is a very good reason why the erotic side of Man has called forth so much more discussion lately than has his appetite for food. The reason is this: that while the urge to eat is a personal matter which concerns no one but the person hungry (or, as the German has it, *der hungrige Mensch*), the sex urge involves, for its true expression, another individual. It is this "other individual" that causes all the trouble.

Though speech input is necessary for speech development, a mere soundtrack is not sufficient. Deaf parents of hearing children were once advised to have the children watch a lot of television. In no case did the children learn English. Without already knowing the language, it is difficult for a child to figure out what the characters in those odd, unresponsive televised worlds are talking about. Live human speakers tend to talk about the here and now in the presence of children; the child can be more of a mind-reader, guessing what the speaker might mean, especially if the child already knows many content words. Indeed, if you are given a translation of the content words in parents' speech to children in some language whose grammar you do not know, it is quite easy to infer what the parents meant. If children can infer parents' meanings, they do not have to be pure cryptographers, trying to crack a code from the statistical structure of the transmissions. They can be a bit more like the archeologists with the Rosetta Stone, who had both a passage from an unknown language and its translation in a known one. For the child, the unknown language is English (or Japanese or Inslekampx or Arabic); the known one is mentalese.

Another reason why television soundtracks might be insufficient is that they are not in Motherese. Compared with conversations among adults, parents' speech to children is slower, more exaggerated in pitch, more directed to the here and now, and more grammatical (it is literally 99 and 44/100ths percent pure, according to one estimate). Surely this makes Motherese easier to learn than the kind of elliptical, fragmentary conversation we saw in the Watergate transcripts. But as we discovered in Chapter 2, Motherese is not an indispensable curriculum of Language-Made-Simple lessons. In some cultures, parents do not talk to their children until the children are capable of keeping up their end of the conversation (though other children might talk to them). Furthermore, Motherese is *not* grammatically simple. That impression is an illusion; grammar is so instinctive that we do not appreciate which constructions are complex until we try to work out the rules behind them. Motherese is riddled with questions containing *who, what,* and *where,* which are among the most complicated constructions in English. For example, to assemble the "simple" question *What did he eat?,* based on *He ate what,* one must move the *what* to the beginning of the sentence, leaving a "trace" that indicates its semantic role of "thing eaten," insert the meaningless auxiliary *do,* make sure that the *do* is in the tense appropriate to the verb, in this case *did,* convert the verb to the infinitive form *eat,* and invert the position of subject and auxiliary from the normal *He did* to the interrogative *Did he.* No mercifully designed language curriculum would use these sentences in Lesson 1, but that is just what mothers do when speaking to their babies.

A better way to think of Motherese is to liken it to the vocalizations that other animals direct to their young. Motherese has interpretable melodies: a rise-and-fall contour for approving, a set of sharp, staccato bursts for prohibiting, a rise pattern for directing attention, and smooth, low legato murmurs for comforting. The psychologist Anne Fernald has shown that these patterns are very widespread across language communities, and may be universal. The melodies attract the child's attention, mark the sounds as speech as opposed to stomach growlings or other noises, distinguish statements, questions, and imperatives, delineate major sentence boundaries, and highlight new words. When given a choice, babies prefer to listen to Motherese than to speech intended for adults.

Surprisingly, though practice is important in training for the gymnastics of speaking, it may be superfluous in learning grammar. For various neurological reasons children are sometimes unable to articulate, but parents report that their comprehension is excellent. Karin Stromswold recently tested one such four-year-old. Though he could not speak, he could understand subtle grammatical differences. He could identify which picture showed "The dog was bitten by the cat" and which showed "The cat was bitten by the dog." He could distinguish pictures that showed "The dogs chase the rabbit" and "The dog chases the rabbit." The boy also responded appropriately when Stromswold asked him, "Show me your room," "Show me your sister's room," "Show me your sister's old room," "Show me your old room," "Show me your new room," "Show me your sister's new room."

In fact, it is not surprising that grammar development does not depend on overt practice, because actually saying something aloud, as opposed to listening to what other people say, does not provide the child with information about the language he or she is trying to learn. The only conceivable information about grammar that speaking could provide would come from feedback from parents on whether the child's utterance was grammatical and meaningful. If a parent punished, corrected, misunderstood, or even reacted differently to a child's ungrammatical sentence, it could in theory inform the child that something in his growing rule system needed to be improved. But parents are remarkably unconcerned about their children's grammar; they care about truthfulness and good behavior. Roger Brown divided the sentences of Adam, Eve, and Sarah into grammatical and ungrammatical lists. For each sentence he checked whether the parent had at the time expressed approval (like "Yes, that's good") or disapproval. The proportion was the same for grammatical sentences and ungrammatical ones, which means that the parent's response had given the child no information about grammar. For example:

> Child: Mamma isn't boy, he a girl.
> Mother: That's right.
>
> Child: And Walt Disney comes on Tuesday.
> Mother: No, he does not.

Brown also checked whether children might learn about the state of their grammars by noticing whether they are being understood. He looked at children's well-formed and badly formed questions and whether their parents seemed to have answered them appropriately (that is, as if they understood them) or with non sequiturs. Again, there was no correlation; *What you can do?* may not be English, but it is perfectly understandable.

Indeed, when fussy parents or meddling experimenters do provide children with feedback, the children tune it out. The psycholinguist Martin Braine once tried for several weeks to stamp out one of his daughter's grammatical errors. Here is the result:

> Child: Want other one spoon, Daddy.
> Father: You mean, you want THE OTHER SPOON.
> Child: Yes, I want other one spoon, please, Daddy.
> Father: Can you say "the other spoon"?
> Child: Other . . . one . . . spoon.
> Father: Say . . . "other."
> Child: Other.
> Father: "Spoon."
> Child: Spoon.
> Father: "Other . . . Spoon."
> Child: Other . . . spoon. Now give me other one spoon?

Braine wrote, "Further tuition is ruled out by her protest, vigorously supported by my wife."

As far as grammar learning goes, the child must be a naturalist, passively observing the speech of others, rather than an experimentalist, manipulating stimuli and recording the results. The implications are profound. Languages are infinite, childhoods finite. To become speakers, children cannot just memorize; they must leap into the linguistic unknown and generalize to an infinite world of as-yet-unspoken sentences. But there are untold numbers of seductive false leaps:

> mind → minded; *but not* find → finded
>
> The ice melted → He melted the ice; *but not* David died → He died David
>
> She seems to be asleep → She seems asleep; *but not* She seems to be sleeping → She seems sleeping
>
> Sheila saw Mary with her best friend's husband → Who did Sheila see Mary with? *but not* Sheila saw Mary and her best friend's husband → Who did Sheila see Mary and?

If children could count on being corrected for making such errors, they could take their chances. But in a world of grammatically oblivious parents, they must be more cautious—if they ever went too far and produced ungrammatical sentences together with the grammatical ones, the world would never tell them they were wrong. They would speak ungrammatically all their lives—though a better way of putting it is that that part of the language, the prohibition against the sentence types that the child was using, would not last beyond a single generation. Thus any no-feedback situation presents a difficult challenge to the design of a learning system, and it is of considerable interest to mathematicians, psychologists, and engineers studying learning in general.

How is the child designed to cope with the problem? A good start would be to build in the basic organization of grammar, so the child would try out only the kinds of generalizations that are possible in the world's languages. Dead ends like *Who did Sheila see Mary and?*, not grammatical in any language, should not even occur to a child, and indeed, no child (or adult) we know of has ever tried it. But this is not enough, because the child also has to figure out how far to leap in the particular language being acquired, and languages vary: some allow many word orders, some only a few; some allow the causative rule to apply freely, others to only a few kinds of verb. Therefore a well-designed child, when faced with several choices in how far to generalize, should, in general, be consecutive: start with the smallest hypothesis about the language that is consistent with what parents say, then expand it outward as the evidence requires. Studies of children's language show that by and large that is how they work. For example, children learning English never leap to the conclusion that it is a free-word-order language and speak in all orders like *give doggie paper; give paper doggie, paper doggie give; doggie paper give,* and so on. Logically speaking, though, that would be consistent with what they hear if they were willing to entertain the possibility that their parents were just taciturn speakers of Korean, Russian, or Swedish, where several orders are possible. But children learning Korean, Russian, and Swedish *do* sometimes err on the side of caution and use only one of the orders allowed in the language, pending further evidence.

Furthermore, in cases where children do make errors and re-cover, their grammars must have some internal checks and balances, so that hearing one kind of sentence can catapult another out of the grammar. For example, if the word-building system is organized so that an irregular form listed in the mental dictionary blocks the appli-cation of the corresponding rule, hearing *held* enough times will even-tually drive out *holded.*

These general conclusions about language learning are interesting, but we would understand them better if we could trace out what actu-ally happens from moment to moment in children's minds as sen-tences come in and they try to distill rules from them. Viewed up close, the problem of learning rules is even harder than it appears from a distance. Imagine a hypothetical child trying to extract patterns from the following sentences, without any innate guidance as to how human grammar works:

Jane eats chicken.
Jane eats fish.
Jane likes fish.

At first glance, patterns jump out. Sentences, the child might conclude, consist of three words: the first must be *Jane,* the second either *eats* or *likes,* the third *chicken* or *fish.* With these micro-rules, the child can already generalize beyond the input, to the brand-new sentence *Jane likes chicken.* So far, so good. But let's say the next two sentences are

Jane eats slowly.
Jane might fish.

The word *might* gets added to the list of words that can appear in second position, and the word *slowly* is added to the list that can appear in third position. But look at the generalizations this would allow:

Jane might slowly.
Jane likes slowly.
Jane might chicken.

Bad start. The same ambiguity that bedevils language parsing in the adult bedevils language acquisition in the child. The moral is that the child must couch rules in grammatical categories like noun, verb, and auxiliary, not in actual words. That way, *fish* as a noun and *fish* as a verb would be kept separate, and the child would not adulterate the noun rule with instances of verbs and vice versa.

How might a child assign words into categories like noun and verb? Clearly, their meanings help. In all languages, words for objects and people are nouns or noun phrases, words for actions and change of state are verbs. (As we saw in Chapter 4, the converse is not true— many nouns, like *destruction,* do not refer to objects and people, and many verbs, like *interest,* do not refer to actions or changes of state.) Similarly, words for kinds of paths and places are prepositions, and words for qualities tend to be adjectives. Recall that children's first words refer to objects, actions, directions, and qualities. This is convenient. If children are willing to guess that words for objects are nouns, words for actions are verbs, and so on, they would have a leg up on the rule-learning problem.

But words are not enough; they must be ordered. Imagine the child trying to figure out what kind of word can occur before the verb *bother*. It can't be done:

> That dog bothers me. [*dog,* a noun]
> What she wears bothers me. [*wears,* a verb]
> Music that is too loud bothers me. [*loud,* an adjective]
> Cheering too loudly bothers me. [*loudly,* an adverb]
> The guy she hangs out with bothers me. [*with,* a preposition]

The problem is obvious. There *is* a certain something that must come before the verb *bother,* but that something is not a kind of word; it is a kind of *phrase,* a noun phrase. A noun phrase always contains a head noun, but that noun can be followed by all kinds of stuff. So it is hopeless to try to learn a language by analyzing sentences word by word. The child must look for phrases.

What does it mean to look for phrases? A phrase is a group of words. For a sentence of four words, there are eight possible ways to group the words into phrases: {That} {dog bothers me}; {That dog} {bothers me}; {That} {dog bothers} {me}, and so on. For a sentence of five words, there are sixteen possible ways; for a sentence of six words, thirty-two ways; for a sentence of $n$ words, $2^{n-1}$—a big number for long sentences. Most of these partitionings would give the child groups of words that would be useless in constructing new sentences, such as *wears bothers* and *cheering too,* but the child, unable to rely on parental feedback, has no way of knowing this. Once again, children cannot attack the language-learning task like a logician free of preconceptions; they need guidance.

This guidance could come from two sources. First, the child could assume that parents' speech respects the basic design of human phrase structure: phrases contain heads; role-players are grouped with heads in the mini-phrases called X-bars; X-bars are grouped with their modifiers inside X-phrases (noun phrase, verb phrase, and so on);

X-phrases can have subjects. To put it crudely, the X-bar theory of phrase structure could be innate. Second, since the meanings of parents' sentences are usually guessable in context, the child could use the meanings to help set up the right phrase structure. Imagine that a parent says *The big dog ate ice cream.* If the child has previously learned the individual words *big, dog, ate,* and *ice cream,* he or she can guess their categories and grow the first twigs of a tree:

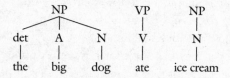

In turn, nouns and verbs must belong to noun phrases and verb phrases, so the child can posit one for each of these words. And if there is a big dog around, the child can guess that *the* and *big* modify *dog,* and connect them properly inside the noun phrase:

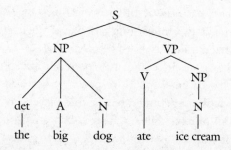

If the child knows that the dog just ate ice cream, he or she can also guess that *ice cream* and *dog* are role-players for the verb *eat. Dog* is a special kind of role-player, because it is the causal agent of the action and the topic of the sentence; hence it is likely to be the subject of the sentence and therefore attaches to the "S." A tree for the sentence has been completed:

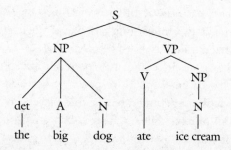

The rules and dictionary entries can be peeled off the tree:

S → NP VP
NP → (det) (A) N
VP → V (NP)
dog: N
ice cream: N
ate: V; eater = subject, thing eaten = object
the: det
big: A

This hypothetical time-lapse photography of the mind of a child at work shows how a child, if suitably equipped, could learn three rules and five words from a single sentence in context.

The use of part-of-speech categories, X-bar phrase structure, and meaning guessed from context is amazingly powerful, but amazing power is what a real-life child needs to learn grammar so quickly, especially without parental feedback. There are many benefits to using a small number of innate categories like N and V to organize incoming speech. By calling both the subject and object phrases "NP," rather than, say, Phrase #1 and Phrase #2, the child automatically can apply hard-won knowledge about nouns in subject position to nouns in object position, and vice versa. For example, our model child can already generalize and use *dog* as an object without having heard an adult do so, and the child tacitly knows that adjectives precede nouns not just in subjects but in objects, again without direct evidence. The child knows that if more than one *dog* is *dogs* in subject position, more than one *dog* is *dogs* in object position. I conservatively estimate that English allows about eight possible phrasemates of a head noun inside a noun phrase, such as *John's* dog; dogs *in the park; big* dogs; dogs *that I like,* and so on. In turn, there are about eight places in a sentence where the whole noun phrase can go, such as *Dog* bites man; Man bites *dog; A dog's* life; Give the boy *a dog;* Talk to *the dog;* and so on. There are three ways to inflect a noun: *dog, dogs, dog's.* And a typical child by the time he or she is in high school has learned something like twenty thousand nouns. If children had to learn all the combinations separately, they would need to listen to about 140 million different sentences. At a rate of a sentence every ten seconds, ten hours a day, it would take over a century. But by unconsciously labeling all nouns as "N" and all noun phrases as "NP," the child has only to hear about twenty-five different kinds of noun phrase and learn the nouns one by one, and the millions of possible combinations become available automatically.

Indeed, if children are blinkered to look for only a small number of phrase types, they automatically gain the ability to produce an infinite number of sentences, one of the quintessential properties of human grammar. Take the phrase *the tree in the park*. If the child mentally labels *the park* as an NP and also labels *the tree in the park* as an NP, the resulting rules generate an NP inside a PP inside an NP—a loop that can be iterated indefinitely, as in *the tree near the ledge by the lake in the park in the city in the east of the state* . . . In contrast, a child who was free to label *in the park* as one kind of phrase and *the tree in the park* as another kind would be deprived of the insight that the phrase contains an example of itself. The child would be limited to reproducing that phrase structure alone. Mental flexibility confines children; innate constraints set them free.

Once a rudimentary but roughly accurate analysis of sentence structure has been set up, the rest of the language can fall into place. Abstract words—nouns that do not refer to objects and people, for example—can be learned by paying attention to where they sit inside a sentence. Since *situation* in *The situation justifies drastic measures* occurs inside a phrase in NP position, it must be a noun. If the language allows phrases to be scrambled around the sentence, like Latin or Warlpiri, the child can discover this feature upon coming across a word that cannot be connected to a tree in the expected place without crossing branches. The child, constrained by Universal Grammar, knows what to focus on in decoding case and agreement inflections: a noun's inflection might depend on whether it is in subject or object position; a verb's might depend on tense, aspect, and the number, person, and gender of its subject and object. If the hypotheses were not confined to this small set, the task of learning inflections would be intractable—logically speaking, an inflection *could* depend on whether the third word in the sentence referred to a reddish or bluish object, whether the last word was long or short, whether the sentence was being uttered indoors or outdoors, and billions of other fruitless possibilities that a grammatically unfettered child would have to test for.

We can now return to the puzzle that opened the chapter: Why aren't babies born talking? We know that part of the answer is that babies have to listen to themselves to learn how to work their articulators, and have to listen to their elders to learn communal phonemes, words, and phrase orders. Some of these acquisitions depend on other ones, forcing development to proceed in a sequence: phonemes before words, words before sentences. But any mental mechanism powerful enough to learn these things could probably do so with a few weeks or months of input. Why does the sequence have to take three years? Could it be any faster?

Perhaps not. Complicated machines take time to assemble, and human infants may be expelled from the womb before their brains are complete. A human, after all, is an animal with a ludicrously large head, and a woman's pelvis, through which it must pass, can be only so big. If human beings stayed in the womb for the proportion of their life cycle that we would expect based on extrapolation from other primates, they would be born at the age of eighteen months. That is the age at which babies in fact begin to put words together. In one sense, then babies *are* born talking!

And we know that babies' brains do change considerably after birth. Before birth, virtually all the neurons (nerve cells) are formed, and they migrate into their proper locations in the brain. But head size, brain weight, and thickness of the cerebral cortex (gray matter), where the synapses (junctions) subserving mental computation are found, continue to increase rapidly in the year after birth. Long-distance connections (white matter) are not complete until nine months, and they continue to grow their speed-inducing myelin insulation throughout childhood. Synapses continue to develop, peaking in number between nine months and two years (depending on the brain region), at which point the child has fifty percent more synapses than the adult! Metabolic activity in the brain reaches adult levels by nine to ten months, and soon exceeds it, peaking around the age of four. The brain is sculpted not only by adding neural material but by chipping it away. Massive numbers of neurons die in utero, and the dying continues during the first two years before leveling off at age seven. Synapses wither from the age of two through the rest of childhood and into adolescence, when the brain's metabolic rate falls back to adult levels. Language development, then, could be on a maturational timetable, like teeth. Perhaps linguistic accomplishments like babbling, first words, and grammar require minimum levels of brain size, long-distance connections, and extra synapses, particularly in the language centers of the brain (which we will explore in the next chapter).

So language seems to develop about as quickly as the growing brain can handle it. What's the rush? Why is language installed so quickly, while the rest of the child's mental development seems to proceed at a more leisurely pace? In a book on evolutionary theory often considered to be one of the most important since Darwin's, the biologist George Williams speculates:

> We might imagine that Hans and Fritz Faustkeil are told on Monday, "Don't go near the water," and that both go wading and are spanked for it. On Tuesday they are told, "Don't play near the fire," and again they disobey and are spanked. On Wednesday they are told, "Don't tease the saber-tooth." This time Hans understands the message, and he bears firmly in mind the consequences of disobedience. He prudently avoids the saber-tooth and escapes the spanking. Poor Fritz escapes the spanking, too, but for a very different reason.
>
> Even today accidental death is an important cause of mortality in early life, and parents who consistently spare the rod in other matters may be moved to violence when a child plays with electric wires or chases a ball into the street. Many of the

accidental deaths of small children would probably have been avoided if the victims had understood and remembered verbal instructions and had been capable of effectively substituting verbal symbols for real experience. This might well have been true also under primitive conditions.

Perhaps it is no coincidence that the vocabulary spurt and beginnings of grammar follow closely on the heels of the baby, quite literally—the ability to walk unaccompanied appears around fifteen months.

Let's complete our exploration of the linguistic life cycle. Everyone knows that it is much more difficult to learn a second language in adulthood than a first language in childhood. Most adults never master a foreign language, especially the phonology—hence the ubiquitous foreign accent. Their development often "fossilizes" into permanent error patterns that no teaching or correction can undo. Of course, there are great individual differences, which depend on effort, attitudes, amount of exposure, quality of teaching, and plain talent, but there seems to be a cap even for the best adults in the best circumstances. The actress Meryl Streep is renowned in the United States for her seemingly convincing accents, but I am told that in England, her British accent in *Plenty* was considered rather awful, and that her Australian accent in the movie about the dingo that ate the baby didn't go over too well down there, either.

Many explanations have been advanced for children's superiority: they exploit Motherese, make errors unself-consciously, are more motivated to communicate, like to conform, are not xenophobic or set in their ways, and have no first language to interfere. But some of these accounts are unlikely, based on what we know about how language acquisition works. For example, children can learn a language without standard Motherese, they make few errors, and they get no feedback for the errors they do make. In any case, recent evidence is calling these social and motivational explanations into doubt. Holding every other factor constant, a key factor stands out: sheer age.

People who immigrate after puberty provide some of the most

compelling examples, even the apparent success stories. A few highly talented and motivated individuals master much of the grammar of a foreign language, but not its sound pattern. Henry Kissinger, who immigrated to the United States as a teenager, retains a frequently satirized German accent; his brother, a few years younger, has no accent. Ukrainian-born Joseph Conrad, whose first language was Polish, is considered one of the best writers in English in this century, but his accent was so thick his friends could barely understand him. Even the adults who succeed at grammar often depend on the conscious exercise of their considerable intellects, unlike children, to whom language acquisition just happens. Vladimir Nabokov, another brilliant writer in English, refused to lecture or be interviewed extemporaneously, insisting on writing out every word beforehand with the help of dictionaries and grammars. As he modestly explained, "I think like a genius, I write like a distinguished author, and I speak like a child." And he had the benefit of being raised in part by an English-speaking nanny.

More systematic evidence comes from the psychologist Elissa Newport and her colleagues. They tested Korean- and Chinese-born students and faculty at the University of Illinois who had spent at least ten years in the United States. The immigrants were given a list of 276 simple English sentences, half of them containing some grammatical error like *The farmer bought two pig* or *The little boy is speak to a policeman.* (The errors were errors with respect to the spoken vernacular, not "proper" written prose.) The immigrants who came to the United States between the ages of three and seven performed identically to American-born students. Those who arrived between the ages of eight and fifteen did increasingly worse the later they arrived, and those who arrived between seventeen and thirty-nine did the worst of all, and showed huge variability unrelated to their age of arrival.

What about acquisition of the mother tongue? Cases in which people make it to puberty without having learned a language are rare, but they all point to the same conclusion. We saw in Chapter 2 that deaf people who are not exposed to sign language until adulthood never do as well as those who learned it as children. Among the wolf-children who are found in the woods or in the homes of psychotic parents after puberty, some develop words, and some, like "Genie," discovered in 1970 at the age of thirteen and a half in a Los Angeles suburb, learn to produce immature, pidgin-like sentences:

Mike paint.
Applesauce buy store.
Neal come happy; Neal not come sad.
Genie have Momma have baby grow up.
I like elephant eat peanut.

But they are permanently incapable of mastering the full grammar of the language. In contrast, one child, Isabelle, was six and a half when she and her mute, brain-damaged mother escaped from the silent imprisonment of her grandfather's house. A year and a half later she had acquired fifteen hundred to two thousand words and produced complex grammatical sentences like

Why does the paste come out if one upsets the jar?
What did Miss Mason say when you told her I cleaned my classroom?
Do you go to Miss Mason's school at the university?

Obviously she was well on her way to learning English as successfully as anyone else; the tender age at which she began made all the difference.

With unsuccessful learners like Genie, there is always a suspicion that the sensory deprivation and emotional scars sustained during the horrific confinement somehow interfered with their ability to learn. But recently a striking case of first language acquisition in a normal adult has surfaced. "Chelsea" was born deaf in a remote town in northern California. A series of inept doctors and clinicians diagnosed her as retarded to emotionally disturbed without recognizing her deafness (a common fate for many deaf children in the past). She grew up shy, dependent, and languageless but otherwise emotionally and neurologically normal, sheltered by a loving family who never believed she was retarded. At the age of thirty-one she was referred to an astonished neurologist, who had her fitted with hearing aids that improved her hearing to near-normal levels. Intensive therapy by a rehabilitative team has brought her to a point where she scores at a ten-year-old level on intelligence tests, knows two thousand words, holds a job in a veterinarian's office, reads, writes, communicates, and has become social and independent. She has only one problem, which becomes apparent as soon as she opens her mouth:

> The small a the hat.
> Richard eat peppers hot.
> Orange Tim car in.
> Banana the eat.
> I Wanda be drive come.
> The boat sits water on.
> Breakfast eating girl.
> Combing hair the boy.
> The woman is bus the going.
> The girl is cone the ice cream shopping buying the man.

Despite intensive training and impressive gains in other spheres, Chelsea's syntax is bizarre.

In sum, acquisition of a normal language is guaranteed for children up to the age of six, is steadily compromised from then until shortly after puberty, and is rare thereafter. Maturational changes in the brain, such as the decline in metabolic rate and number of neurons during the early school-age years, and the bottoming out of the number of synapses and metabolic rate around puberty, are plausible causes. We do know that the language-learning circuitry of the brain is more plastic in childhood; children learn or recover language when the left hemisphere of the brain is damaged or even surgically removed (though not quite at normal levels), but comparable damage in an adult usually leads to permanent aphasia.

"Critical periods" for specific kinds of learning are common in the animal kingdom. There are windows in development in which ducklings learn to follow large moving objects, kittens' visual neurons become tuned to vertical, horizontal, and oblique lines, and white-crowned sparrows duplicate their fathers' songs. But why should learning ever decline and fall? Why throw away such a useful skill?

Critical periods seem paradoxical, but only because most of us have an incorrect understanding of the biology of organisms' life histories. We tend to think that genes are like the blueprints in a factory and organisms are like the appliances that the factory turns out. Our picture is that during gestation, when the organism is built, it is permanently fitted with the parts it will carry throughout its lifetime. Children and teenagers and adults and old people have arms and legs and a heart because arms and legs and a heart were part of the infant's factory-installed equipment. When a part vanishes for no reason, we are puzzled.

But now try to think of the life cycle in a different way. Imagine that what the genes control is not a factory sending appliances into the world, but a machine shop in a thrifty theater company to which props and sets and materials periodically return to be dismantled and reassembled for the next production. At any point, different contraptions can come out of the shop, depending on current need. The most obvious biological illustration is metamorphosis. In insects, the genes build an eating machine, let it grow, build a container around it, dissolve it into a puddle of nutrients, and recycle them into a breeding machine. Even in humans, the sucking reflex disappears, teeth erupt twice, and a suite of secondary sexual characteristics emerge in a maturational schedule. Now complete the mental backflip. Think of metamorphosis and maturational emergence not as the exception but as the rule. The genes, shaped by natural selection, control bodies throughout the life span; designs hang around during the times of life that they are useful, not before or after. The reason that we have arms at age sixty is not because they have stuck around since birth, but because arms are as useful to a sixty-year-old as they were to a baby.

This inversion (an exaggeration, but a useful one) flips the critical-period question with it. The question is no longer "Why does a learning ability disappear?" but "When is the learning ability needed?" We have already noted that the answer might be "As early as possible," to allow the benefits of language to be enjoyed for as much of life as possible. Now note that learning a language—as opposed to *using* a language—is perfectly useful as a one-shot skill. Once the details of the local language have been acquired from the surrounding adults, any further ability to learn (aside from vocabulary) is superfluous. It is like borrowing a floppy disk drive to load a new computer with the software you will need, or borrowing a turntable to copy your old collection of LP's onto tape; once you are done, the machines can be returned. So language-acquisition circuitry is not needed once it has been used; it should be dismantled if keeping it around incurs any costs. And it probably does incur costs. Metabolically, the brain is a pig. It consumes a fifth of the body's oxygen and similarly large portions of its calories and phospholipids. Greedy neural tissue lying around beyond its point of usefulness is a good candidate for the recycling bin. James Hurford, the world's only computational evolutionary linguist, has put these kinds of assumptions into a computer simulation of evolving humans, and finds that a critical period for language acquisition centered in early childhood is the inevitable outcome.

Even if there is some utility to our learning a second language as adults, the critical period for language acquisition may have evolved as part of a larger fact of life: the increasing feebleness and vulnerability with advancing age that biologists call "senescence." Common sense says that the body, like all machines, must wear out with use, but this is another misleading implication of the appliance metaphor. Organisms are self-replenishing, self-repairing systems, and there is no physical reason why we should not be biologically immortal, as in fact lineages of cancer cells used in laboratory research are. That would not mean that we would *actually* be immortal. Every day there is a certain probability that we will fall off a cliff, catch a virulent disease, be struck by lightning, or be murdered by a rival, and sooner or later one of those lightning bolts or bullets will have our name on it. The question is, is every day a lottery in which the odds of drawing a fatal ticket are the same, or do the odds get worse and worse the longer we play? Senescence is the bad news that the odds do change; elderly people are killed by falls and flus that their grandchildren easily survive. A major question in modern evolutionary biology is why this should be true, given that selection operates at every point of an organism's life history. Why aren't we built to be equally hale and hearty every day of our lives, so that we can pump out copies of ourselves indefinitely?

The solution, from George Williams and P. B. Medawar, is ingenious. As natural selection designed organisms, it must have been faced with countless choices among features that involved different tradeoffs of costs and benefits at different ages. Some materials might be strong and light but wear out quickly, whereas others might be heavier but more durable. Some biochemical processes might deliver excellent products but leave a legacy of accumulating pollution within the body. There might be a metabolically expensive cellular repair mechanism that comes in most useful late in life when wear and tear have accumulated. What does natural selection do when faced with these tradeoffs? In general, it will favor an option with benefits to the young organism and costs to the old one over an option with the same average benefit spread out evenly over the life span. This asymmetry is rooted in the inherent asymmetry of death. If a lightning bolt kills a forty-year-old, there will be no fifty-year-old or sixty-year-old to worry about, but there will have been a twenty-year-old and a thirty-year-old. Any bodily feature designed for the benefit of the potential over-forty incarnations, at the expense of the under-forty incarnations, will have gone to waste. And the logic is the same for unforeseeable death at any age: the brute mathematical fact is that all things being equal, there is a better chance of being a young person than being an old person. So genes that strengthen young organisms at the expense of old organisms have the odds in their favor and will tend to accumulate over evolutionary timespans, whatever the bodily system, and the result is overall senescence.

Thus language acquisition might be like other biological functions. The linguistic clumsiness of tourists and students might be the price we pay for the linguistic genius we displayed as babies, just as the decrepitude of age is the price we pay for the vigor of youth.

# 10

# Language Organs and Grammar Genes

*"Ability to Learn Grammar Laid to Gene by Researcher."* This 1992 headline appeared not in a supermarket tabloid but in an Associated Press news story, based on a report at the annual meeting of the principal scientific association in the United States. The report had summarized evidence that Specific Language Impairment runs in families, focusing on the British family we met in Chapter 2 in which the inheritance pattern is particularly clear. The syndicated columnists James J. Kilpatrick and Erma Bombeck were incredulous. Kilpatrick's column began:

### BETTER GRAMMAR THROUGH GENETICS

Researchers made a stunning announcement the other day at a meeting of the American Association for the Advancement of Science. Are you ready? Genetic biologists have identified the grammar gene.

Yes! It appears from a news account that Steven Pinker of MIT and Myrna Gopnik of McGill University have solved a puzzle that has baffled teachers of English for years. Some pupils master grammar with no more than a few moans of protest. Others, given the same instruction, persist in saying that

Susie invited her and I to the party. It is all a matter of heredity. This we can handle.

A single dominant gene, the biologists believe, controls the ability to learn grammar. A child who says "them marbles is mine" is not necessarily stupid. He has all his marbles. The child is simply a little short on chromosomes.

It boggles the mind. Before long the researchers will isolate the gene that controls spelling . . . [the column continues] . . . neatness. . . . The read-a-book gene . . . a gene to turn down the boom box . . . another to turn off the TV . . . politeness . . . chores . . . homework . . .

Bombeck wrote:

#### POOR GRAMMAR? IT ARE IN THE GENES

It was not much of a surprise to read that kids who are unable to learn grammar are missing a dominant gene. . . . At one time in his career, my husband taught high school English. He had 37 grammar-gene deficients in his class at one time. What do you think the odds of that happening are? They didn't have a clue where they were. A comma could have been a petroglyph. A subjective complement was something you said to a friend when her hair came out right. A dangling participle was not their problem. . . .

Where is that class of young people today, you ask? They are all major sports figures, rock stars and television personalities who make millions spewing out words such as "bummer," "radical" and "awesome" and thinking they are complete sentences.

The syndicated columns, third-hand newspaper stories, editorial cartoons, and radio shows following the symposium gave me a quick education about how scientific discoveries get addled by journalists working under deadline pressure. To set the record straight: the discovery of the family with the inherited language disorder belongs to Gopnik; the reporter who generously shared the credit with me was confused by the fact that I chaired the session and thus introduced Gopnik to the audience. No grammar gene was identified; a defective gene was inferred, from the way the syndrome runs in the family. A single gene is thought to *disrupt* grammar, but that does not mean a single gene *controls* grammar. (Removing the distributor wire prevents a car from moving, but that does not mean a car is controlled by its distributor wire.) And of course, what is disrupted is the ability to converse normally in everyday English, not the ability to learn the standard written dialect in school.

But even when they know the facts, many people share the columnists' incredulity. Could there really be a gene tied to something as specific as grammar? The very idea is an assault on the deeply rooted belief that the brain is a general-purpose learning device, void and without form prior to experience of the surrounding culture. And if there are grammar genes, what do they do? Build the grammar organ, presumably—a metaphor, from Chomsky, that many find just as preposterous.

But if there is a language instinct, it has to be embodied somewhere in the brain, and those brain circuits must have been prepared for their role by the genes that built them. What kind of evidence could show that there are genes that build parts of brains that control grammar? The ever-expanding toolkit of the geneticist and neurobiologist is mostly useless. Most people do not want their brains impaled by electrodes, injected with chemicals, rearranged by surgery, or removed for slicing and staining. (As Woody Allen said, "The brain is my second-favorite organ.") So the biology of language remains poorly understood. But accidents of nature and ingenious indirect techniques have allowed neurolinguists to learn a surprising amount. Let's try to home in on the putative grammar gene, beginning with a bird's-eye view of the brain and zooming in on smaller and smaller components.

We can narrow down our search at the outset by throwing away half the brain. In 1861 the French physician Paul Broca dissected the brain of an aphasic patient who had been nicknamed "Tan" by hospital workers because that was the only syllable he uttered. Broca discovered a large cyst producing a lesion in Tan's left hemisphere. The next eight cases of aphasia he observed also had left-hemisphere lesions, too many to be attributed to chance. Broca concluded that "the faculty for articulate language" resides in the left hemisphere.

In the 130 years since, Broca's conclusion has been confirmed by many kinds of evidence. Some of it comes from the convenient fact that the right half of the body and of perceptual space is controlled by the left hemisphere of the brain and vice versa. Many people with aphasia suffer weakness or paralysis on the right side, including Tan and the recovered aphasic of Chapter 2, who awoke thinking that he had slept on his right arm. The link is summed up in Psalms 137:5–6:

> If I forget thee, O Jerusalem, let my right hand forget her cunning.
> If I do not remember thee, let my tongue cleave to the roof of my mouth.

Normal people recognize words more accurately when the words are flashed to the right side of their visual field than when they are flashed to the left, even when the language is Hebrew, which is written from right to left. When different words are presented simultaneously to the two ears, the person can make out the word coming into the right ear better. In some cases of otherwise incurable epilepsy, surgeons disconnect the two cerebral hemispheres by cutting the bundle of fibers running between them. After surgery the patients live completely normal lives, except for a subtlety discovered by the neuroscientist Michael Gazzaniga: when the patients are kept still, they can describe events taking place in their right visual field and can name objects in their right hand, but cannot describe events taking place in their left visual field or name objects placed in their left hand (though the right hemisphere can display its awareness of those events by nonverbal means like gesturing and pointing). The left half of their world has been disconnected from their language center.

When neuroscientists look directly at the brain, using a variety of techniques, they can actually see language in action in the left hemisphere. The anatomy of the normal brain—its bulges and creases—is slightly asymmetrical. In some of the regions associated with language, the differences are large enough to be seen with the naked eye. Computerized Axial Tomography (CT or CAT) and Magnetic Resonance Imaging (MRI) use a computer algorithm to reconstruct a picture of the living brain in cross-section. Aphasics' brains almost always show lesions in the left hemisphere. Neurologists can temporarily paralyze one hemisphere by injecting sodium amytal into the carotid artery. A patient with a sleeping right hemisphere can talk; a patient with a sleeping left hemisphere cannot. During brain surgery, patients can remain conscious under local anesthetic because the brain has no pain receptors. The neurosurgeon Wilder Penfield found that small electric shocks to certain parts of the left hemisphere could silence the patient in mid-sentence. (Neurosurgeons do these manipulations not out of curiosity but to be sure that they are not cutting out vital parts of the brain along with the diseased ones.) In a technique used on normal research subjects, electrodes are pasted all over the scalp, and the subjects' electroencephalograms (EEG's) are recorded as they read or hear words. There are recognizable jumps in the electrical signal that are synchronized with each word, and they are more prominent in the electrodes pasted on the left side of the skull than in those on the right (though this finding is tricky to interpret, because an electrical signal generated deep in one part of the brain can radiate out of another part).

In a new technique called Positron Emission Tomography (PET), a volunteer is injected with mildly radioactive glucose or water, or inhales a radioactive gas, comparable in dosage to a chest X-ray, and puts his head inside a ring of gamma-ray detectors. The parts of the brain that are more active burn more glucose and have more oxygenated blood sent their way. Computer algorithms can reconstruct which parts of the brain are working harder from the pattern of radiation that emanates from the head. An actual picture of metabolic activity within a slice of the brain can be displayed in a computer-generated photograph, with the more active areas showing up in bright reds and yellows, the quiet areas in dark indigos. By subtracting an image of the brain when its owner is watching meaningless patterns or listening to meaningless sounds from an image when the owner is understanding words or speech, one can see which areas of the brain "light up" during language processing. The hot spots, as expected, are on the left side.

What exactly is engaging the left hemisphere? It is not merely speechlike sounds, or wordlike shapes, or movements of the mouth, but abstract *language*. Most aphasic people—Mr. Ford from Chapter 2, for example—can blow out candles and suck on straws, but their writing suffers as much as their speech; this shows that it is not mouth control but language control that is damaged. Some aphasics remain fine singers, and many are superb at swearing. In perception, it has long been known that tones are discriminated better when they are played to the left ear, which is connected most strongly to the right hemisphere. But this is only true if the tones are perceived as musical sounds like hums; when the ears are Chinese or Thai and the same tones are features of phonemes, the advantage is to the right ear and the left hemisphere it feeds.

If a person is asked to shadow someone else's speech (repeat it as the talker is talking) and, simultaneously, to tap a finger to the right or the left hand, the person has a harder time tapping with the right finger than with the left, because the right finger competes with language for the resources of the left hemisphere. Remarkably, the psychologist Ursula Bellugi and her colleagues have shown that the same thing happens when deaf people shadow one-handed signs in American Sign Language: they find it harder to tap with their right finger than with their left finger. The gestures must be tying up the left hemispheres, but it is not because they are gestures; it is because they are *linguistic* gestures. When a person (either a signer or a speaker) has to shadow a goodbye wave, a thumbs-up sign, or a meaningless gesticulation, the fingers of the right hand and the left hand are slowed down equally.

The study of aphasia in the deaf leads to a similar conclusion. Deaf signers with damage to their left hemispheres suffer from forms of sign aphasia that are virtually identical to the aphasia of hearing victims with similar lesions. For example, Mr. Ford's sign-language counterparts are unimpaired at nonlinguistic tasks that place similar demands on the eyes and hands, such as gesturing, pantomiming, recognizing faces, and copying designs. Injuries to the right hemisphere of deaf signers produce the opposite pattern: they remain flawless at signing but have difficulty performing visuospatial tasks, just like hearing patients with injured right hemispheres. It is a fascinating discovery. The right hemisphere is known to specialize in visuospatial abilities, so one might have expected that sign language, which depends on visuospatial abilities, would be computed in the right hemisphere. Bellugi's findings show that language, whether by ear and mouth or by eye and hand, is controlled by the left hemisphere. The left hemisphere must be handling the abstract rules and trees underlying language, the grammar and the dictionary and the anatomy of words, and not merely the sounds and the mouthings at the surface.

Why is language so lopsided? A better question is, why is the rest of a person so symmetrical? Symmetry is an inherently improbable arrangement of matter. If you were to fill in the squares of an 8 × 8 checkerboard at random, the odds are less than one in a billion that the pattern would be bilaterally symmetrical. The molecules of life are asymmetrical, as are most plants and many animals. Making a body bilaterally symmetrical is difficult and expensive. Symmetry is so demanding that among animals with a symmetrical design, any disease or weakness can disrupt it. As a result, organisms from scorpion flies to barn swallows to human beings find symmetry sexy (a sign of a fit potential mate) and gross asymmetry a sign of deformity. There must be something in an animal's lifestyle that makes a symmetrical design worth its price. The crucial lifestyle feature is mobility: the species with bilaterally symmetrical body plans are the ones that are designed to move in straight lines. The reasons are obvious. A creature with an asymmetrical body would veer off in circles, and a creature with asymmetrical sense organs would eccentrically monitor one side of its body even though equally interesting things can happen on either side.

Though locomoting organisms are symmetrical side-to-side, they are not (apart from Dr. Dolittle's Push-mi-pull-you) symmetrical front-and-back. Thrusters apply force best in one direction, so it is easier to build a vehicle that can move in one direction and turn than a vehicle that can move equally well in forward and reverse (or that can scoot off in any direction at all, like a flying saucer). Organisms are not symmetrical up-and-down because gravity makes up different from down.

The symmetry in sensory and motor organs is reflected in the brain, most of which, at least in nonhumans, is dedicated to processing sensation and programming action. The brain is divided into maps of visual, auditory, and motor space that literally reproduce the structure of real space: if you move over a small amount in the brain, you find neurons that correspond to a neighboring region of the world as the animal senses it. So a symmetrical body and a symmetrical perceptual world is controlled by a brain that is itself almost perfectly symmetrical.

No biologist has explained why the left brain controls right space and vice versa. It took a psycholinguist, Marcel Kinsbourne, to come up with the only speculation that is even remotely plausible. All bilaterally symmetrical invertebrates (worms, insects, and so on) have the more straightforward arrangement in which the left side of the central nervous system controls the left side of the body and the right side controls the right side. Most likely, the invertebrate that was the ancestor of the chordates (animals with a stiffening rod around their spinal cords, including fish, amphibians, birds, reptiles, and mammals) had this arrangement as well. But all the chordates have "contralateral" control: right brain controls left body and left brain controls right body. What could have led to the rewiring? Here is Kinsbourne's idea. Imagine that you are a creature with the left-brain-left-body arrangement. Now turn your head around to look behind you, a full 180 degrees back, like an owl. (Stop at 180 degrees; don't go around and around like the girl in *The Exorcist.*) Now imagine that your head is stuck in that position. Your nerve cables have been given a half-twist, so the left brain would control your right body and vice versa.

Now, Kinsbourne is not suggesting that some primordial rubbernecker literally got its head stuck, but that changes in the genetic instructions for building the creature resulted in the half-twist during embryonic development—a torsion that one can actually see happening during the development of snails and some flies. This may sound like a perverse way to build an organism, but evolution does it all the time, because it never works from a fresh drawing board but has to tinker with what is already around. For example, our sadistically designed S-shaped spines are the product of bending and straightening the arched backbones of our quadrupedal forebears. The Picassoesque face of the flounder was the product of warping the head of a kind of fish that had opted to cling sideways to the ocean floor, bringing around the eye that had been staring uselessly into the sand. Since Kinsbourne's hypothetical creature left no fossils and has been extinct for over half a billion years, no one knows why it would have undergone the rotation. (Perhaps one of *its* ancestors had changed its posture, like the flounder, and subsequently righted itself. Evolution, which has no foresight, may have put its head back into alignment with its body by giving the head another quarter-twist in the same direction, rather than by the more sensible route of undoing the original quarter-twist.) But it does not really matter; Kinsbourne is only proposing that such a rotation must have taken place; he is not claiming he can reconstruct why it happened. (In the case of the snail, where the rotation is accompanied by a bending, like one of the arms of a pretzel, scientists are more knowledgeable. As my old biology textbook explains, "While the head and foot remain stationary, the visceral mass is rotated through an angle of 180°, so that the anus . . . is carried upward and finally comes to lie [above] the head. . . . The advantages of this arrangement are clear enough in an animal that lives in a shell with only one opening.")

In support of the theory, Kinsbourne notes that invertebrates have their main neural cables laid along their bellies and their hearts in their backs, whereas chordates have their neural cables laid along their backs and their hearts in their chests. This is exactly what one would expect from a 180-degree head-to-body turn in the transition from one group to the other, and Kinsbourne could not find any reports of an animal that has only one or two out of the three reversals that his theory says must have happened together. Major changes in body architecture affect the entire design of the animal and can be very difficult to undo. We are the descendants of that twisted creature, and half a billion years later, a stroke in the left hemisphere leaves the right arm tingly.

The benefits of a symmetrical body plan all have to do with sensing and moving in the bilaterally indifferent environment. For body systems that do not interact directly with the environment, the symmetrical blueprint can be overridden. Internal organs such as the heart, liver, and stomach are good examples; they are not in contact with the layout of the external world, and they are grossly asymmetrical. The same thing happens on a much smaller scale in the microscopic circuitry of the brain.

Think about the act of deliberately manipulating some captive object. The actions are not being keyed to the environment; the manipulator is putting the object anywhere it wants. So the organism's forelimbs, and the brain centers controlling them, do not have to be symmetrical in order to react to events appearing unpredictably on one side or the other; they can be tailored to whatever configuration is most efficient to carry out the action. Manipulating an object often benefits from a division of labor between the limbs, one holding the object, the other acting on it. The result is the asymmetrical claws of lobsters, and the asymmetrical brains that control paws and hands in a variety of species. Humans are by far the most adept manipulators in the animal kingdom, and we are the species that displays the strongest and most consistent limb preference. Ninety percent of people in all societies and periods in history are right-handed, and most are thought to possess one or two copies of a dominant gene that imposes the right-hand (left-brain) bias. Possessors of two copies of the recessive version of the gene develop without this strong right-hand bias; they turn into the rest of the right-handers and into the left-handers and ambidextrics.

Processing information that is spread out over time but not space is another function where symmetry serves no purpose. Given a certain amount of neural tissue necessary to perform such a function, it makes more sense to put it all in one place with short interconnections, rather than have half of it communicate with the other half over a slow, noisy, long-distance connection between the hemispheres. Thus the control of song is strongly lateralized in the left hemispheres of many birds, and the production and recognition of calls and squeaks is somewhat lateralized in monkeys, dolphins, and mice.

Human language may have been concentrated in one hemisphere because it, too, is coordinated in time but not environmental space: words are strung together in order but do not have to be aimed in various directions. Possibly, the hemisphere that already contained computational microcircuitry necessary for control of the fine, deliberate, sequential manipulation of captive objects was the most natural place in which to put language, which also requires sequential control. In the lineage leading to humans, that happened to be the left hemisphere. Many cognitive psychologists believe that a variety of mental processes requiring sequential coordination and arrangement of parts co-reside in the left hemisphere, such as recognizing and imagining multipart objects and engaging in step-by-step logical reasoning. Gazzaniga, testing the two hemispheres of a split-brain patient separately, found that the newly isolated left hemisphere had the same IQ as the entire connected brain before surgery!

Linguistically, most left-handers are not mirror images of the righty majority. The left hemisphere controls language in virtually all right-handers (97%), but the right hemisphere controls language in a minority of left-handers, only about 19%. The rest have language in the left hemisphere (68%) or redundantly in both. In all of these lefties, language is more evenly distributed between the hemispheres than it is in righties, and thus the lefties are more likely to withstand a stroke on one side of the brain without suffering from aphasia. There is some evidence that left-handers, though better at mathematical, spatial, and artistic activities, are more susceptible to language impairment, dyslexia, and stuttering. Even righties with left-handed relatives (presumably, those righties possessing only one copy of the dominant right-bias gene) appear to parse sentences in subtly different ways than pure righties.

Language, of course, does not use up the entire left half of the brain. Broca observed that Tan's brain was mushy and deformed in the regions immediately above the Sylvian fissure—the huge cleavage that separates the distinctively human temporal lobe from the rest of the brain. The area in which Tan's damage began is now called Broca's area, and several other anatomical regions hugging both sides of the Sylvian fissure affect language when they are damaged. The most prominent are shown as the large gray blobs in the diagram (see page 314). In about 98% of the cases where brain damage leads to language problems, the damage is somewhere on the banks of the Sylvian fissure of the left hemisphere. Penfield found that most of the spots that disrupted language when he stimulated them were there, too. Though the language areas appear to be separated by large gulfs, this may be an illusion. The cerebral cortex (gray matter) is a large sheet of two-dimensional tissue that has been wadded up to fit inside the spherical skull. Just as crumpling a newspaper can appear to scramble the pictures and text, a side view of a brain is a misleading picture of which regions are adjacent. Gazzaniga's coworkers have developed a technique that uses MRI pictures of brain slices to reconstruct what the person's cortex would look like if somehow it could be unwrinkled into a flat sheet. They found that all the areas that have been implicated in language are adjacent in one continuous territory. This region of the cortex, the left perisylvian region, can be considered to be the language organ.

Let us zoom in closer. Tan and Mr. Ford, in whom Broca's area was damaged, suffered from a syndrome of slow, labored, ungrammatical speech called Broca's aphasia. Here is another example, from a man called Peter Hogan. In the first passage he describes what brought him into the hospital; in the second, his former job in a paper mill:

> Yes . . . ah . . . Monday . . . ah . . . Dad and Peter Hogan, and
> Dad . . . ah . . . hospital . . . and ah . . . Wednesday . . . Wednes-

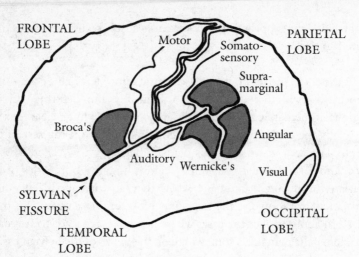

day nine o'clock and ah Thursday . . . ten o'clock ah doctors
. . . two . . . two . . . an doctors and . . . ah . . . teeth . . . yah
. . . And a doctor an girl . . . and gums, an I.

Lower Falls . . . Maine . . . Paper. Four hundred tons a day!
And ah . . . sulphur machines, and ah . . . wood . . . Two weeks
and eight hours. Eight hours . . . no! Twelve hours, fifteen
hours . . . workin . . . workin . . . workin! Yes, and ah . . .
sulphur. Sulphur and . . . Ah wood. Ah . . . handlin! And ah
sick, four years ago.

Broca's area is adjacent to the part of the motor-control strip
dedicated to the jaws, lip, and tongue, and it was once thought that
Broca's area is involved in the production of language (though obviously not speech per se, because writing and signing are just as
affected). But the area seems to be implicated in grammatical processing in general. A defect in grammar will be most obvious in the output, because any slip will lead to a sentence that is conspicuously
defective. Comprehension, on the other hand, can often exploit the
redundancy in speech to come up with sensible interpretations with
little in the way of actual parsing. For example, one can understand
*The dog bit the man* or *The apple that the boy is eating is red* just by

knowing that dogs bite men, boys eat apples, and apples are red. Even *The car pushes the truck* can be guessed at because the cause is mentioned before the effect. For a century, Broca's aphasics fooled neurologists by using shortcuts. Their trickery was finally unmasked when psycholinguists asked them to act out sentences that could be understood only by their syntax, like *The car is pushed by the truck* or *The girl whom the boy is pushing is tall.* The patients gave the correct interpretation half the time and its opposite half the time—a mental coin flip.

There are other reasons to believe that the front portion of the perisylvian cortex, where Broca's area is found, is involved in grammatical processing. When people read a sentence, electrodes pasted over the front of their left hemispheres pick up distinctive patterns of electrical activity at the point in the sentence at which it becomes ungrammatical. Those electrodes also pick up changes during the portions of a sentence in which a moved phrase must be held in memory while the reader awaits its trace, like *What did you say* (trace) *to John?* Several studies using PET and other techniques to measure blood flow have shown that this region lights up when people listen to speech in a language they know, tell stories, or understand complex sentences. Various control tasks and subtractions confirm that it is processing the structure of sentences, not just thinking about their content, that engages this general area. A recent and very carefully designed experiment by Karin Stromswold and the neurologists David Caplan and Nat Alpert obtained an even more precise picture; it showed one circumscribed *part* of Broca's area lighting up.

So is Broca's area the grammar organ? Not really. Damage to Broca's area alone usually does not produce long-lasting severe aphasia; the surrounding areas and underlying white matter (which connects Broca's area to other brain regions) must be damaged as well. Sometimes symptoms of Broca's aphasia can be produced by a stroke or Parkinson's disease that damages the basal ganglia, complex neural centers buried inside the frontal lobes that are otherwise needed for skilled movement. The labored speech output of Broca's aphasics may be distinct from the lack of grammar in their speech, and may impli-

cate not Broca's area but hidden parts of the cortex nearby that tend
to be damaged by the same lesions. And, most surprisingly of all, some
kinds of grammatical abilities seem to survive damage to Broca's area.
When asked to distinguish grammatical from ungrammatical sentences, some Broca's aphasics can detect even subtle violations of the
rules of syntax, as in pairs like these:

> John was finally kissed Louise.
> John was finally kissed by Louise.
>
> I want you will go to the store now.
> I want you to go to the store now.
>
> Did the old man enjoying the view?
> Did the old man enjoy the view?

Still, aphasics do not detect all ungrammaticalities, nor do all aphasics
detect them, so the role of Broca's area in language is maddeningly
unclear. Perhaps the area underlies grammatical processing by converting messages in mentalese into grammatical structures and vice
versa, in part by communicating via the basal ganglia with the prefron-
tal lobes, which subserve abstract reasoning and knowledge.

Broca's area is also connected by a band of fibers to a second
language organ, Wernicke's area. Damage to Wernicke's area produces
a very different syndrome of aphasia. Howard Gardner describes his
encounter with a Mr. Gorgan:

> "What brings you to the hospital?" I asked the 72-year-
> old retired butcher four weeks after his admission to the hos-
> pital.
>
> "Boy, I'm sweating, I'm awful nervous, you know, once
> in a while I get caught up, I can't mention the tarripoi, a month
> ago, quite a little, I've done a lot well, I impose a lot, while, on
> the other hand, you know what I mean, I have to run around,
> look it over, trebbin and all that sort of stuff."
>
> I attempted several times to break in, but was unable to
> do so against this relentlessly steady and rapid outflow. Finally,
> I put up my hand, rested it on Gorgan's shoulder, and was able
> to gain a moment's reprieve.

"Thank you, Mr. Gorgan. I want to ask you a few—"

"Oh sure, go ahead, any old think you want. If I could I would. Oh, I'm taking the word the wrong way to say, all of the barbers here whenever they stop you it's going around and around, if you know what I mean, that is tying and tying for repucer, repuceration, well, we were trying the best that we could while another time it was with the beds over there the same thing . . ."

Wernicke's aphasia is in some ways the complement of Broca's. Patients utter fluent streams of more-or-less grammatical phrases, but their speech makes no sense and is filled with neologisms and word substitutions. Unlike many Broca's patients, Wernicke's patients have consistent difficulty naming objects; they come up with related words or distortions of the sound of the correct one:

table: "chair"
elbow: "knee"
clip: "plick"
butter: "tubber"
ceiling: "leasing"
ankle: "ankley, no mankle, no kankle"
comb: "close, saw it, cit it, cut, the comb, the came"
paper: "piece of handkerchief, pauper, hand pepper, piece of
   hand paper"
fork: "tonsil, teller, tongue, fung"

A striking symptom of Wernicke's aphasia is that the patients show few signs of comprehending the speech around them. In a third kind of aphasia, the connection between Wernicke's area and Broca's is damaged, and these patients are unable to repeat sentences. In a fourth kind, Broca's and Wernicke's and the link between them are intact but they are an island cut off from the rest of the cortex, and these patients eerily repeat what they hear without understanding it or ever speaking spontaneously. For these reasons, and because Wernicke's area is adjacent to the part of the cortex that processes sound, the area was once thought to underlie language comprehension. But that would not explain why the speech of these patients sounds so psychotic. Wernicke's area seems to have a role in looking up words and funneling them to other areas, notably Broca's, that assemble or parse them syntactically. Wernicke's aphasia, perhaps, is the product of an intact Broca's area madly churning out phrases without the intended message and intended words that Wernicke's area ordinarily supplies. But to be honest, no one really knows what either Broca's area or Wernicke's area is for.

Wernicke's area, together with the two shaded areas adjacent to it in the diagram (the angular and supramarginal gyri), sit at the cross-roads of three lobes of the brain, and hence are ideally suited to integrating streams of information about visual shapes, sounds, bodily sensations (from the "somatosensory" strip), and spatial relations (from the parietal lobe). It would be a logical place to store links between the sounds of words and the appearance and geometry of what they refer to. Indeed, damage to this general vicinity often causes a syndrome that is called *anomia*, though a more mnemonic label might be "no-name-ia," which is literally what it means. The neuropsychologist Kathleen Baynes describes "HW," a business executive who suffered a stroke in this general area. He is highly intelligent, articulate, and conversationally adept but finds it virtually impossible to retrieve nouns from his mental dictionary, though he can understand them. Here is how he responded when Baynes asked him to describe a picture of a boy falling from a stool as he reaches into a jar on a shelf and hands a cookie to his sister:

> First of all this is falling down, just about, and is gonna fall down and they're both getting something to eat . . . but the trouble is this is gonna let go and they're both gonna fall down . . . I can't see well enough but I believe that either she or will have some food that's not good for you and she's to get some for her, too . . . and that you get it there because they shouldn't go up there and get it unless you tell them that they could have it. And so this is falling down and for sure there's one they're

going to have for food and, and this didn't come out right, the, uh, the stuff that's uh, good for, it's not good for you but it, but you love, um mum mum [smacks lips] . . . and that so they've . . . see that, I can't see whether it's in there or not . . . I think she's saying, I want two or three, I want one, I think, I think so, and so, so she's gonna get this one for sure it's gonna fall down there or whatever, she's gonna get that one and, and there, he's gonna get one himself or more, it all depends with this when they fall down . . . and when it falls down there's no problem, all they got to do is fix it and go right back up and get some more.

HW uses noun phrases perfectly but cannot retrive the nouns to put inside them: he uses pronouns, gerunds like *falling down,* and a few generic nouns like *food* and *stuff,* referring to particular objects with convoluted circumlocutions. Verbs tend to pose less of a problem for anomics; they are much harder for Broca's aphasics, presumably because verbs are intimately linked to syntax.

There are other indications that these regions in the rear of the perisylvian are implicated in storing and retrieving words. When people read perfectly grammatical sentences and come across a word that makes no sense, like *The boys heard Joe's orange about Africa,* electrodes pasted near the back of the skull pick up a change in their EEG's (although, as I have mentioned, it is only a guess that the blips are coming from below the electrodes). When people put their heads in the PET scanner, this general part of the brain lights up when they hear words (and pseudo-words, like *tweal*) and even when they read words on a screen and have to decide whether the words rhyme—a task requiring them to imagine the word's sounds.

A very gross anatomy of the language sub-organs within the perisylvian might be: front of the perisylvian (including Broca's area), grammatical processing; rear of the perisylvian (including Wernicke's and the three-lobe junction), the sounds of words, especially nouns, and some aspects of their meaning. Can we zoom in still closer, and locate smaller areas of brain that carry out more circumscribed language tasks? The answer is no and yes. No, there are no smaller patches of brain that one can draw a line around and label as some linguistic module—at least, not today. But yes, there must be portions of cortex that carry out circumscribed tasks, because brain damage can lead to language deficits that are startlingly specific. It is an intriguing paradox.

Here are some examples. Although impairments of what I have been calling the sixth sense, speech perception, can arise from damage to most areas of the left perisylvian (and speech perception causes several parts of the perisylvian to light up in PET studies), there is a specific syndrome called Pure Word Deafness that is exactly what it sounds like: the patients can read and speak, and can recognize environmental sounds like music, slamming doors, and animal cries, but cannot recognize spoken words; words are as meaningless as if they were from a foreign language. Among patients with problems in grammar, some do not display the halting articulation of Broca's aphasia but produce fluent ungrammatical speech. Some aphasics leave out verbs, inflections, and function words; others use the wrong ones. Some cannot comprehend complicated sentences involving traces (like *The man who the woman kissed* (trace) *hugged the child*) but can comprehend complex sentences involving reflexives (like *The girl said that the woman washed herself*). Other patients do the reverse. There are Italian patients who mangle their language's inflectional suffixes (similar to the *-ing, -s,* and *-ed* of English) but are almost flawless with its derivational suffixes (similar to *-able, -ness,* and *-er*).

The mental thesaurus, in particular, is sometimes torn into pieces with clean edges. Among anomic patients (those who have trouble using nouns), different patients have problems with different kinds of nouns. Some can use concrete nouns but not abstract nouns. Some can use abstract nouns but not concrete nouns. Some can use nouns for nonliving things but have trouble with nouns for living things; others can use nouns for living things but have trouble with nouns for nonliving things. Some can name animals and vegetables but not foods, body parts, clothing, vehicles, or furniture. There are patients who have trouble with nouns for anything but animals, patients who cannot name body parts, patients who cannot name objects typically found indoors, patients who cannot name colors, and patients who have trouble with proper names. One patient could not name fruits or vegetables: he could name an abacus and a sphinx but not an apple or a peach. The psychologist Edgar Zurif, jesting the neurologist's habit of giving a fancy name to every syndrome, has suggested that it be called anomia for bananas, or "banananomia."

Does this mean that the brain has a produce section? No one has found one, nor centers for inflections, traces, phonology, and so on. Pinning brain areas to mental functions has been frustrating. Frequently one finds two patients with lesions in the same general area but with different kinds of impairment, or two patients with the same impairment but lesions in different areas. Sometimes a circumscribed impairment, like the inability to name animals, can be caused by massive lesions, brain-wide degeneration, or a blow to the head. And about ten percent of the time a patient with a lesion in the general vicinity of Wernicke's area can have a Broca-like aphasia, and a patient with lesions near Broca's area can have a Wernicke-like aphasia.

Why has it been so hard to draw an atlas of the brain with areas for different parts of language? According to one school of thought, it is because there aren't any; the brain is a meatloaf. Except for sensation and movement, mental processes are patterns of neuronal activity that are widely distributed, hologram-style, all over the brain. But the meatloaf theory is hard to reconcile with the amazingly specific deficits of many brain-damaged patients, and it is becoming obsolete in this "decade of the brain." Using tools that are getting more sophisticated each month, neurobiologists are charting vast territories that once bore the unhelpful label "association cortex" in the old textbooks, and are delineating dozens of new regions with their own functions or styles of processing, like visual areas specializing in object shape, spatial layout, color, 3D stereo-vision, simple motion, and complex motion.

For all we know, the brain might have regions dedicated to processes as specific as noun phrases and metrical trees; our methods for studying the human brain are still so crude that we would be unable to find them. Perhaps the regions look like little polka dots or blobs or stripes scattered around the general language areas of the brain. They might be irregularly shaped squiggles, like gerrymandered political districts. In different people, the regions might be pulled and stretched onto different bulges and folds of the brain. (All of these arrangements are found in brain systems we understand better, like the visual system.) If so, the enormous bomb craters that we call brain lesions, and the blurry snapshots we call PET scans, would leave their whereabouts unknown.

There is already some evidence that the linguistic brain might be organized in this tortuous way. The neurosurgeon George Ojemann, following up on Penfield's methods, electrically stimulated different sites in conscious, exposed brains. He found that stimulating within a site no more than a few millimeters across could disrupt a single function, like repeating or completing a sentence, naming an object, or reading a word. But these dots were scattered over the brain (largely, but not exclusively, in the perisylvian regions) and were found in different places in different individuals.

From the standpoint of what the brain is designed to do, it would not be surprising if language subcenters are idiosyncratically tangled or scattered over the cortex. The brain is a special kind of organ, the organ of computation, and unlike an organ that moves stuff around in the physical world such as the hip or the heart, the brain does not need its functional parts to have nice cohesive shapes. As long as the connectivity of the neural microcircuitry is preserved, its parts can be put in different places and do the same thing, just as the wires connecting a set of electrical components can be haphazardly stuffed into a cabinet, or the headquarters of a corporation can be located anywhere if it has good communication links to its plants and warehouses. This seems especially true of words: lesions or electrical stimulation over wide areas of the brain can cause naming difficulties. A word is a bundle of different kinds of information. Perhaps each word is like a hub that can be positioned anywhere in a large region, as long as its spokes extend to the parts of the brain storing its sound, its syntax, its logic, and the appearance of the things it stands for.

The developing brain may take advantage of the disembodied nature of computation to position language circuits with some degree of flexibility. Say a variety of brain areas have the potential to grow the precise wiring diagrams for language components. An initial bias causes the circuits to be laid down in their typical sites; the alternative sites are then suppressed. But if those first sites get damaged within a certain critical period, the circuits can grow elsewhere. Many neurologists believe that this is why the language centers are located in unexpected places in a significant minority of people. Birth is traumatic, and not just for the familiar psychological reasons. The birth canal squeezes the baby's head like a lemon, and newborns frequently suffer small strokes and other brain insults. Adults with anomalous language areas may be the recovered victims of these primal injuries. Now that MRI machines are common in brain research centers, visiting journalists and philosophers are sometimes given pictures of their brains to take home as a souvenir. Occasionally the picture will reveal a walnut-sized dent, which, aside from some teasing from friends who say they knew it all along, bespeaks no ill effects.

There are other reasons why language functions have been so hard to pin down in the brain. Some kinds of linguistic knowledge might be stored in multiple copies, some of higher quality than others, in several places. Also, by the time stroke victims can be tested systematically, they have often recovered some of their facility with language, in part by compensating with general reasoning abilities. And neurologists are not like electronics technicians who can wiggle a probe into the input or output line of some component to isolate its function. They must tap the whole patient via his or her eyes and ears and mouth and hands, and there are many computational waystations between the stimulus they present and the response they observe. For example, naming an object involves recognizing it, looking up its entry in the mental dictionary, accessing its pronunciation, articulating it, and perhaps also monitoring the output for errors by listening to it. A naming problem could arise if any of these processes tripped up.

There is some hope that we will have better localization of mental processes soon, because more precise brain-imaging technologies are rapidly being developed. One example is Functional MRI, which can measure—with much more precision than PET—how hard the different parts of the brain are working during different kinds of mental activity. Another is Magneto-Encephalography, which is like EEG but can pinpoint the part of the brain that an electromagnetic signal is coming from.

We will never understand language organs and grammar genes by looking only for postage-stamp-sized blobs of brain. The computations underlying mental life are caused by the wiring of the intricate networks that make up the cortex, networks with millions of neurons, each neuron connected to thousands of others, operating in thousandths of a second. What would we see if we could crank up the microscope and peer into the microcircuitry of the language areas? No one knows, but I would like to give you an educated guess. Ironically, this is both the aspect of the language instinct that we know the least about and the aspect that is the most important, because it is there that the actual causes of speaking and understanding lie. I will present you with a dramatization of what grammatical information processing might be like from a neuron's-eye view. It is not something that you should take particularly seriously; it is simply a demonstration that the language instinct is compatible in principle with the billiard-ball causality of the physical universe, not just mysticism dressed up in a biological metaphor.

Neural network modeling is based on a simplified toy neuron. This neuron can do just a few things. It can be active or inactive. When active, it sends a signal down its axon (output wire) to the other cells it is connected to; the connections are called synapses. Synapses can be excitatory or inhibitory and can have various degrees of strength. The neuron at the receiving end adds up any signals coming in from excitatory synapses, subtracts any signals coming in from inhibitory synapses, and if the sum exceeds a threshold, the receiving neuron becomes active itself.

A network of these toy neurons, if large enough, can serve as a computer, calculating the answer to any problem that can be specified precisely, just like the page-crawling Turing machine in Chapter 3 that could deduce that Socrates is mortal. That is because toy neurons can be wired together in a few simple ways that turn them into "logic gates," devices that can compute the logical relations "and," "or," and "not" that underlie deduction. The meaning of the logical relation "and" is that the statement "A and B" is true if A is true and if B is true. An AND gate that computes that relation would be one that turns itself on if all of its inputs are on. If we assume that the threshold for our toy neurons is .5, then a set of incoming synapses whose weights are each less than .5 but that sum to greater than .5, say .4 and .4, will function as an AND gate, such as the one on the left here:

AND          OR          NOT

The meaning of the logical relation "or" is that a statement "A or B" is true if A is true or if B is true. Thus an OR gate must turn on if at least one of its inputs is on. To implement it, each synaptic weight must be greater than the neuron's threshold, say .6, like the middle circuit in the diagram. Finally, the meaning of the logical relation "not" is that a statement "Not A" is true if A is false, and vice versa. Thus a NOT gate should turn its output off if its input is on, and vice versa. It is implemented by an inhibitory synapse, shown on the right, whose negative weight is sufficient to turn off an output neuron that is otherwise always on.

Here is how a network of neurons might compute a moderately complex grammatical rule. The English inflection -s as in *Bill walks* is a suffix that should be applied under the following conditions: when the subject is in the third person AND singular AND the action is in the present tense AND is done habitually (this is its "aspect," in lingo)—

but NOT if the verb is irregular like *do, have, say,* or *be* (for example, we say *Bill is,* not *Bill be's*). A network of neural gates that computes these logical relations looks like this:

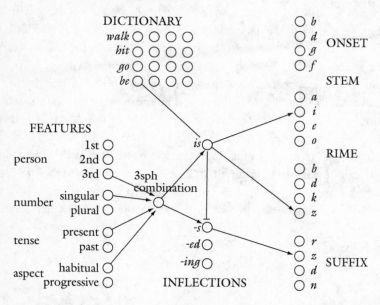

First, there is a bank of neurons standing for inflectional features on the lower left. The relevant ones are connected via an AND gate to a neuron that stands for the combination third person, singular number, present tense, and habitual aspect (labeled "3sph"). That neuron excites a neuron corresponding to the *-s* inflection, which in turn excites the neuron corresponding to the phoneme *z* in a bank of neurons that represent the pronunciations of suffixes. If the verb is regular, this is all the computation that is needed for the suffix; the pronunciation of the stem, as specified in the mental dictionary, is simply copied over verbatim to the stem neurons by connections I have not drawn in. (That is, the form for *to hit* is just *hit + s*; the form for *to wug* is just *wug + s*.) For irregular verbs like *be,* this process must be blocked, or else the neural network would produce the incorrect *be's*. So the 3sph combination neuron also sends a signal to a neuron that stands for the entire irregular form *is*. If the person whose brain we are modeling is intending to use the verb *be*, a neuron stand-ing for the verb *be* is already active, and it, too, sends activation to the *is* neuron. Because the two inputs to *is* are connected as an AND gate, both must be on to activate *is*. That is, if and only if the person is thinking of *be* and third-person-singular-present-habitual at the same time, the *is* neuron is activated. The *is* neuron inhibits the -*s* inflection via a NOT gate formed by an inhibitory synapse, preventing *ises* or *be's*, but activates the vowel *i* and the consonant *z* in the bank of neurons standing for the stem. (Obviously I have omitted many neurons and many connections to the rest of the brain.)

I have hand-wired this network, but the connections are specific to English and in a real brain would have to have been learned. Con-tinuing our neural network fantasy for a while, try to imagine what this network might look like in a baby. Pretend that each of the pools of neurons is innately there. But wherever I have drawn an arrow from a single neuron in one pool to a single neuron in another, imagine a suite of arrows, from every neuron in one pool to every neuron in another. This corresponds to the child innately "expecting" there to be, say, suffixes for persons, numbers, tenses, and aspects, as well as possible irregular words for those combinations, but not knowing exactly which combinations, suffixes, or irregulars are found in the particular language. Learning them corresponds to strengthening some of the synapses at the arrowheads (the ones I happen to have drawn in) and letting the others stay invisible. This could work as follows. Imagine that when the infant hears a word with a *z* in its suffix, the *z* neuron in the suffix pool at the right edge of the diagram gets activated, and when the infant thinks of third person, singular number, present tense, and habitual aspect (parts of his construal of the event), those four neurons at the left edge get activated, too. If the activation spreads backwards as well as forwards, and if a synapse gets strengthened every time it is activated at the same time that its output neuron is already active, then all the synapses lining the paths between "3rd," "singular," "present," "habitual" at one end, and "z" at the other end, get strengthened. Repeat the experience enough times, and the partly specified neonate network gets tuned into the adult one I have pictured.

Let's zoom in even closer. What primal solderer laid down the pools of neurons and the innate potential connections among them? This is one of the hottest topics in contemporary neuroscience, and we are beginning to get the glimmerings of how embryonic brains get wired. Not the language areas of humans, of course, but the eyeballs of fruit flies and the thalamuses of ferrets and the visual cortexes of cats and monkeys. Neurons destined for particular cortical areas are born in specific areas along the walls of the ventricles, the fluid-filled cavities at the center of the cerebral hemispheres. They then creep outward toward the skull into their final resting place in the cortex along guy wires formed by the glial cells (the support cells that, together with neurons, constitute the bulk of the brain). The connections between neurons in different regions of the cortex are often laid down when the intended target area releases some chemical, and the axons growing every which way from the source area "sniff out" that chemical and follow the direction in which its concentration increases, like plant roots growing toward sources of moisture and fertilizer. The axons also sense the presence of specific molecules on the glial surfaces on which they creep, and can steer themselves like Hansel and Gretel following the trail of bread crumbs. Once the axons reach the general vicinity of their target, more precise synaptic connections can be formed because the growing axons and the target neurons bear certain molecules on their surfaces that match each other like a lock and key and adhere in place. These initial connections are often quite sloppy, though, with neurons exuberantly sending out axons that grow toward, and connect to, all kinds of inappropriate targets. The inappropriate ones die off, either because their targets fail to provide some chemical necessary for their survival, or because the connections they form are not used enough once the brain turns on in fetal development.

Try to stay with me in this neuro-mythological quest: we are beginning to approach the "grammar genes." The molecules that guide, connect, and preserve neurons are proteins. A protein is specified by a gene, and a gene is a sequence of bases in the DNA string found in a chromosome. A gene is turned on by "transcription factors" and other regulatory molecules—gadgets that latch on to a sequence of bases somewhere on a DNA molecule and unzip a neighboring stretch, allowing that gene to be transcribed into RNA, which is then translated into protein. Generally these regulatory factors are themselves proteins, so the process of building an organism is an intricate cascade of DNA making proteins, some of which interact with other DNA to make more proteins, and so on. Small differences in the timing or amount of some protein can have large effects on the organism being built.

Thus a single gene rarely specifies some identifiable part of an organism. Instead, it specifies the release of some protein at specific times in development, an ingredient of an unfathomably complex recipe, usually having some effect in molding a suite of parts that are also affected by many other genes. Brain wiring in particular has a complex relationship to the genes that lay it down. A surface molecule may not be used in a single circuit but in many circuits, each guided by a specific combination. For example, if there are three proteins, X, Y, and Z, that can sit on a membrane, one axon might glue itself to a surface that has X and Y and not Z, and another might glue itself to a surface that has Y and Z but not X. Neuroscientists estimate that about thirty thousand genes, the majority of the human genome, are used to build the brain and nervous sytem.

And it all begins with a single cell, the fertilized egg. It contains two copies of each chromosome, one from the mother, one from the father. Each parental chromosome was originally assembled in the parents' gonads by randomly splicing together parts of the chromosomes of the two grandparents.

We have arrived at a point at which we can define what grammar genes would be. The grammar genes would be stretches of DNA that code for proteins, or trigger the transcription of proteins, in certain times and places in the brain, that guide, attract, or glue neurons into networks that, in combination with the synaptic tuning that takes place during learning, are necessary to compute the solution to some grammatical problem (like choosing an affix or a word).

So do grammar genes really exist, or is the whole idea just loopy? Can we expect the scenario in the 1990 editorial cartoon by Brian Duffy? A pig, standing upright, asks a farmer, "What's for dinner? Not me, I hope." The farmer says to his companion, "That's the one that received the human gene implant."

For any grammar gene that exists in every human being, there is currently no way to verify its existence directly. As in many cases in biology, genes are easiest to identify when they correlate with some difference between individuals, often a difference implicated in some pathology.

We certainly know that there is something in the sperm and egg that affects the language abilities of the child that grows out of their union. Stuttering, dyslexia (a difficulty in reading that is often related to a difficulty in mentally snipping syllables into their phonemes), and Specific Language Impairment (SLI) all run in families. This does not prove that they are genetic (recipes and wealth also run in families), but these three syndromes probably are. In each case there is no plausible environmental agent that could act on afflicted family members while sparing the normal ones. And the syndromes are far more likely to affect both members of a pair of identical twins, who share an environment and all their DNA, than both members of a pair of fraternal twins, who share an environment and only half of their DNA. For example, identical four-year-old twins tend to mispronounce the same words more often than fraternal twins, and if a child has Specific Language Impairment, there is an eighty percent chance that an identical twin will have it too, but only a thirty-five percent chance that a fraternal twin will have it. It would be interesting to see whether adopted children resemble their biological family members, who share their DNA but not their environments. I am unaware of any adoption study that tests for SLI or dyslexia, but one study has found that a measure of early language ability in the first year of life (a measure that combines vocabulary, vocal imitation, word combinations, jabbering, and word comprehension) was correlated with the general cognitive ability and memory of the birth mother, but not of the adoptive mother or father.

The K family, three generations of SLI sufferers, whose members say things like *Carol is cry in the church* and can not deduce the plural of *wug*, is currently one of the most dramatic demonstrations that defects in grammatical abilities might be inherited. The attention-grabbing hypothesis about a single dominant autosomal gene is based on the following Mendelian reasoning. The syndrome is suspected of being genetic because there is no plausible environmental cause that would single out some family members and spare their agemates (in one case, one fraternal twin was affected, the other not), and because the syndrome has struck fifty-three percent of the family members but strikes no more than about three percent of the population at large. (In principle, the family could just have been unlucky; after all, they were not randomly selected from the population but came to the geneticists' attention only *because* of the high concentration of the syndrome. But it is unlikely.) A single gene is thought to be responsible because if several genes were responsible, each eroding language ability by a bit, there would be several degrees of disability among the family members, depending on how many of the damaging genes they inherited. But the syndrome seems to be all-or-none: the school system and family members all agree on who does and who does not have the impairment, and in most of Gopnik's tests the impaired members cluster together at the low end of the scale while the normal members cluster at the high end, with no overlap. The gene is thought to be autosomal (not on the X chromosome) and dominant because the syndrome struck males and females with equal frequency, and in all cases the spouse of an impaired parent, whether husband or wife, was normal. If the gene were recessive and autosomal, it would be necessary to have two impaired parents to inherit the syndrome. If it were recessive and on the X chromosome, only males would have it; females would be carriers. And if it were dominant and on the X chromosome, an impaired father would pass it on to all of his daughters and none of his sons, because sons get their X chromosome from their mother, and daughters get one from each parent. But one of the daughters of an impaired man was normal.

This single gene is not, repeat not, responsible for all the circuitry underlying grammar, contrary to the Associated Press, James Kilpatrick, et al. Remember that a single defective component can bring a complex machine to a halt even when the machine needs many properly functioning parts to work. In fact, it is possible that the normal version of the gene does not build grammar circuitry at all. Maybe the defective version manufactures a protein that gets in the way of some chemical process necessary for laying down the language circuits. Maybe it causes some adjacent area in the brain to overgrow its own territory and spill into the territory ordinarily allotted to language.

But the discovery is still quite interesting. Most of the language-impaired family members were average in intelligence, and there are sufferers in other families who are way above average; one boy studied by Gopnik was tops in his math class. So the syndrome shows that there must be some pattern of genetically guided events in the development in the brain (namely, the events disrupted in this syndrome) that is specialized for the wiring in of linguistic computation. And these construction sites seem to involve circuitry necessary for the processing of grammar in the mind, not just the articulation of speech sounds by the mouth or the perception of speech sounds by the ear. Though the afflicted family members as children suffered from difficulties in articulating speech and developed language late, most of them outgrew the articulation problems, and their lasting deficits involve grammar. For example, although the impaired family members often leave off the *-ed* and *-s* suffixes, it is not because they cannot hear or say those sounds; they easily discriminate between *car* and *card*, and never pronounce *nose* as *no*. In other words, they treat a sound differently when it is a permanent part of a word and when it is added to a word by a rule of grammar.

Equally interestingly, the impairment does not wipe out any part of grammar completely, nor does it compromise all parts equally. Though the impaired family members had trouble changing the tense of test sentences and applying suffixes in their spontaneous speech, they were not hopeless; they just performed far less accurately than their unimpaired relatives. These probabilistic deficits seemed to be concentrated in morphology and the features it manipulates, like tense, person, and number; other aspects of grammar were less affected. The impaired members could, for example, detect verb phrase violations in sentences like *The nice girl gives* and *The girl eats a cookie to the boy,* and could act out many complex commands. The lack of an exact correspondence between a gene and a single function is exactly what we would expect, knowing how genes work.

So for now there is suggestive evidence for grammar genes, in the sense of genes whose effects seem most specific to the development of the circuits underlying parts of grammar. The chromosomal locus of the putative gene is completely unknown, as is its effect on the structure of the brain. But blood samples are being drawn from the family for genetic analysis, and MRI scans of brains from other individuals with Specific Language Impairment have already been found to lack the asymmetry in the perisylvian areas that we find in linguistically normal brains. Other researchers on language disorders, some excited by Gopnik's claims, others skeptical of them, have begun to screen their patients with careful tests of their grammatical abilities and their family histories. They are seeking to determine how commonly Specific Language Impairment is inherited and how many distinct syndromes of the impairment there might be. You can expect to read about some interesting discoveries about the neurology and genetics of language in the next few years.

In modern biology, it is hard to discuss genes without discussing genetic variation. Aside from identical twins, no two people—in fact, no two sexually reproducing organisms—are genetically identical. If this were not true, evolution as we know it could not have happened. If there are language genes, then, shouldn't normal people be innately different from one another in their linguistic abilities? Are they? Must I qualify everything I have said about language and its development, because no two people have the same language instinct?

It is easy to get carried away with the geneticists' discovery that many of our genes are as distinctive as our fingerprints. After all, you can open up any page of *Grey's Anatomy* and expect to find a depiction of organs and their parts and arrangements that will be true of any normal person. (Everyone has a heart with four chambers, a liver, and so on.) The biological anthropologist John Tooby and the cognitive psychologist Leda Cosmides have resolved the apparent paradox.

Tooby and Cosmides argue that differences between people must be minor quantitative variations, not qualitatively different designs. The reason is sex. Imagine that two people were really built from fundamentally different designs: either physical designs, like the structure of the lungs, or neurological designs, like the circuitry underlying some cognitive process. Complex machines require many finely meshing parts, which in turn require many genes to build them. But the chromosomes are randomly snipped, spliced, and shuffled during the formation of sex cells, and then are paired with other chimeras at fertilization. If two people really had different designs, their offspring would inherit a mishmash of fragments from the genetic blueprints of each—as if the plans for two cars were cut up with scissors and the pieces taped back together without our caring about which scrap originally came from which car. If the cars are of different designs, like a Ferrari and a jeep, the resulting contraption, if it could be built at all, would certainly not get anywhere. Only if the two designs were extremely similar to begin with could the new pastiche work.

That is why the variation that geneticists tell us about is microscopic—differences in the exact sequence of molecules in proteins whose overall shape and function are basically the same, kept within narrow limits of variation by natural selection. That variation is there for a purpose: by shuffling the genes each generation, lineages of organisms can stay one step ahead of the microscopic, rapidly evolving disease parasites that fine-tune themselves to infiltrate the chemical environments of their hosts. But above the germ's-eye view, at the macroscopic level of functioning biological machinery visible to an anatomist or psychologist, variation from one individual to another must be quantitative and minor; thanks to natural selection, all normal people must be qualitatively the same.

But this does not mean that individual differences are boring. Genetic variation can open our eyes to the degree of structure and complexity that the genes ordinarily give to the mind. If genes just equipped a mind with a few general information-processing devices like a short-term memory and a correlation detector, some people might be better than others at holding things in memory or learning contingencies, and that would be about it. But if the genes built a mind with many elaborate parts dedicated to particular tasks, the unique genetic hand that is dealt to each person would give rise to an unprecedented profile of innate cognitive quirks.

I quote from a recent article in *Science*:

> When Oskar Stöhr and Jack Yufe arrived in Minnesota to participate in University of Minnesota psychologist Thomas J. Bouchard, Jr.'s study of identical twins reared apart, they were both sporting blue double-breasted epauletted shirts, mustaches, and wire-rimmed glasses. Identical twins separated at birth, the two men, in their late 40s, had met once before two decades earlier. Nonetheless, Oskar, raised as a Catholic in Germany, and Jack, reared by his Jewish father in Trinidad, proved to have much in common in their tastes and personalities— including hasty tempers and idiosyncratic senses of humor (both enjoyed surprising people by sneezing in elevators).

And both flushed the toilet both before and after using it, kept rubber bands around their wrists, and dipped buttered toast in their coffee.

Many people are skeptical of such anecdotes. Are the parallels just coincidences, the overlap that is inevitable when two biographies are scrutinized in enough detail? Clearly not. Bouchard and his behavior geneticist colleagues D. Lykken, M. McGue, and A. Tellegen are repeatedly astonished by the spooky similarities they discover in their identical twins reared apart but that never appear in their fraternal twins reared apart. Another pair of identical twins meeting for the first time discovered that they both used Vademecum toothpaste, Canoe shaving lotion, Vitalis hair tonic, and Lucky Strike cigarettes. After the meeting they sent each other identical birthday presents that crossed in the mail. One pair of women habitually wore seven rings. Another pair of men pointed out (correctly) that a wheel bearing in Bouchard's car needed replacing. And quantitative research corroborates the hundreds of anecdotes. Not only are very general traits like IQ, extroversion, and neuroticism partly heritable, but so are specific ones like degree of religious feeling, vocational interests, and opinions about the death penalty, disarmament, and computer music.

Could there really be a gene for sneezing in elevators? Presumably not, but there does not have to be. Identical twins share all their genes, not just one of them. So there are fifty thousand genes for sneezing in elevators—which are also fifty thousand genes for liking blue double-breasted epauletted shirts, using Vitalis hair tonic, wearing seven rings, and all the rest. The reason is that the relationship between particular genes and particular psychological traits is doubly indirect. First, a single gene does not build a single brain module; the brain is a delicately layered soufflé in which each gene product is an ingredient with a complex effect on many properties of many circuits. Second, a single brain module does not produce a single behavioral trait. Most of the traits that capture our attention emerge out of unique combinations of kinks in many different modules. Here is an analogy. Becoming an all-star basketball player requires many physical advantages, like height, large hands, excellent aim, good peripheral vision, lots of fast-twitch muscle tissue, efficient lungs, and springy tendons. Though these traits are probably genetic to a large degree, there does not have to be a basketball gene; those men for whom the genetic slot machine stopped at three cherries play in the NBA, while the more numerous seven-foot klutzes and five-foot sharpshooters go into some other line of work. No doubt the same is true of any interesting behavioral trait like sneezing in elevators (which is no odder than an aptitude for shooting a ball through a hoop with someone's hand in your face). Perhaps the sneezing-in-elevators gene complex is the one that specifies just the right combination of thresholds and cross-connections among the modules governing humor, reactions to enclosed spaces, sensitivity to the mental states of others such as their anxiety and boredom, and the sneezing reflux.

No one has ever studied heritable variation in language, but I have a strong suspicion of what it is like. I would expect the basic design of language, from X-bar syntax to phonological rules and vocabulary structure, to be uniform across the species; how else could children learn to talk and adults understand one another? But the complexity of language circuitry leaves plenty of scope for quantitative variation to combine into unique linguistic profiles. Some module might be relatively stunted or hypertrophied. Some normally unconscious representation of sound or meaning or grammatical structure might be more accessible to the rest of the brain. Some connection between language circuitry and the intellect or emotions might be faster or slower.

Thus I predict that there are idiosyncratic combinations of genes (detectable in identical twins reared apart) behind the raconteur, the punster, the accidental poet, the sweet-talker, the rapier-like wit, the sesquipedalian, the word-juggler, the owner of the gift of gab, the Reverend Spooner, the Mrs. Malaprop, the Alexander Haig, the woman (and her teenage son!) I once tested who can talk backwards, and the student at the back of every linguistics classroom who objects that *Who do you believe the claim that John saw?* doesn't sound so bad. Between 1988 and 1992, many people suspected that the chief executive of the United States and his second-in-command were not playing with a full linguistic deck:

> I am less interested in what the definition is. You might argue technically, are we in a recession or not. But when there's this kind of sluggishness and concern—definitions, heck with it.
>
> I'm all for Lawrence Welk. Lawrence Welk is a wonderful man. He used to be, or was, or—wherever he is now, bless him.
>
> —George Bush

Hawaii has always been a very pivotal role in the Pacific. It is IN the Pacific. It is a part of the United States that is an island that is right here.

[Speaking to the United Negro College Fund, whose motto is

"A mind is a terrible thing to waste":] What a terrible thing to
have lost one's mind. Or not to have a mind at all. How true
that is.

—Dan Quayle

And who knows what unrepeatable amalgam of genes creates the linguistic genius?

If people don't want to come out to the ballpark, nobody's
going to stop them.
You can observe a lot just by watching.
In baseball, you don't know nothing.
Nobody goes there anymore. It's too crowded.
It ain't over till it's over.
It gets late early this time of year.

—Yogi Berra

And NUH is the letter I use to spell Nutches
Who live in small caves, known as Nitches, for hutches.
These Nutches have troubles, the biggest of which is
The fact that there are many more Nutches than Nitches.
Each Nutch in a Nitch knows that some other Nutch
Would like to move into his Nitch very much.
So each Nutch in a Nitch has to watch that small Nitch
Or Nutches who haven't got Nitches will snitch.

—Dr. Seuss

Lolita, light of my life, fire of my loins. My sin, my soul. Lo-
lee-ta: the tip of the tongue taking a trip of three steps down
the palate to tap, at three, on the teeth. Lo. Lee. Ta.

—Valdimir Nabokov

I have a dream that one day this nation will rise up and
live out the true meaning of its creed: "We hold these truths to
be self-evident, that all men are created equal."

I have a dream that one day on the red hills of Georgia
the sons of former slaves and the sons of former slaveowners
will be able to sit down together at the table of brotherhood.

I have a dream that one day even the state of Mississippi, a state sweltering with the people's injustice, sweltering with the heat of oppression, will be transformed into an oasis of freedom and justice.

I have a dream that my four little children will one day live in a nation where they will not be judged by the color of their skin but by the content of their character.

—Martin Luther King, Jr.

This goodly frame, the earth, seems to me a sterile promontory, this most excellent canopy, the air, look you, this brave o'er-hanging firmament, this majestical roof fretted with golden fire, why, it appears no other thing to me than a foul and pestilent congregation of vapours. What a piece of work is a man! how noble in reason! how infinite in faculty! in form and moving how express and admirable! in action how like an angel! in apprehension how like a god! the beauty of the world! the paragon of animals! And yet, to me, what is this quintessence of dust?

—William Shakespeare

# 11

⊰⊱

# The Big Bang

*The elephant's trunk is six feet long and one foot thick and contains sixty* thousand muscles. Elephants can use their trunks to uproot trees, stack timber, or carefully place huge logs in position when recruited to build bridges. An elephant can curl its trunk around a pencil and draw characters on letter-size paper. With the two muscular extensions at the tip, it can remove a thorn, pick up a pin or a dime, uncork a bottle, slide the bolt off a cage door and hide it on a ledge, or grip a cup so firmly, without breaking it, that only another elephant can pull it away. The tip is sensitive enough for a blindfolded elephant to ascertain the shape and texture of objects. In the wild, elephants use their trunks to pull up clumps of grass and tap them against their knees to knock off the dirt, to shake coconuts out of palm trees, and to powder their bodies with dust. They use their trunks to probe the ground as they walk, avoiding pit traps, and to dig wells and siphon water from them. Elephants can walk underwater on the beds of deep rivers or swim like submarines for miles, using their trunks as snorkels. They communicate through their trunks by trumpeting, humming, roaring, piping, purring, rumbling, and making a crumpling-metal sound by rapping the trunk against the ground. The trunk is lined with chemoreceptors that allow the elephant to smell python hidden in the grass or food a mile away.

Elephants are the only living animals that possess this extraordinary organ. Their closest living terrestrial relative is the hyrax, a mammal that you would probably not be able to tell from a large guinea pig. Until now you have probably not given the uniqueness of the elephant's trunk a moment's thought. Certainly no biologist has made a fuss about it. But now imagine what might happen if some biologists were elephants. Obsessed with the unique place of the trunk in nature, they might ask how it could have evolved, given that no other organism has a trunk or anything like it. One school might try to think up ways to narrow the gap. They would first point out that the elephant and the hyrax share about 90% of their DNA and thus could not be all that different. They might say that the trunk must not be as complex as everyone thought; perhaps the number of muscles had been miscounted. They might further note that the hyrax really does have a trunk, but somehow it has been overlooked; after all, the hyrax does have nostrils. Though their attempts to train hyraxes to pick up objects with their nostrils have failed, some might trumpet their success at training the hyraxes to push toothpicks around with their tongues, noting that stacking tree trunks or drawing on blackboards differ from it only in degree. The opposite school, maintaining the uniqueness of the trunk, might insist that it appeared all at once in the offspring of a particular trunkless elephant ancestor, the product of a single dramatic mutation. Or they might say that the trunk somehow arose as an automatic by-product of the elephant's having evolved a large head. They might add another paradox for trunk evolution: the trunk is absurdly more intricate and well coordinated than any ancestral elephant would have needed.

These arguments might strike us as peculiar, but every one of them has been made by scientists of a different species about a complex organ that that species alone possesses, language. As we shall see in this chapter, Chomsky and some of his fiercest opponents agree on one thing: that a uniquely human language instinct seems to be incompatible with the modern Darwinian theory of evolution, in which complex biological systems arise by the gradual accumulation over generations of random genetic mutations that enhance reproductive success. ~~Either there is no language instinct, or it must~~ have ~~evolved by other means.~~ Since I have been trying to convince you that there is a language instinct but would certainly forgive you if you would rather believe Darwin than believe me, I would also like to convince you that you need not make that choice. Though we know few details about how the language instinct evolved, there is no reason to doubt that the principal explanation is the same as for any other complex instinct or organ, Darwin's theory of natural selection.

Language is obviously as different from other animals' communication systems as the elephant's trunk is different from other animals' nostrils. Nonhuman communication systems are based on one of three designs: a finite repertory of calls (one for warnings of predators, one for claims to territory, and so on), a continuous analog signal that registers the magnitude of some state (the livelier the dance of the bee, the richer the food source that it is telling its hivemates about), or a series of random variations on a theme (a birdsong repeated with a new twist each time: Charlie Parker with feathers). As we have seen, human language has a very different design. The discrete combinatorial system called "grammar" makes human language infinite (there is no limit to the number of complex words or sentence in a language), digital (this infinity is achieved by rearranging discrete elements in particular orders and combinations, not by varying some signal along a continuum like the mercury in a thermometer), and compositional (each of the infinite combinations has a different meaning predictable from the meanings of its parts and the rules and principles arranging them).

Even the seat of human language in the brain is special. The vocal calls of primates are controlled not by their cerebral cortex but by phylogenetically older neural structures in the brain stem and limbic systems, structures that are heavily involved in emotion. Human vocalizations other than language, like sobbing, laughing, moaning, and shouting in pain, are also controlled subcortically. Subcortical structures even control the swearing that follows the arrival of a hammer on a thumb, that emerges as an involuntary tic in Tourette's syndrome, and that can survive as Broca's aphasics' only speech. Genuine language, as we saw in the preceding chapter, is seated in the cerebral cortex, primarily the left perisylvian region.

Some psychologists believe that changes in the vocal organs and in the neural circuitry that produces and perceives speech sounds are the *only* aspects of language that evolved in our species. On this view, there are a few general learning abilities found throughout the animal kingdom, and they work most efficiently in humans. At some point in history language was invented and refined, and we have been learning it ever since. The idea that species-specific behavior is caused by anatomy and general intelligence is captured in the Gary Larson *Far Side* cartoon in which two bears hide behind a tree near a human couple relaxing on a blanket. One says: "C'mon! Look at these fangs! . . . Look at these claws! . . . You think we're supposed to eat just honey and berries?"

According to this view, chimpanzees are the second-best learners in the animal kingdom, so they should be able to acquire a language too, albeit a simpler one. All it takes is a teacher. In the 1930s and 1940s two psychologist couples adopted baby chimpanzees. The chimps became part of the family and learned to dress, use the toilet, brush their teeth, and wash the dishes. One of them, Gua, was raised alongside a boy of the same age but never spoke a word. The other, Viki, was given arduous training in speech, mainly by the foster parents' moulding the puzzled chimp's lips and tongue into the right shapes. With a lot of practice, and often with the help of her own hands, Viki learned to make three utterances that charitable listeners could hear as *papa, mama,* and *cup,* though she often confused them when she got excited. She could respond to some stereotyped formulas, like *Kiss me* and *Bring me the dog,* but stared blankly when asked to act out a novel combination like *Kiss the dog.*

But Gua and Viki were at a disadvantage: they were forced to use their vocal apparatus, which was not designed for speech and which they could not voluntarily control. Beginning in the late 1960s, several famous projects claimed to have taught language to baby chimpanzees with the help of more user-friendly media. (Baby chimps are used because the adults are not the hairy clowns in overalls you see on television, but strong, vicious wild animals who have bitten fingers off several well-known psychologists.) Sarah learned to string magnetized plastic shapes on a board. Lana and Kanzi learned to press buttons with symbols on a large computer console or point to them on a portable tablet. Washoe and Koko (a gorilla) were said to have acquired American Sign Language. According to their trainers, these apes learned hundreds of words, strung them together in meaningful sentence, and coined new phrases, like *water bird* for a swan and *cookie rock* for a stale Danish. "Language is no longer the exclusive domain of man," said Koko's trainer, Francine (Penny) Patterson.

These claims quickly captured the public's imagination and were played up in popular science books and magazines and television programs like *National Geographic, Nova, Sixty Minutes,* and *20/20.* Not only did the projects seem to consummate our age-old yearning to talk to the animals, but the photo opportunities of attractive women communing with apes, evocative of the beauty-and-the-beast archetype, were not lost on the popular media. Some of the projects were covered by *People, Life,* and *Penthouse* magazines, and they were fictionalized in a bad movie starring Holly Hunter called *Animal Behavior* and in a famous Pepsi commercial.

Many scientists have also been captivated, seeing the projects as a healthy deflation of our species' arrogant chauvinism. I have seen popular-science columns that list the acquisition of language by chimpanzees as one of the major scientific discoveries of the century. In a recent, widely excerpted book, Carl Sagan and Ann Druyan have used the ape language experiments as part of a call for us to reassess our place in nature:

> A sharp distinction between human beings and "animals" is essential if we are to bend them to our will, make them work for us, wear them, eat them—without any disquieting tinges of guilt or regret. With untroubled consciences, we can render whole species extinct—as we do today to the tune of 100 speces a day. Their loss is of little import: Those beings, we tell

ourselves, are not like us. An unbridgeable gap has thus a practical role to play beyond the mere stroking of human egos. Isn't
there much to be proud of in the lives of monkeys and apes?
Shouldn't we be glad to acknowledge a connection with
Leakey, Imo, or Kanzi? Remember those macaques who would
rather go hungry then profit from harming their fellows; might
we have a more optimistic view of the human future if we were
sure our ethics were up to their standards? And, viewed from
this perspective, how shall we judge our treatment of monkeys
and apes?

This well-meaning but misguided reasoning could only have come
from writers who are not biologists. Is it really "humility" for us to
save species from extinction because we think they are like us? Or
because they seem like a bunch of nice guys? What about all the
creepy, nasty, selfish animals who do not remind us of ourselves, or
our image of what we would like to be—can we go ahead and wipe
them out? And Sagan and Druyan are no friends of the apes if they
think the reason we should treat the apes fairly is that they can be
taught human language. Like many other writers, Sagan and Druyan
are far too credulous about the claims of the chimpanzee trainers.

People who spend a lot of time with animals are prone to developing indulgent attitudes about their powers of communication. My
great-aunt Bella insisted in all sincerity that her Siamese cat Rusty
understood English. Many of the claims of the ape trainers were not
much more scientific. Most of the trainers were schooled in the behaviorist tradition of B. F. Skinner and are ignorant of the study of language; they latched on to the most tenuous resemblance between
chimp and child and proclaimed that their abilities are fundamentally
the same. The more enthusiastic trainers went over the heads of scientists and made their engaging case directly to the public on the
*Tonight Show* and *National Geographic*. Patterson in particular has
found ways to excuse Koko's performance on the grounds that the
gorilla is fond of puns, jokes, metaphors, and mischievous lies. Generally the stronger the claims about the animal's abilities, the skimpier

the data made available to the scientific community for evaluation. Most of the trainers have refused all requests to share their raw data, and Washoe's trainers, Beatrice and Alan Gardner, threatened to sue another researcher because he used frames of one of their films (the only raw data available to him) in a critical scientific article. That researcher, Herbert Terrance, together with the psychologists Lara Ann Petitto, Richard Sanders, and Tom Bever, had tried to teach ASL to one of Washoe's relatives, whom they named Nim Chimpsky. They carefully tabulated and analyzed his signs, and Petitto, with the psychologist Mark Seidenberg, also scrutinized the videotapes and what published data there were on the other signing apes, whose abilities were similar to Nim's. More recently, Joel Wallman has written a history of the topic called *Aping Language*. The moral of their investigation is: Don't believe everything you hear on the *Tonight Show*.

To begin with, the apes did *not* "learn American Sign Language." This preposterous claim is based on the myth that ASL is a crude system of pantomimes and gestures rather than a full language with complex phonology, morphology, and syntax. In fact the apes had not learned *any* true ASL signs. The one deaf native signer on the Washoe team later made these candid remarks:

> Every time the chimp made a sign, we were supposed to write it down in the log. . . . They were always complaining because my log didn't show enough signs. All the hearing people turned in logs with long lists of signs. They always saw more signs than I did. . . . I watched really carefully. The chimp's hands were moving constantly. Maybe I missed something, but I don't think so. I just wasn't seeing any signs. The hearing people were logging every movement the chimp made as a sign. Every time the chimp put his finger in his mouth, they'd say "Oh, he's making the sign for *drink*," and they'd give him some milk. . . . When the chimp scratched itself, they'd record it as the sign for *scratch*. . . . When [the chimps] want something, they reach. Sometimes [the trainers would] say, "Oh, amazing, look at that, it's exactly like the ASL sign for *give!*" It wasn't.

To arrive at their vocabulary counts in the hundreds, the investigators would also "translate" the chimps' pointing as a sign for *you*, their hugging as a sign for *hug*, their picking, tickling, and kissing as signs for *pick, tickle,* and *kiss*. Often the same movement would be credited to the chimps as different "words," depending on what the observers thought the appropriate word would be in the context. In the experiments in which the chimps interacted with a computer console, the key that the chimp had to press to initialize the computer was translated as the word *please*. Petitto estimates that with more standard criteria the true vocabulary count would be closer to 25 than 125.

Actually, what the chimps were really doing was more interesting than what they were claimed to be doing. Jane Goodall, visiting the project, remarked to Terrace and Petitto that every one of Nim's so-called signs was familiar to her from her observations of chimps in the wild. The chimps were relying heavily on the gestures in their natural repertoire, rather than learning true arbitrary ASL signs with their combinatorial phonological structure of hand shapes, motions, locations, and orientations. Such backsliding is common when humans train animals. Two enterprising students of B. F. Skinner, Keller and Marian Breland, took his principles for shaping the behavior of rats and pigeons with schedules of reward and turned them into a lucrative career of training circus animals. They recounted their experiences in a famous article called "The Misbehavior of Organisms," a play on Skinner's book *The Behavior of Organisms*. In some of their acts the animals were trained to insert poker chips in little juke boxes and vending machines for a food reward. Though the training schedules were the same for the various animals, their species-specific instincts bled through. The chickens spontaneously pecked at the chips, the pigs tossed and rooted them with their snouts, and the raccoons rubbed and washed them.

The chimp's abilities at anything one would want to call grammar were next to nil. Signs were not coordinated into the well-defined motion contours of ASL and were not inflected for aspect, agreement, and so on—a striking omission, since inflection is the primary means in ASL of conveying who did what to whom and many other kinds of information. The trainers frequently claim that the chimps have syntax, because pairs of signs are sometimes placed in one order more often than chance would predict, and because the brighter chimps can act out sequences like *Would you please carry the cooler to Penny.* But remember from the Loebner Prize competition (for the most convincing computer simulation of a conversational partner) how easy it is to fool people into thinking that their interlocutors have humanlike talents. To understand the request, the chimp could ignore the symbols *would, you, please, carry, the,* and *to;* all the chimp had to notice was the order of the two nouns (and in most of the tests, not even that, because it is more natural to carry a cooler to a person than a person to a cooler). True, some of the chimps can carry out these commands more reliably than a two-year-old child, but this says more about temperament than about grammar: the chimps are highly trained animal acts, and a two-year-old is a two-year-old.

As far as spontaneous output is concerned, there is no comparison. Over several years of intensive training, the average length of the chimps' "sentences" remains constant. With nothing more than exposure to speakers, the average length of a child's sentences shoots off like a rocket. Recall that typical sentences from a two-year-old child are *Look at that train Ursula brought* and *We going turn light on so you can't see.* Typical sentences from a language-trained chimp are:

> Nim eat Nim eat.
> Drink eat me Nim.
> Me gum me gum.
> Tickle me Nim play.
> Me eat me eat.
> Me banana you banana me you give.
> You me banana me banana you.
> Banana me me me eat.
> Give orange me give eat orange me eat orange give me eat
>   orange give me you.

These jumbles bear scant resemblance to children's sentences. (By watching long enough, of course, one is bound to find random combinations in the chimps' gesturing that can be given sensible interpretations, like *water bird*). But the strings *do* resemble animal behavior in the wild. The zoologist E. O. Wilson, summing up a survey of animal communication, remarked on its most striking property: animals, he said, are "repetitious to the point of inanity."

Even putting aside vocabulary, phonology, morphology, and syntax, what impresses one the most about chimpanzee signing is that fundamentally, deep down, chimps just don't "get it." They know that the trainers like them to sign and that signing often gets them what they want, but they never seem to feel in their bones what language is and how to use it. They do not take turns in conversation but instead blithely sign simultaneously with their partner, frequently off to the side or under a table rather than in the standardized signing space in front of the body. (Chimps also like to sign with their feet, but no one blames them for taking advantage of this anatomical gift.) The chimps seldom sign spontaneously; they have to be molded, drilled, and coerced. Many of their "sentences," especially the ones showing systematic ordering, are direct imitations of what the trainer has just signed, or minor variants of a small number of formulas that they have been trained on thousands of times. They do not even clearly get the idea that a particular sign might refer to a kind of object. Most of the chimps' object signs can refer to any aspect of the situation with which an object is typically associated. *Toothbrush* can mean "toothbrush," "toothpaste," "brushing teeth," "I want my toothbrush," or "It's time for bed." *Juice* can mean "juice," "where juice is usually kept," or "Take me to where the juice is kept." Recall from Ellen Markman's experiments in Chapter 5 that children use these "thematic" associations when sorting pictures into groups, but they ignore them when learning word meanings: to them, a *dax* is a dog or another dog, not a dog or its bone. Also, the chimps rarely make statements that comment on interesting objects or actions; virtually all their signs are demands for something they want, usually food or tickling. I cannot help but think of a moment with my two-year-old niece Eva that captures how different are the minds of child and chimp. One night the family was driving on an expressway, and when the adult conversation died down, a tiny voice from the back seat said, "Pink." I followed her gaze, and on the horizon several miles away I could make out a pink neon sign. She was commenting on its color, just for the sake of commenting on its color.

Within the field of psychology, most of the ambitious claims about chimpanzee language are a thing of the past. Nim's trainer Herbert Terrace, as mentioned, turned from enthusiast to whistle-blower. David Premack, Sarah's trainer, does not claim that what she acquired is comparable to human language; he uses the symbol system as a tool to do chimpanzee cognitive psychology. The Gardners and Patterson have distanced themselves from the community of scientific discourse for over a decade. Only one team is currently making claims about language. Sue Savage-Rumbaugh and Duane Rumbaugh concede that the chimps they trained at the computer console did not learn much. But they are now claiming that a different variety of chimpanzee does much better. Chimpanzees come from some half a dozen mutually isolated "islands" of forest in the west African continent, and the groups have diverged over the past million years to the point where some of the groups are sometimes classified as belonging to different species. Most of the trained chimps were "common chimps"; Kanzi is a "pygmy chimp" or "bonobo," and he learned to bang on visual symbols on a portable tablet. Kanzi, says Savage-Rumbaugh, does substantially better at learning symbols (and at understanding spoken language) than common chimps. Why he would be expected to do so much better than members of his sibling species is not clear; contrary to some reports in the press, pygmy chimps are no more closely related to humans than common chimps are. Kanzi is said to have learned his graphic symbols without having been laboriously trained on them—but he was at his mother's side watching while *she* was laboriously trained on them (unsuccessfully). He is said to use the symbols for purposes other than requesting—but at best only four percent of the time. He is said to use three-symbol "sentences"—but they are really fixed formulas with no internal structure and are not even three symbols long. The so-called sentences are all chains like the symbol for chase followed by the symbol for hide followed by a point to the person Kanzi wants to do the chasing and hiding. Kanzi's language abilities, if one is being charitable, are above those of his common cousins by a just-noticeable difference, but no more.

What an irony it is that the supposed attempt to bring *Homo sapiens* down a few notches in the natural order has taken the form of us humans hectoring another species into emulating our instinctive form of communication, or some artificial form we have invented, as if that were the measure of biological worth. The chimpanzees' resistance is no shame on them; a human would surely do no better if trained to hoot and shriek like a chimp, a symmetrical project that makes about as much scientific sense. In fact, the idea that some species needs our intervention before its members can display a useful skill, like some bird that could not fly until given a human education, is far from humble!

So human language differs dramatically from natural and artificial animal communication. What of it? Some people, recalling Darwin's insistence on the gradualness of evolutionary change, seem to believe that a detailed examination of chimps' behavior is unnecessary: they must have some form of language, as a matter of principle. Elizabeth Bates, a vociferous critic of Chomskyan approaches to language, writes:

> If the basic structural principles of language cannot be learned (bottom up) or derived (top down), there are only two possible explanations for their existence: either Universal Grammar was endowed to us directly by the Creator, or else our species has undergone a mutation of unprecedented magnitude, a cognitive equivalent of the Big Bang. . . . We have to abandon any strong version of the discontinuity claim that has characterized generative grammar for thirty years. We have to find some way to ground symbols and syntax in the mental material that we share with other species.

But, in fact, if human language is unique in the modern animal kingdom, as it appears to be, the implications for a Darwinian account of its evolution would be as follows: none. A language instinct unique to modern humans poses no more of a paradox than a trunk unique to modern elephants. No contradiction, no Creator, no big bang.

Modern evolutionary biologists are alternately amused and annoyed by a curious fact. Though most educated people profess to believe in Darwin's theory, what they really believe in is a modified version of the ancient theological notion of the Great Chain of Being: that all species are arrayed in a linear hierarchy with humans at the top. Darwin's contribution, according to this belief, was showing that each species on the ladder evolved from the species one rung down, instead of being allotted its rung by God. Dimly remembering their high school biology classes that took them on a tour of the phyla from "primitive" to "modern," people think roughly as follows: amoebas begat sponges which begat jellyfish which begat flatworms which begat trout which begat frogs which begat lizards which begat dinosaurs which begat anteaters which begat monkeys which begat chimpanzees which begat us. (I have skipped a few steps for the sake of brevity.)

**The Wrong Theory**

Amoebas
|
Sponges
|
Jellyfish
|
Flatworms
|
Trout
|
Frogs
|
Lizards
|
Dinosaurs
|
Anteaters
|
Monkeys
|
Chimpanzees
|
*Homo sapiens*

Hence the paradox: humans enjoy language while their neighbors on the adjacent rung have nothing of the kind. We expect a fade-in, but we see a big bang.

But evolution did not make a ladder; it made a bush. We did not evolve from chimpanzees. We and chimpanzees evolved from a common ancestor, now extinct. The human-chimp ancestor evolved not from monkeys but from an even older ancestor of the two, also extinct. And so on, back to our single-celled forebears. Paleontologists like to say that to a first approximation, all species are extinct (ninety-nine percent is the usual estimate). The organisms we see around us are distant cousins, not great-grandparents; they are a few scattered twig-tips of an enormous tree whose branches and trunk are no longer with us. Simplifying a lot:

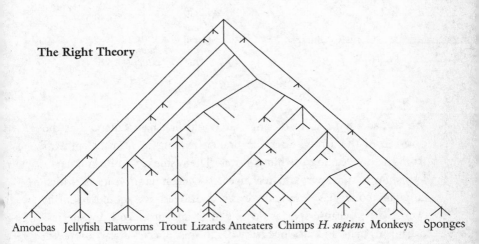

**The Right Theory**

Amoebas  Jellyfish  Flatworms  Trout  Lizards  Anteaters  Chimps  *H. sapiens*  Monkeys  Sponges

Zooming in on our branch, we see chimpanzees off on a separate sub-branch, not sitting on top of us.

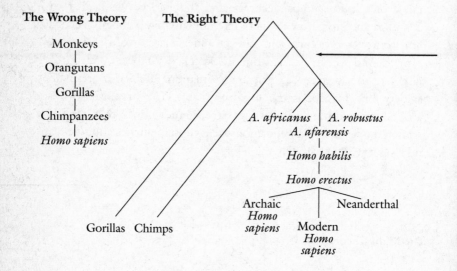

We also see that a form of language could first have emerged at the position of the arrow, after the branch leading to humans split off from the one leading to chimpanzees. The result would be language-less chimps and approximately five to seven million years in which language could have gradually evolved. Indeed, we should zoom in even closer, because species do not mate and produce baby species; organisms mate and produce baby organisms. Species are an abbreviation for chunks of a vast family tree composed of individuals, such as the *particular* gorilla, chimp, australopithecine, *erectus,* archaic *sapiens,* Neanderthal, and modern *sapiens* I have named in this family tree:

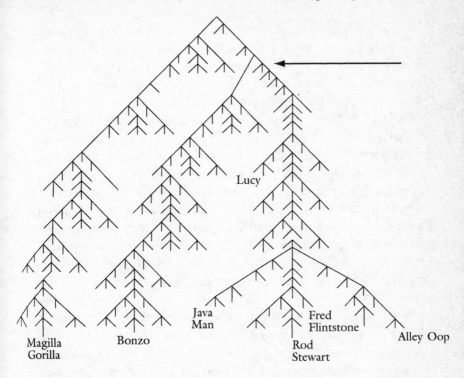

So if the first trace of a proto-language ability appeared in the ancestor at the arrow, there could have been on the order of 350,000 generations between then and now for the ability to have been elaborated and fine-tuned to the Universal Grammar we see today. For all we know, language could have had a gradual fade-in, even if no extant species, not even our closest living relatives the chimpanzees, have it. There were plenty of organisms with intermediate language abilities, but they are all dead.

Here is another way to think about it. People see chimpanzees, the living species closest to us, and are tempted to conclude that they, at the very least, must have some ability that is ancestral to language. But because the evolutionary tree is a tree of individuals, not species, "the living species closest to us" has no special status; what that species is depends on the accidents of extinction. Try the following thought experiment. Imagine that anthropologists discover a relict population of *Homo habilis* in some remote highland. *Habilis* would now be our closest living relatives. Would that take the pressure off chimps, so it is not so important that they have something like language after all? Or do it the other way around. Imagine that some epidemic wiped out all the apes several thousand years ago. Would Darwin be in danger unless we showed that monkeys had language? If you are inclined to answer yes, just push the thought experiment one branch up: imagine that in the past some extraterrestrials developed a craze for primate fur coats, and hunted and trapped all the primates to extinction except hairless us. Would insectivores like anteaters have to shoulder the proto-language burden? What if the aliens went for mammals in general? Or developed a taste for vertebrate flesh, sparing us because they like the sitcom reruns that we inadvertently broadcast into space? Would we then have to look for talking starfish? Or ground syntax in the mental material we share with sea cucumbers?

Obviously not. Our brains, and chimpanzee brains, and anteater brains, have whatever wiring they have; the wiring cannot change depending on which other species a continent away happen to survive or go extinct. The point of these thought experiments is that the gradualness that Darwin made so much about applies to lineages of individual organisms in a bushy family tree, not to entire living species in a great chain. For reasons that we will cover soon, an ancestral ape with nothing but hoots and grunts is unlikely to have given birth to a baby who could learn English or Kivunjo. But it did not have to; there was a chain of several hundred thousand generations of grandchildren in which such abilities could gradually blossom. To determine when in fact language began, we have to look at people, and look at animals, and note what we see; we cannot use the idea of phyletic continuity to legislate the answer from the armchair.

The difference between bush and ladder also allows us to put a lid on a fruitless and boring debate. That debate is over what qualifies as True Language. One side lists some qualities that human language has but that no animal has yet demonstrated: reference, use of symbols displaced in time and space from their referents, creativity, categorical speech perception, consistent ordering, hierarchical structure, infinity, recursion, and so on. The other side finds some counterexample in the animal kingdom (perhaps budgies can discriminate speech sounds, or dolphins or parrots can attend to word order when carrying out commands, or some songbird can improvise indefinitely without repeating itself) and then gloats that the citadel of human uniqueness has been breached. The Human Uniqueness team relinquishes that criterion but emphasizes others or adds new ones to the list, provoking angry objections that they are moving the goalposts. To see how silly this all is, imagine a debate over whether flatworms have True Vision or houseflies have True Hands. Is an iris critical? Eyelashes? Fingernails? Who cares? This is a debate for dictionary-writers, not scientists. Plato and Diogenes were not doing biology when Plato defined man as a "featherless biped" and Diogenes refuted him with a plucked chicken.

The fallacy in all this is that there is some line to be drawn across the ladder, the species on the rungs above it being credited with some glorious trait, those below lacking it. In the tree of life, traits like eyes or hands or infinite vocalizations can arise on any branch, or several times on different branches, some leading to humans, some not. There is an important scientific issue at stake, but it is not whether some species possesses the true version of a trait as opposed to some pale imitation or vile impostor. The issue is which traits are *homologous* to which other ones.

Biologists distinguish two kinds of similarity. "Analogous" traits are ones that have a common function but arose on different branches of the evolutionary tree and are in an important sense not "the same" organ. The wings of birds and the wings of bees are a textbook example; they are both used for flight and are similar in some ways because anything used for flight has to be built in those ways, but they arose independently in evolution and have nothing in common beyond their use in flight. "Homologous" traits, in contrast, may or may not have a common function, but they descended from a common ancestor and hence have some common structure that bespeaks their being "the same" organ. The wing of a bat, the front leg of a horse, the flipper of a seal, the claw of a mole, and the hand of a human have very different functions, but they are all modifications of the forelimb of the ancestor of all mammals, and as a result they share nonfunctional traits like the number of bones and the ways they are connected. To distinguish analogy from homology, biologists usually look at the overall architecture of the organs and focus on their most useless properties—the useful ones could have arisen independently in two lineages *because* they are useful (a nuisance to taxonomists called convergent evolution). We deduce that bat wings are really hands because we can see the wrist and count the joints in the fingers, and because that is not the only way that nature could have built a wing.

The interesting question is whether human language is homologous to—biologically "the same thing" as—anything in the modern animal kingdom. Discovering a similarity like sequential ordering is pointless, especially when it is found on a remote branch that is surely not ancestral to humans (birds, for example). Here primates are relevant, but the ape-trainers and their fans are playing by the wrong rules. Imagine that their wildest dreams are realized and some chimpanzee can be taught to produce real signs, to group and order them consistently to convey meaning, to use them spontaneously to describe events, and so on. Does that show that the human ability to learn language evolved from the chimp ability to learn the artificial sign system? Of course not, any more than a seagull's wings show that it evolved from mosquitos. Any resemblance between the chimps' symbol system and human language would not be a legacy of their common ancestor; the features of the symbol system were deliberately designed by the scientists and acquired by the chimps because it was useful to them then and there. To check for homology, one would have to find some signature trait that reliably emerges both in ape symbol systems and in human language, and that is not so indispensable to communication that it was likely to have emerged twice, once in the course of human evolution and once in the lab meetings of the psychologists as they contrived the system to teach their apes. One could look for such signatures in development, checking the apes for some echo of the standard human sequence from syllable babbling to jargon babbling to first words to two-word sequences to a grammar explosion. One could look at the developed grammar, seeing if apes invent or favor some specimen of nouns and verbs, inflections, X-bar syntax, roots and stems, auxiliaries in second position inverting to form questions, or other distinctive aspects of universal human grammar. (These structures are not so abstract as to be undetectable; they leapt out of the data when linguists first looked at American Sign Language and creoles, for example.) And one could look at neuroanatomy, checking for control by the left perisylvian regions of the cortex, with grammar more anterior, dictionary more posterior. This line of questioning, routine in biology since the nineteenth century, has never been applied to chimp signing, though one can make a good prediction of what the answers would be.

How plausible is it that the ancestor to language first appeared after the branch leading to humans split off from the branch leading to chimps? Not very, says Philip Lieberman, one of the scientists who believe that vocal tract anatomy and speech control are the only things that were modified in evolution, not a grammar module: "Since Darwinian natural selection involves small incremental steps that enhance the present function of the specialized module, the evolution of a 'new' module is logically impossible." Now, something has gone seriously awry in this argument. Humans evolved from single-celled ancestors. Single-celled ancestors had no arms, legs, heart, eyes, liver, and so on. Therefore eyes and livers are logically impossible.

The point that the argument misses is that although natural selection involves incremental steps that enhance functioning, the enhancements do not have to be an existing module. They can slowly build a module out of some previously nondescript stretch of anatomy, or out of the nooks and crannies between existing modules, which the biologists Stephen Jay Gould and Richard Lewontin call "spandrels," from the architectural term for the space between two arches. An example of a new module is the eye, which has arisen de novo some forty separate times in animal evolution. It can begin in an eyeless organism with a patch of skin whose cells are sensitive to light.

The patch can deepen into a pit, cinch up into a sphere with a hole in front, grow a translucent cover over the hole, and so on, each step allowing the owner to detect events a bit better. An example of a module growing out of bits that were not originally a module is the elephant's trunk. It is a brand-new organ, but homologies suggest that it evolved from a fusion of the nostrils and some of the upper lip muscles of the extinct elephant-hyrax common ancestor, followed by radical complications and refinements.

Language could have arisen, and probably did arise, in a similar way: by a revamping of primate brain circuits that originally had no role in vocal communication, and by the addition of some new ones. The neuroanatomists Al Galaburda and Terrence Deacon have discovered areas in monkey brains that correspond in location, input-output cabling, and cellular composition to the human language areas. For example, there are homologues to Wernicke's and Broca's areas and a band of fibers connecting the two, just as in humans. The regions are not involved in producing the monkeys' calls, nor are they involved in producing their gestures. The monkey seems to use the regions corresponding to Wernicke's area and its neighbors to recognize sound sequences and to discriminate the calls of other monkeys from its own calls. The Broca's homologues are involved in control over the muscles of the face, mouth, tongue, and larynx, and various subregions of these homologues receive inputs from the parts of the brain dedicated to hearing, the sense of touch in the mouth, tongue, and larynx, and areas in which streams of information from all the senses converge. No one knows exactly why this arrangement is found in monkeys and, presumably, their common ancestor with humans, but the arrangement would have given evolution some parts it could tinker with to produce the human language circuitry, perhaps exploiting the confluence of vocal, auditory, and other signals there.

Brand-new circuits in this general territory could have arisen, too. Neuroscientists charting the cortex with electrodes have occasionally found mutant monkeys who have one extra visual map in their brains compared to standard monkeys (visual maps are the postage-stamp-sized brain areas that are a bit like internal graphics buffers, registering the contours and motions of the visible world in a distorted picture). A sequence of genetic changes that duplicate a brain map or circuit, reroute its inputs and outputs, and frob, twiddle, and tweak its internal connections could manufacture a genuinely new brain module.

Brains can be rewired only if the genes that control their wiring have changed. This brings up another bad argument about why chimp signing must be like human language. The argument is based on the finding that chimpanzees and humans share 98% to 99% of their DNA, a factoid that has become as widely circulated as the supposed four hundred Eskimo words for snow (the comic strip *Zippy* recently quoted the figure as "99.9%"). The implication is that we must be 99% similar to chimpanzees.

But geneticists are appalled at such reasoning and take pains to stifle it in the same breath that they report their results. The recipe for the embryological soufflé is so baroque that small genetic changes can have enormous effects on the final product. And a 1% difference is not even so small. In terms of the information content in the DNA it is 10 megabytes, big enough for the Universal Grammar with lots of room left over for the rest of the instructions on how to turn a chimp into a human. Indeed, a 1% difference in total DNA does not even mean that only 1% of human and chimpanzee genes are different. It could, in theory, mean that 100% of human and chimpanzee genes are different, each by 1%. DNA is a discrete combinatorial code, so a 1% difference in the DNA for a gene can be as significant as a 100% difference, just as changing one bit in every byte, or one letter in every word, can result in a new text that is 100% different, not 10% or 20% different. The reason, for DNA, is that even a single amino-acid substitution can change the shape of a protein enough to alter its function completely; this is what happens in many fatal genetic diseases. Data on genetic similarity are useful in figuring out how to connect up a family tree (for example, whether gorillas branched off from a common ancestor of humans and chimps or humans branched off from a common ancestor of chimps and gorillas) and perhaps even to date the divergences using a "molecular clock." But they say nothing about how similar the organisms' brains and bodies are.

The ancestral brain could have been rewired only if the new circuits had some effect on perception and behavior. The first steps toward human language are a mystery. This did not stop philosophers in the nineteenth century from offering fanciful speculations, such as that speech arose as imitations of animal sounds or as oral gestures that resembled the objects they represented, and linguists subsequently gave these speculations pejorative names like the bow-wow theory and the ding-dong theory. Sign language has frequently been suggested as an intermediate, but that was before scientists discovered that sign language was every bit as complex as speech. Also, signing seems to depend on Broca's and Wernicke's areas, which are in close proximity to vocal and auditory areas on the cortex, respectively. To the extent that brain areas for abstract computation are placed near the centers that process their inputs and outputs, this would suggest that speech is more basic. If I were forced to think about intermediate steps, I might ponder the vervet monkey alarm calls studied by Cheney and Seyfarth, one of which warns of eagles, one of snakes, and one of leopards. Perhaps a set of quasi-referential calls like these came under the voluntary control of the cerebral cortex, and came to be produced in combination for complicated events; the ability to analyze combinations of calls was then applied to the parts of each call. But I admit that this idea has no more evidence in its favor than the ding-dong theory (or than Lily Tomlin's suggestion that the first human sentence was "What a hairy back!").

Also unknown is when, in the lineage beginning at the chimp-human common ancestor, proto-language first evolved, or the rate at which it developed into the modern language instinct. In the tradition of the drunk looking for his keys under the lamppost because that is where the light is best, many archaeologists have tried to infer our extinct ancestors' language abilities from their tangible remnants such as stone tools and dwellings. Complex artifacts are thought to reflect a complex mind which could benefit from complex language. Regional variation in tools is thought to suggest cultural transmission, which depends in turn on generation-to-generation communication, perhaps via language. However, I suspect that any investigation that depends on what an ancient group left behind will seriously underestimate the antiquity of language. There are many modern hunter-gatherer peoples with sophisticated language and technology, but their baskets, clothing, baby slings, boomerangs, tents, traps, bows and arrows, and poisoned spears are not made of stone and would rot into nothing quickly after their departure, obscuring their linguistic competence from future archaeologists.

Thus the first traces of language could have appeared as early as *Australopithecus afarensis* (first discovered as the famous "Lucy" fossil), at 4 million years old our most ancient fossilized ancestor. Or perhaps even earlier; there are few fossils from the time between the human-chimp split 5 to 7 million years ago and *A. afarensis*. Evidence for a lifestyle into which language could plausibly be woven gets better with later species. *Homo habilis*, which lived about 2.5 to 2 million years ago, left behind caches of stone tools that may have been home bases or local butchering stations; in either case they suggest some degree of cooperation and acquired technology. *Habilis* was also considerate enough to have left us some of their skulls, which bear faint imprints of the wrinkle patterns of their brains. Broca's area is large and prominent enough to be visible, as are the supramarginal and angular gyri (the language areas shown in the brain diagram in Chapter 10), and these areas are larger in the left hemisphere. We do not, however, know whether habilines used them for language; remember that even monkeys have a small homologue to Broca's area. *Homo erectus*, which spread from Africa across much of the old world from 1.5 million to 500,000 years ago (all the way to China and Indonesia), controlled fire and almost everywhere used the same symmetrical, well-crafted stone hand-axes. It is easy to imagine some form of language contributing to such successes, though again we cannot be sure.

Modern *Homo sapiens*, which is thought to have appeared about 200,000 years ago and to have spread out of Africa 100,000 years ago, had skulls like ours and much more elegant and complex tools, showing considerable regional variation. It is hard to believe that they lacked language, given that biologically they *were* us, and all biologically modern humans have language. This elementary fact, by the way, demolishes the date most commonly given in magazine articles and textbooks for the origin of language: 30,000 years ago, the age of the gorgeous cave art and decorated artifacts of Cro-Magnon humans in the Upper Paleolithic. The major branches of humanity diverged well before then, and all their descendants have identical language abilities; therefore the language instinct was probably in place well before the cultural fads of the Upper Paleolithic emerged in Europe. Indeed, the logic used by archaeologists (who are largely unaware of psycholinguistics) to pin language to that date is faulty. It depends on there being a single "symbolic" capacity underlying art, religion, decorated tools, and language, which we now know is false (just think of linguistic idiot savants like Denyse and Crystal from Chapter 2, or, for that matter, any normal three-year-old).

One other ingenious bit of evidence has been applied to language origins. Newborn babies, like other mammals, have a larynx that can rise up and engage the rear opening of the nasal cavity, allowing air to pass from nose to lungs avoiding the mouth and throat. Babies become human at three months when their larynx descends to a position low in their throats. This gives the tongue the space to move both up and down and back and forth, changing the shape of two resonant cavities and defining a large number of possible vowels. But it comes at a price. In *The Origin of Species* Darwin noted "the strange fact that every particle of food and drink which we swallow has to pass over the orifice of the trachea, with some risk of falling into the lungs." Until the recent invention of the Heimlich maneuver, choking on food was the sixth leading cause of accidental death in the United States, claiming six thousand victims a year. The positioning of the larynx deep in the throat, and the tongue far enough low and back to articulate a range of vowels, also compromised breathing and chewing. Presumably the communicative benefits outweighed the physiological costs.

Lieberman and his colleagues have tried to reconstruct the vocal tracts of extinct hominids by deducing where the larynx and its associated muscles could have fit into the space at the base of their fossilized skulls. They argue that all species prior to modern *Homo sapiens,* including Neanderthals, had a standard mammalian airway with its reduced space of possible vowels. Lieberman suggests that until modern *Homo sapiens,* language must have been quite rudimentary. But Neanderthals have their loyal defenders and Lieberman's claim remains controversial. In any case, e lengeege weth e smell nember ef vewels cen remeen quete expresseve, so we cannot conclude that a hominid with a restricted vowel space had little language.

So far I have talked about when and how the language instinct might have evolved, but not why. In a chapter of *The Origin of Species,* Darwin painstakingly argued that his theory of natural selection could account for the evolution of instincts as well as bodies. If language is like other instincts, presumably it evolved by natural selection, the only successful scientific explanation of complex biological traits.

Chomsky, one might think, would have everything to gain by grounding his controversial theory about a language origin in the firm foundation of evolutionary theory, and in some of his writings he has hinted at a connection. But more often he is skeptical:

> It is perfectly safe to attribute this development [of innate mental structure] to "natural selection," so long as we realize that there is no substance to this assertion, that it amounts to nothing more than a belief that there is some naturalistic explanation for these phenomena. . . . In studying the evolution of mind, we cannot guess to what extent there are physically possible alternatives to, say, transformational generative grammar, for an organism meeting certain other physical conditions characteristic of humans. Conceivably, there are none—or very few—in which case talk about evolution of the language capacity is beside the point.

> Can the problem [the evolution of language] be addressed today? In fact, little is known about these matters. Evolutionary

> theory is informative about many things, but it has little to say, as of now, about questions of this nature. The answers may well lie not so much in the theory of natural selection as in molecular biology, in the study of what kinds of physical systems can develop under the conditions of life on earth and why, ultimately because of physical principles. It surely cannot be assumed that every trait is specifically selected. In the case of such systems as language . . . it is not easy even to imagine a course of selection that might have given rise to them.

What could he possibly mean? Could there be a language organ that evolved by a process different from the one we have always been told is responsible for the other organs? Many psychologists, impatient with arguments that cannot be fit into a slogan, pounce on such statements and ridicule Chomsky as a crypto-creationist. They are wrong, though I think Chomsky is wrong too.

To understand the issues, we first must understand the logic of Darwin's theory of natural selection. Evolution and natural selection are not the same thing. Evolution, the fact that species change over time because of what Darwin called "descent with modification," was already widely accepted in Darwin's time but was attributed to many now-discredited processes such as Lamarck's inheritance of acquired characteristics and some internal urge or drive to develop in a direction of increasing complexity culminating in humans. What Darwin and Alfred Wallace discovered and emphasized was a particular cause of evolution, natural selection. Natural selection applies to any set of entities with the properties of *multiplication, variation,* and *heredity.* Multiplication means that the entities copy themselves, that the copies are also capable of copying themselves, and so on. Variation means that the copying is not perfect; errors crop up from time to time, and these errors may give an entity traits that enable it to copy itself at higher or lower rates relative to other entities. Heredity means that a variant trait produced by a copying error reappears in subsequent copies, so the trait is perpetuated in the lineage. Natural selection is the mathematically necessary outcome that any traits that foster superior replication will tend to spread through the population over many generations. As a result, the entities will come to have traits that appear to have been designed for effective replication, including traits that are means to this end, like the ability to gather energy and materials from the environment and to safeguard them from competitors. These replicating entities are what we recognize as "organisms," and the replication-enhancing traits they accumulated by this process are called "adaptations."

At this point many people feel proud of themselves for spotting what they think is a fatal flaw. "Aha! The theory is circular! All it says is that traits that lead to effective replication lead to effective replication. Natural selection is 'the survival of the fittest' and the definition of 'the fittest' is 'those who survive.' " Not!! The power of the theory of natural selection is that it connects two independent and very different ideas. The first idea is the appearance of design. By "appearance of design" I mean something that an engineer could look at and surmise that its parts are shaped and arranged so as to carry out some function. Give an optical engineer an eyeball from an unknown species, and the engineer could immediately tell that it is designed for forming an image of the surroundings: it is built like a camera, with a transparent lens, contractable diaphragm, and so on. Moreover, an image-forming device is not just any old piece of bric-a-brac but a tool that is useful for finding food and mates, escaping from enemies, and so on. Natural selection explains how this design came to be, using a *second* idea: the actuarial statistics of reproduction in the organism's ancestors. Take a good look at the two ideas:

1. A part of an organism appears to have been engineered to enhance its reproduction.
2. That organism's ancestors reproduced more effectively than their competitors.

Note that (1) and (2) are logically independent. They are about different things: engineering design, and birth and death rates. They are about different organisms: the one you're interested in, and its ancestors. You can say that an organism has good vision and that good vision should help it reproduce (1), without knowing how well that organism, or any organism, in fact reproduces (2). Since "design" merely implies an enhanced *probability* of reproduction, a particular organism with well-designed vision may, in fact, not reproduce at all. Maybe it will be struck by lightning. Conversely, it may have a myopic sibling that in fact reproduces better, if, for instance, the same lightning bolt killed a predator who had the sibling in its sights. The theory of natural selection says that (2), the ancestors' birth and death rates, is the explanation for (1), the organism's engineering design—so it is not circular in the least.

This means that Chomsky was too flip when he dismissed natural selection as having no substance, as nothing more than a belief that there is some naturalistic explanation for a trait. In fact, it is not so easy to show that a trait is a product of selection. The trait has to be hereditary. It has to enhance the probability of reproduction of the organism, relative to organisms without the trait, in an environment like the one its ancestors lived in. There has to have been a sufficiently long lineage of similar organisms in the past. And because natural selection has no foresight, each intermediate stage in the evolution of an organ must have conferred some reproductive advantage on its possessor. Darwin noted that his theory made strong predictions and could easily be falsified. All it would take is the discovery of a trait that showed signs of design but that appeared somewhere other than at the end of a linage of replicators that could have used it to help in their replication. One example would be the existence of a trait designed only for the beauty of nature, such as a beautiful but cumbersome peacock tail evolving in moles, whose potential mates are too blind to be attracted to it. Another would be a complex organ that can exist in no useful intermediate form, such as a part-wing that could not have been useful for anything until it was one hundred percent of its current size and shape. A third would be an organism that was not produced by an entity that can replicate, such as some insect that spontaneously grew out of rocks, like a crystal. A fourth would be a trait designed to benefit an organism other than the one that caused the trait to appear, such as horses evolving saddles. In the comic strip *Li'l Abner,* the cartoonist Al Capp featured selfless organisms called shmoos that laid chocolate cakes instead of eggs and that cheerfully barbecued themselves so that people could enjoy their delicious boneless meat. The discovery of a real-life shmoo would instantly refute Darwin.

Hasty dismissals aside, Chomsky raises a real issue when he brings up alternatives to natural selection. Thoughtful evolutionary theorists since Darwin have been adamant that not every beneficial trait is an adaptation to be explained by natural selection. When a flying fish leaves the water, it is extremely adaptive for it to reenter the water. But we do not need natural selection to explain this happy event; gravity will do just fine. Other traits, too, need an explanation different from selection. Sometimes a trait is not an adaptation in itself but a consequence of something else that is an adaptation. There is no advantage to our bones being white instead of green, but there is an advantage to our bones being rigid; building them out of calcium is one way to make them rigid, and calcium happens to be white. Sometimes a trait is constrained by its history, like the S-bend in our spine that we inherited when four legs became bad and two legs good. Many traits may just be impossible to grow within the constraints of a body plan and the way the genes build the body. The biologist J.B.S. Haldane once said that there are two reasons why humans do not turn into angels: moral imperfection and a body plan that cannot accommodate both arms and wings. And sometimes a trait comes about by dumb luck. If enough time passes in a small population of organisms, all kinds of coincidences will be preserved in it, a process called genetic drift. For example, in a particular generation all the stripeless organisms might be hit by lightning or die without issue; stripedness will reign thereafter, whatever its advantages or disadvantages.

Stephen Jay Gould and Richard Lewontin have accused biologists (unfairly, most believe) of ignoring these alternative forces and putting too much stock in natural selection. They ridicule such explanations as "just-so stories," an allusion to Kipling's whimsical tales of how various animals got their body parts. Gould and Lewontin's essays have been influential in the cognitive sciences, and Chomsky's skepticism that natural selection can explain human language is in the spirit of their critique.

But Gould and Lewontin's potshots do not provide a useful model of how to reason about the evolution of a complex trait. One of their goals was to undermine theories of human behavior that they envisioned as having right-wing political implications. The critiques also reflect their day-to-day professional concerns. Gould is a paleontologist, and paleontologists study organisms after they have turned into rocks. They look more at grand patterns in the history of life than at the workings of an individual's long-defunct organs. When they discover, for example, that the dinosaurs were extinguished by an asteroid slamming into the earth and blacking out the sun, small differences in reproductive advantages understandably seem beside the point. Lewontin is a geneticist, and geneticists tend to look at the raw code of the genes and their statistical variation in a population, rather than the complex organs they build. Adaptation can seem like a minor force to them, just as someone examining the 1's and 0's of a computer program in machine language without knowing what the program does might conclude that the patterns are without design. The mainstream in modern evolutionary biology is better represented by biologists like George Williams, John Maynard Smith, and Ernst Mayr, who are concerned with the design of whole living organisms. Their consensus is that natural selection has a very special place in evolution, and that the existence of alternatives does *not* mean that the explanation of a biological trait is up for grabs, depending only on the taste of the explainer.

The biologist Richard Dawkins has explained this reasoning lucidly in his book *The Blind Watchmaker*. Dawkins notes that the fundamental problem of biology is to explain "complex design." The problem was appreciated well before Darwin. The theologian William Paley wrote:

In crossing a heath, suppose I pitched my foot against a *stone,* and were asked how the stone came to be there; I might possibly answer, that, for anything I knew to the contrary, it had lain there for ever: nor would it perhaps be very easy to show the absurdity of this answer. But suppose I had found a *watch* upon the ground, and it should be inquired how the watch happened to be in that place; I should hardly think of the answer which I had before given, that for anything I knew, the watch might have always been there.

Paley noted that a watch has a delicate arrangement of tiny gears and springs that function together to indicate the time. Bits of rock do not spontaneously exude metal which forms itself into gears and springs which then hop into an arrangement that keeps time. We are forced to conclude that the watch had an artificer who designed the watch with the goal of timekeeping in mind. But an organ like an eye is even more complexly and purposefully designed than a watch. The eye has a transparent protective cornea, a focusing lens, a light-sensitive retina at the focal plane of the lens, an iris whose diameter changes with the illumination, muscles that move one eye in tandem with the other, and neural circuits that detect edges, color, motion, and depth. It is impossible to make sense of the eye without noting that it appears to have been designed for seeing—if for no other reason than that it displays an uncanny resemblance to the man-made camera. If a watch entails a watchmaker and a camera entails a cameramaker, then an eye entails an eyemaker, namely God. Biologists today do not disagree with Paley's laying out of the problem. They disagree only with his solution. Darwin is history's most important biologist because he showed how such "organs of extreme perfection and complication" could arise from the purely physical process of natural selection.

And here is the key point. Natural selection is not just a scientifically respectable alternative to divine creation. It is the *only* alternative that can explain the evolution of a complex organ like the eye. The reason that the choice is so stark—God or natural selection—is that structures that can do what the eye does are extremely low-probability arrangements of matter. By an unimaginably large margin, most objects thrown together out of generic stuff, even generic animal stuff, cannot bring an image into focus, modulate incoming light, and detect edges and depth boundaries. The animal stuff in an eye seems to have been assembled with the goal of seeing in mind—but in whose mind, if not God's? How else could the mere *goal* of seeing well *cause* something to see well? The very special power of natural selection is to remove the paradox. What causes eyes to see well now is that they descended from a long line of ancestors that saw a bit better than their rivals, which allowed them to out-reproduce those rivals. The small random improvements in seeing were retained and combined and concentrated over the eons, leading to better and better eyes. The ability of *many* ancestors to see a *bit* better in the *past* causes a *single* organism to see *extremely* well *now*.

Another way of putting it is that natural selection is the only process that can steer a lineage of organisms along the path in the astronomically vast space of possible bodies leading from a body with no eye to a body with a functioning eye. The alternatives to natural selection can, in contrast, only grope randomly. The odds that the coincidences of genetic drift would result in just the right genes coming together to build a functioning eye are infinitesimally small. Gravity alone may make a flying fish fall into the ocean, a nice big target, but gravity alone cannot make bits of a flying fish embryo fall into place to make a flying fish eye. When one organ develops, a bulge of tissue or some nook or cranny can come along for free, the way an S-bend accompanies an upright spine. But you can bet that such a cranny will not just happen to have a functioning lens and a diaphragm and a retina all perfectly arranged for seeing. It would be like the proverbial hurricane that blows through a junkyard and assembles a Boeing 747. For these reasons, Dawkins argues that natural selection is not only the correct explanation for life on earth but is bound to be the correct explanation for anything we would be willing to call "life" anywhere in the universe.

And adaptive complexity, by the way, is also the reason that the evolution of complex organs tends to be slow and gradual. It is not that large mutations and rapid change violate some law of evolution. It is only that complex engineering requires precise arrangements of delicate parts, and if the engineering is accomplished by accumulating random changes, those changes had better be small. Complex organs evolve by small steps for the same reason that a watchmaker does not use a sledgehammer and a surgeon does not use a meat cleaver.

So we now know which biological traits to credit to natural selection and which ones to other evolutionary processes. What about language? In my mind, the conclusion is inescapable. Every discussion in this book has underscored the adaptive complexity of the language instinct. It is composed of many parts: syntax, with its discrete combinatorial system building phrase structures; morphology, a second combinatorial system building words; a capacious lexicon; a revamped vocal tract; phonological rules and structures; speech perception; parsing algorithms; learning algorithms. Those parts are physically realized as intricately structured neural circuits, laid down by a cascade of precisely timed genetic events. What these circuits make possible is an extraordinary gift: the ability to dispatch an infinite number of precisely structured thoughts from head to head by modulating exhaled breath. The gift is obviously useful for reproduction—think of Williams' parable of little Hans and Fritz being ordered to stay away from the fire and not to play with the saber-tooth. Randomly jigger a neural network or mangle a vocal tract, and you will not end up with a system with these capabilities. The language instinct, like the eye, is an example of what Darwin called "that perfection of structure and co-adaptation which justly excites our admiration," and as such it bears the unmistakable stamp of nature's designer, natural selection.

If Chomsky maintains that grammar shows signs of complex design but is skeptical that natural selection manufactured it, what alternative does he have in mind? What he repeatedly mentions is physical law. Just as the flying fish is compelled to return to the water and calcium-filled bones are compelled to be white, human brains might, for all we know, be compelled to contain circuits for Universal Grammar. He writes:

> These skills [for example, learning a grammar] may well have arisen as a concomitant of structural properties of the brain that

developed for other reasons. Suppose that there was selection for bigger brains, more cortical surface, hemispheric specialization for analytic processing, or many other structural properties that can be imagined. The brain that evolved might well have all sorts of special properties that are not individually selected; there would be no miracle in this, but only the normal workings of evolution. We have no idea, at present, how physical laws apply when $10^{10}$ neurons are placed in an object the size of a basketball, under the special conditions that arose during human evolution.

We may not, just as we don't know how physical laws apply under the special conditions of hurricanes sweeping through junkyards, but the possibility that there is an undiscovered corollary of the laws of physics that causes brains of human size and shape to develop the circuitry for Universal Grammar seems unlikely for many reasons.

At the microscopic level, what set of physical laws could cause a surface molecule guiding an axon along a thicket of glial cells to cooperate with millions of other such molecules to solder together just the kinds of circuits that would compute something as useful to an intelligent social species as grammatical language? The vast majority of the astronomical number of ways of wiring together a large neural network would surely lead to something else: bat sonar, or nest-building, or go-go dancing, or, most likely of all, random neural noise.

At the level of the whole brain, the remark that there has been selection for bigger brains is, to be sure, common in writings about human evolution (especially from paleoanthropologists). Given that premise, one might naturally think that all kinds of computational abilities might come as a by-product. But if you think about it for a minute, you should quickly see that the premise has it backwards. Why would evolution ever have selected for sheer bigness of brain, that bulbous, metabolically greedy organ? A large-brained creature is sentenced to a life that combines all the disadvantages of balancing a watermelon on a broomstick, running in place in a down jacket, and, for women, passing a large kidney stone every few years. Any selection on brain size itself would surely have favored the pinhead. Selection for more powerful computational abilities (language, perception, reasoning, and so on) must have given us a big brain as a by-product, not the other way around!

But even given a big brain, language does not fall out the way that flying fish fall out of the air. We see language in dwarfs whose heads are much smaller than a basketball. We also see it in hydrocephalics whose cerebral hemispheres have been squashed into grotesque shapes, sometimes a thin layer lining the skull like the flesh of a coconut, but who are intellectually and linguistically normal. Conversely, there are Specific Language Impairment victims with brains of normal size and shape and with intact analytic processing (recall that one of Gopnik's subjects was fine with math and computers). All the evidence suggests that it is the precise wiring of the brain's microcircuitry that makes language happen, not gross size, shape, or neuron packing. The pitiless laws of physics are unlikely to have done us the favor of hooking up that circuitry so that we could communicate with one another in words.

Incidentally, to attribute the basic design of the language instinct to natural selection is not to indulge in just-so storytelling that can spuriously "explain" any trait. The neuroscientist William Calvin, in his book *The Throwing Madonna*, explains the left-brain specialization for hand control, and consequently for language, as follows. Female hominids held their baby on their left side so the baby would be calmed by their heartbeat. This forced the mothers to use their right arm for throwing stones at small prey. Therefore the race became right-handed and left-brained. Now, this really *is* a just-so story. In all human societies that hunt, it is the men who do the hunting, not the women. Moreover, as a former boy I can attest that hitting an animal with a rock is not so easy. Calvin's throwing madonna is about as likely as Roger Clemens hurling split-fingered fastballs over the plate with a squirming infant on his lap. In the second edition to his book Calvin had to explain to readers that he only meant it as a joke; he was trying to show that such stories are no less plausible than serious adaptationist explanations. But such blunt-edged satire misses the point almost as much as if it had been intended as serious. The throwing madonna is qualitatively different from genuine adaptationist explanations, for not only is it instantly falsified by empirical and engineering considerations, but it is a nonstarter for a key theoretical reason: natural selection is an explanation for the extremely improbable. If brains are lateralized at all, lateralization on the left is not extremely improbable—its chances are exactly fifty percent! We do not need a circuitous tracing of left brains to anything else, for here the alternatives to selection are perfectly satisfying. It is a good illustration of how the logic of natural selection allows us to distinguish legitimate selectionist accounts from just-so stories.

To be fair, there are genuine problems in reconstructing how the language faculty might have evolved by natural selection, though the psychologist Paul Bloom and I have argued that the problems are all resolvable. As P. B. Medawar noted, language could not have begun in the form it supposedly took in the first recorded utterance of the infant Lord Macaulay, who after having been scalded with hot tea allegedly said to his hostess, "Thank you, madam, the agony is sensibly abated." If language evolved gradually, there must have been a sequence of intermediate forms, each useful to its possessor, and this raises several questions.

First, if language involves, for its true expression, another individual, who did the first grammar mutant talk to? One answer might be: the fifty percent of the brothers and sisters and sons and daughters who shared the new gene by common inheritance. But a more general answer is that the neighbors could have partly understood what the mutant was saying even if they lacked the new-fangled circuitry, just using overall intelligence. Though we cannot parse strings like *skid crash hospital,* we can figure out what they probably mean, and English speakers can often do a reasonably good job understanding Italian newspaper stories based on similar words and background knowledge. If a grammar mutant is making important distinctions that can be decoded by others only with uncertainty and great mental effort, it could set up a pressure for them to evolve the matching system that allows those distinctions to be recovered reliably by an automatic, unconscious parsing process. As I mentioned in Chapter 8, natural selection can take skills that are acquired with effort and uncertainty and hardwire them into the brain. Selection could have ratcheted up language abilities by favoring the speakers in each generation that the hearers could best decode, and the hearers who could best decode the speakers.

A second problem is what an intermediate grammar would have looked like. Bates asks:

> What protoform can we possibly envision that could have given birth to constraints on the extraction of noun phrases from an embedded clause? What could it conceivably mean for an organism to possess half a symbol, or three quarters of a rule? . . . monadic symbols, absolute rules and modular systems must be acquired as a whole, on a yes-or-no basis—a process that cries out for a Creationist explanation.

The question is rather odd, because it assumes that Darwin literally meant that organs must evolve in successively larger fractions (half, three quarters, and so on). Bates' rhetorical question is like asking what it could conceivably mean for an organism to possess half a head or three quarters of an elbow. Darwin's real claim, of course, is that organs evolve in successively more complex forms. Grammars of intermediate *complexity* are easy to imagine; they could have symbols with a narrower range, rules that are less reliably applied, modules with fewer rules, and so on. In a recent book Derek Bickerton answers Bates even more concretely. He gives the term "protolanguage" to chimp signing, pidgins, child language in the two-word stage, and the unsuccessful partial language acquired after the critical period by Genie and other wolf-children. Bickerton suggests that *Homo erectus* spoke in protolanguage. Obviously there is still a huge gulf between these relatively crude systems and the modern adult language instinct, and here Bickerton makes the jaw-dropping additional suggestion that a single mutation in a single woman, African Eve, simultaneously wired in syntax, resized and reshaped the skull, and reworked the vocal tract. But we can extend the first half of Bickerton's argument without accepting the second half, which is reminiscent of hurricanes assembling jetliners. The languages of children, pidgin speakers, immigrants, tourists, aphasics, telegrams, and headlines show that there is a vast continuum of viable language systems varying in efficiency and expressive power, exactly what the theory of natural selection requires.

A third problem is that each step in the evolution of a language instinct, up to and including the most recent ones, must enhance fitness. David Premack writes:

> I challenge the reader to reconstruct the scenario that would confer selective fitness on recursiveness. Language evolved, it is conjectured, at a time when humans or protohumans were hunting mastodons. . . . Would it be a great advantage for one of our ancestors squatting alongside the embers, to be able to remark: "Beware of the short beast whose front hoof Bob cracked when, having forgotten his own spear back at camp, he got in a glancing blow with the dull spear he borrowed from Jack"?
>
>   Human language is an embarrassment for evolutionary theory because it is vastly more powerful than one can account for in terms of selective fitness. A semantic language with simple mapping rules, of a kind one might suppose that the chimpanzee would have, appears to confer all the advantages one normally associates with discussions of mastodon hunting or the like. For discussions of that kind, syntactic classes, structure-dependent rules, recursion and the rest, are overly powerful devices, absurdly so.

I am reminded of a Yiddish expression, "What's the matter, is the bride too beautiful?" The objection is a bit like saying that the cheetah is much faster than it has to be, or that the eagle does not need such good vision, or that the elephant's trunk is an overly powerful device, absurdly so. But it is worth taking up the challenge.

First, bear in mind that selection does not need great advantages.

Given the vastness of time, tiny advantages will do. Imagine a mouse that was subjected to a minuscule selection pressure for increased size—say, a one percent reproductive advantage for offspring that were one percent bigger. Some arithmetic shows that the mouse's descendants would evolve to the size of an elephant in a few thousand generations, an evolutionary eyeblink.

Second, if contemporary hunter-gatherers are any guide, our ancestors were not grunting cave men with little more to talk about than which mastodon to avoid. Hunter-gatherers are accomplished toolmakers and superb amateur biologists with detailed knowledge of the life cycles, ecology, and behavior of the plants and animals they depend on. Language would surely have been useful in anything resembling such a lifestyle. It is possible to imagine a superintelligent species whose isolated members cleverly negotiated their environment without communicating with one another, but what a waste! There is a fantastic payoff in trading hard-won knowledge with kin and friends, and language is obviously a major means of doing so.

And grammatical devices designed for communicating precise information about time, space, objects, and who did what to whom are not like the proverbial thermonuclear fly-swatter. Recursion in particular is extremely useful; it is not, as Premack implies, confined to phrases with tortuous syntax. Without recursion you can't say *the man's hat* or *I think he left*. Recall that all you need for recursion is an ability to embed a noun phrase inside another noun phrase or a clause within a clause, which falls out of rules as simple as "NP → det N PP" and "PP → P NP." With this ability a speaker can pick out an object to an arbitrarily fine level of precision. These abilities can make a big difference. It makes a difference whether a far-off region is reached by taking the trail that is in front of the large tree or the trail that the large tree is in front of. It makes a difference whether that region has animals that you can eat or animals that can eat you. It makes a difference whether it has fruit that is ripe or fruit that was ripe or fruit that will be ripe. It makes a difference whether you can get there if you walk for three days or whether you can get there and walk for three days.

Third, people everywhere depend on cooperative efforts for survival, forming alliances by exchanging information and commitments. This too puts complex grammar to good use. It makes a difference whether you understand me as saying that if you give me some of your fruit I will share meat that I will get, or that you should give me some fruit because I shared meat that I got, or that if you don't give me some fruit I will take back the meat that I got. And once again, recursion is far from being an absurdly powerful device. Recursion allows sentences like *He knows that she thinks that he is flirting with Mary* and other means of conveying gossip, an apparently universal human vice.

But could these exchanges really produce the rococo complexity of human grammar? Perhaps. Evolution often produces spectacular abilities when adversaries get locked into an "arms race," like the struggle between cheetahs and gazelles. Some anthropologists believe that human brain evolution was propelled more by a cognitive arms race among social competitors than by mastery of technology and the physical environment. After all, it doesn't take that much brain power to master the ins and outs of a rock or to get the better of a berry. But outwitting and second-guessing an organism of approximately equal mental abilities with non-overlapping interests, at best, and malevolent intentions, at worst, makes formidable and ever-escalating demands on cognition. And a cognitive arms race clearly could propel a linguistic one. In all cultures, social interactions are mediated by persuasion and argument. How a choice is framed plays a large role in determining which alternative people choose. Thus there could easily have been selection for any edge in the ability to frame an offer so that it appears to present maximal benefit and minimal cost to the negotiating partner, and in the ability to see through such attempts and to formulate attractive counterproposals.

Finally, anthropologists have noted that tribal chiefs are often both gifted orators and highly polygynous—a splendid prod to any imagination that cannot conceive of how linguistic skills could make a Darwinian difference. I suspect that evolving humans lived in a world in which language was woven into the intrigues of politics, economics, technology, family, sex, and friendship that played key roles in individual reproductive success. They could no more live with a Me-Tarzan-you-Jane level of grammar than we could.

The brouhaha raised by the uniqueness of language has many ironies. The spectacle of humans trying to ennoble animals by forcing them to mimic human forms of communication is one. The pains that have been taken to portray language as innate, complex, and useful but not a product of the one force in nature that can make innate complex useful things is another. Why should language be considered such a big deal? It has allowed humans to spread out over the planet and wreak large changes, but is that any more extraordinary than coral that build islands, earthworms that shape the landscape by building soil, or the photosynthesizing bacteria that first released corrosive oxygen into the atmosphere, an ecological catastrophe of its time? Why should talking humans be considered any weirder than elephants, penguins, beavers, camels, rattlesnakes, hummingbirds, electric eels, leaf-mimicking insects, giant sequoias, Venus flytraps, echolocating bats, or deep-sea fish with lanterns growing out of their heads? Some of these creatures have traits unique to their species, others do not, depending only on the accidents of which of their relatives have become extinct. Darwin emphasized the genealogical connectedness of all living things, but evolution is descent *with modification,* and natural selection has shaped the raw materials of bodies and brains to fit them into countless differentiated niches. For Darwin, such is the "grandeur in this view of life": "that whilst this planet has gone cycling on according to the fixed law of gravity, from so simple a beginning endless forms most beautiful and wonderful have been, and are being, evolved."

# 12

# The Language Mavens

*Imagine that you are watching a nature documentary. The video shows* the usual gorgeous footage of animals in their natural habitats. But the voiceover reports some troubling facts. Dolphins do not execute their swimming strokes properly. White-crowned sparrows carelessly debase their calls. Chickadees' nests are incorrectly constructed, pandas hold bamboo in the wrong paw, the song of the humpback whale contains several well-known errors, and monkeys' cries have been in a state of chaos and degeneration for hundreds of years. Your reaction would probably be, What on earth could it mean for the song of the humpback whale to contain an "error"? Isn't the song of the humpback whale whatever the humpback whale decides to sing? Who is this announcer, anyway?

But for human language, most people think that the same pronouncements not only are meaningful but are cause for alarm. Johnny can't construct a grammatical sentence. As educational standards decline and pop culture disseminates the inarticulate ravings and unintelligible patois of surfers, jocks, and valley girls, we are turning into a nation of functional illiterates: misusing *hopefully*, confusing *lie* and *lay*, treating *data* as a singular noun, letting our participles dangle.

English itself will steadily decay unless we get back to basics and start to respect our language again.

To a linguist or psycholinguist, of course, language is like the song of the humpback whale. The way to determine whether a construction is "grammatical" is to find people who speak the language and ask them. So when people are accused of speaking "ungrammatically" in their own language, or of consistently violating a "rule," there must be some different sense of "grammatical" and "rule" in the air. In fact, the pervasive belief that people do not know their own language is a nuisance in doing linguistic research. A linguist's question to an informant about some form in his or her speech (say, whether the person uses *sneaked* or *snuck*) is often lobbed back with the ingenuous counterquestion "Gee, I better not take a chance; which is correct?"

In this chapter I had better resolve this contradiction for you. Recall columnist Erma Bombeck, incredulous at the very idea of a grammar gene because her husband taught thirty-seven high school students who thought that "bummer" was a sentence. You, too, might be wondering: if language is as instinctive as spinning a web, if every three-year-old is a grammatical genius, if the design of syntax is coded in our DNA and wired into our brains, why is the English language in such a mess? Why does the average American sound like a gibbering fool every time he opens his mouth or puts pen to paper?

The contradiction begins in the fact that the words "rule," "grammatical," and "ungrammatical," have very different meanings to a scientist and to a layperson. The rules people learn (or, more likely, fail to learn) in school are called *prescriptive* rules, prescribing how one "ought" to talk. Scientists studying language propose *descriptive* rules, describing how people *do* talk. They are completely different things, and there is a good reason that scientists focus on descriptive rules.

To a scientist, the fundamental fact of human language is its sheer improbability. Most objects in the universe—lakes, rocks, trees, worms, cows, cars—cannot talk. Even in humans, the utterances in a language are an infinitesimal fraction of the noises people's mouths are capable of making. I can arrange a combination of words that explains how octopuses make love or how to remove cherry stains; rearrange the words in even the most minor way, and the result is a sentence with a different meaning or, most likely of all, word salad. How are we to account for this miracle? What would it take to build a device that could duplicate human language?

Obviously, you need to build in some kind of rules, but what kind? Prescriptive rules? Imagine trying to build a talking machine by designing it to obey rules like "Don't split infinitives" or "Never begin a sentence with *because*." It would just sit there. In fact, we already have machines that don't split infinitives; they're called screwdrivers, bathtubs, cappuccino-makers, and so on. Prescriptive rules are useless without the much more fundamental rules that create the sentences and define the infinitives and list the word *because* to begin with, the rules of Chapters 4 and 5. These rules are never mentioned in style manuals or school grammars because the authors correctly assume that anyone capable of reading the manuals must already have the rules. No one, not even a valley girl, has to be told not to say *Apples eat the boy* or *The child seems sleeping* or *Who did you meet John and?* or the vast, vast majority of the millions of trillions of mathematically possible combinations of words. So when a scientist considers all the high-tech mental machinery needed to arrange words into ordinary sentences, prescriptive rules are, at best, inconsequential little decorations. The very fact that they have to be drilled shows that they are alien to the natural workings of the language system. One can choose to obsess over prescriptive rules, but they have no more to do with human language than the criteria for judging cats at a cat show have to do with mammalian biology.

So there is no contradiction in saying that every normal person can speak grammatically (in the sense of systematically) and ungrammatically (in the sense of nonprescriptively), just as there is no contradiction in saying that a taxi obeys the laws of physics but breaks the laws of Massachusetts. But this raises a question. Someone, somewhere, must be making decisions about "correct English" for the rest of us. Who? There is no English Language Academy, and this is just as well; the purpose of the Académie Française is to amuse journalists from other countries with bitterly argued decisions that the French gaily ignore. Nor were there any Founding Fathers at some English Language Constitutional Conference at the beginning of time. The legislators of "correct English," in fact, are an informal network of copy-editors, dictionary usage panelists, style manual and handbook writers, English teachers, essayists, columnists, and pundits. Their authority, they claim, comes from their dedication to implementing standards that have served the language well in the past, especially in the prose of its finest writers, and that maximize its clarity, logic, consistency, conciseness, elegance, continuity, precision, stability, integrity, and expressive range. (Some of them go further and say that they are actually safeguarding the ability to *think* clearly and logically. This radical Whorfianism is common among language pundits, not surprisingly; who would settle for being a schoolmarm when one can be an upholder of rationality itself?) William Safire, who writes the weekly column "On Language" for *The New York Time Magazine,* calls himself a "language maven," from the Yiddish word meaning expert, and this gives us a convenient label for the entire group.

To whom I say: Maven, shmaven! *Kibbitzers* and *nudniks* is more like it. For here are the remarkable facts. Most of the prescriptive rules of the language mavens make no sense on any level. They are bits of folklore that originated for screwball reasons several hundred years ago and have perpetuated themselves ever since. For as long as they have existed, speakers have flouted them, spawning identical plaints about the imminent decline of the language century after century. All the best writers in English at all periods, incuding Shakespeare and most of the mavens themselves, have been among the flagrant flout-ers. The rules conform neither to logic nor to tradition, and if they were ever followed they would force writers into fuzzy, clumsy, wordy, ambiguous, incomprehensible prose, in which certain thoughts are not expressible at all. Indeed, most of the "ignorant errors" these rules are supposed to correct display an elegant logic and an acute sensitivity to the grammatical texture of the language, to which the mavens are oblivious.

The scandal of the language mavens began in the eighteenth century. London had become the political and financial center of England, and England had become the center of a powerful empire. The London dialect was suddenly an important world language. Scholars began to criticize it as they would any artistic or civil institution, in part to question the customs, hence authority, of court and aristocracy. Latin was still considered the language of enlightenment and learning (not to mention the language of a comparably vast empire), and it was offered as an ideal of precision and logic to which English should aspire. The period also saw unprecedented social mobility, and anyone who desired education and self-improvement and who wanted to distinguish himself as cultivated had to master the best version of English. These trends created a demand for handbooks and style manuals, which were soon shaped by market forces. Casting English grammar into the mold of Latin grammar made the books useful as a way of helping young students learn Latin. And as the competition became cutthroat, the manuals tried to outdo one another by including greater numbers of increasingly fastidious rules that no refined person could afford to ignore. Most of the hobgoblins of a contemporary prescriptive grammar (don't split infinitives, don't end a sentence with a preposition) can be traced back to these eighteenth-century fads.

Of course, forcing modern speakers of English to not—whoops, not to split an infinitive because it isn't done in Latin makes about as much sense as forcing modern residents of England to wear laurels and togas. Julius Caesar could not have split an infinitive if he had wanted to. In Latin the infinitive is a single word like *facere* or *dicere,* a syntactic atom. English is a different kind of language. It is an "isolating" language, building sentences around many simple words instead of a few complicated ones. The infinitive is composed of two words—a complementizer, *to,* and a verb, like *go.* Words, by definition, are rearrangeable units, and there is no conceivable reason why an adverb should not come between them:

Space—the final frontier . . . These are the voyages of the starship *Enterprise.* Its five-year mission: to explore strange new worlds, to seek out new life and new civilizations, to boldly go where no man has gone before.

To *go boldly* where no man has gone before? Beam me up, Scotty; there's no intelligent life down here. As for outlawing sentences that end with a preposition (impossible in Latin for good reasons having to do with its case-marking system, reasons that are irrelevant in case-poor English)—as Winston Churchill would have said, it is a rule up with which we should not put.

But once introduced, a prescriptive rule is very hard to eradicate, no matter how ridiculous. Inside the educational and writing establishments, the rules survive by the same dynamic that perpetuates ritual genital mutilations and college fraternity hazing: I had to go through it and am none the worse, so why should you have it any easier? Anyone daring to overturn a rule by example must always worry that readers will think he or she is ignorant of the rule, rather than challenging it. (I confess that this has deterred me from splitting some splitworthy infinitives.) Perhaps most importantly, since prescriptive rules are so psychologically unnatural that only those with access to the right schooling can abide by them, they serve as shibboleths, differentiating the elite from the rabble.

The concept of shibboleth (Hebrew for "torrent") comes from the Bible:

And the Gileadites took the passages of Jordan before the Ephraimites: and it was so, that when those Ephraimites which were escaped said, Let me go over; that the men of Gilead said unto him, Art thou an Ephraimite? If he said, Nay; Then said they unto him, Say now Shibboleth: and he said Sibboleth: for he could not frame to pronounce it right. Then they took him, and slew him at the passages of the Jordan: and there fell at that time of the Ephraimites forty and two thousand. (Judges 12:5–6)

This is the kind of terror that has driven the prescriptive grammar market in the United States during the past century. Throughout the country people have spoken a dialect of English, some of whose features date to the early modern English period, that H. L. Mencken called The American Language. It had the misfortune of not becoming the standard of government and education, and large parts of the "grammar" curriculum in American schools have been dedicated to stigmatizing it as ungrammatical, sloppy speech. Familiar examples are *aks a question, workin', ain't, I don't see no birds, he don't, them boys, we was,* and past-tense forms like *drug, seen, clumb, drownded,* and *growed.* For ambitious adults who had been unable to complete school, there were full-page magazine ads for correspondence courses, containing lists of examples under screaming headlines like "DO YOU MAKE ANY OF THESE EMBARRASSING MISTAKES?"

Frequently the language mavens claim that nonstandard American English is not just different but less sophisticated and logical. The case, they would have to admit, is hard to make for nonstandard irregular verbs like *drag–drug* (and even more so for regularizations like *feeled* and *growed*). After all, in "correct" English, Richard Lederer notes, "Today we speak, but first we spoke; some faucets leak, but never loke. Today we write, but first we wrote; we bite our tongues, but never bote." At first glance, the mavens would seem to have a better argument when it comes to the leveling of inflectional distinctions in *He don't* and *We was.* But then, this has been the trend in Standard English for centuries. No one gets upset that we no longer distinguish tne second person singular form of verbs, like *sayest.* And by this criterion it is the nonstandard dialects that are superior, because they provide their speakers with second person plural pronouns like *y'all* and *youse,* and Standard English does not.

At this point, defenders of the standard are likely to pull out the notorious double negative, as in *I can't get no satisfaction.* Logically speaking, the two negatives cancel each other out, they teach; Mr. Jagger is actually saying that he is satisfied. The song should be entitled "I Can't Get *Any* Satisfaction." But this reasoning is not satisfactory. Hundreds of languages require their speakers to use a negative element somewhere within the "scope," as linguists call it, of a negated verb. The so-called double negative, far from being a corruption, was the norm in Chaucer's Middle English, and negation in standard French—as in *Je ne sais pas,* where *ne* and *pas* are both negative—is a familiar contemporary example. Come to think of it, Standard English is really no different. What do *any, even* and *at all* mean in the following sentences?

> I didn't buy any lottery tickets.
> I didn't eat even a single French fry.
> I didn't eat fried food at all today.

Clearly, not much: you can't use them alone, as the following strange sentences show:

> I bought any lottery tickets.
> I ate even a single French fry.
> I ate fried food at all today.

What these words are doing is exactly what *no* is doing in nonstandard American English, such as in the equivalent *I didn't buy no lottery tickets*—agreeing with the negated verb. The slim difference is that nonstandard English co-opted the word *no* as the agreement element, whereas Standard English co-opted the word *any;* aside from that, they are pretty much translations. And one more point has to be made. In the grammar of standard English, a double negative does *not* assert the corresponding affirmative. No one would dream of saying *I can't get no satisfaction* out of the blue to boast that he easily attains contentment. There are circumstances in which one might use the construction to deny a preceding negation in the discourse, but denying a negation is not the same as asserting an affirmative, and even then one could probably only use it by putting heavy stress on the negative element, as in the following contrived example:

> As hard as I try not to be smug about the misfortunes of my adversaries, I must admit that I can't get *no* satisfaction out of his tenure denial.

So the implication that use of the nonstandard form would lead to confusion is pure pedantry.

A tin ear for prosody (stress and intonation) and an obliviousness to the principles of discourse and rhetoric are important tools of the trade for the language maven. Consider an alleged atrocity committed by today's youth: the expression *I could care less.* The teenagers are trying to express disdain, the adults note, in which case they should be saying *I couldn't care less.* If they could care less than they do, that means that they really do care, the opposite of what they are trying to say. But if these dudes would stop ragging on teenagers and scope out the construction, they would see that their argument is bogus. Listen to how the two versions are pronounced:

```
COULDN'T care                    I

              LE               CARE
i             ESS.                   LE
                        could            ESS.
```

The melodies and stresses are completely different, and for a good reason. The second version is not illogical, it's *sarcastic.* The point of sarcasm is that by making an assertion that is manifestly false or accompanied by ostentatiously mannered intonation, one deliberately implies its opposite. A good paraphrase is, "Oh yeah, as if there was something in the world that I care less about."

Sometimes an alleged grammatical "error" is logical not only in the sense of "rational" but in the sense of respecting distinctions made by the formal logician. Consider this alleged barbarism, brought up by nearly every language maven:

Everyone returned to their seats.
Anyone who thinks a Yonex racquet has improved their game,
    raise your hand.
If anyone calls, tell them I can't come to the phone.
Someone dropped by but they didn't say what they wanted.
No one should have to sell their home to pay for medical care.
He's one of those guys who's always patting themself on the

back. [an actual quote from Holden Caulfield in J. D. Salinger's *Catcher in the Rye*]

They explain: *everyone* means *every one,* a singular subject, which may not serve as the antecedent of a plural pronoun like *them* later in the sentence. "Everyone returned to *his* seat," they insist. "If anyone calls, tell *him* I can't come to the phone."

If you were the target of these lessons, at this point you might be getting a bit uncomfortable. *Everyone returned to his seat* makes it sound like Bruce Springsteen was discovered during intermission to be in the audience, and everyone rushed back and converged on his seat to await an autograph. If there is a good chance that a caller may be female, it is odd to ask one's roommate to tell *him* anything (even if you are not among the people who are concerned about "sexist language"). Such feelings of disquiet—a red flag to any serious linguist—are well founded in this case. The next time you get corrected for this sin, ask Mr. Smartypants how you should fix the following:

Mary saw everyone before John noticed them.

Now watch him squirm as he mulls over the downright unintelligible "improvement," *Mary saw everyone before John noticed him.*

The logical point that you, Holden Caulfield, and everyone but the language mavens intuitively grasp is that *everyone* and *they* are not an "antecedent" and a "pronoun" referring to the same person in the world, which would force them to agree in number. They are a "quantifier" and a "bound variable," a different logical relationship. *Everyone returned to their seats* means "For all X, X returned to X's seat." The "X" does not refer to any particular person or group of people; it is simply a placeholder that keeps track of the roles that players play across different relationships. In this case, the X that comes back to a seat is the same X that owns the seat that X comes back to. The *their* there does not, in fact, have plural number, because it refers neither to one thing nor to many things; it does not refer at all. The same goes for the hypothetical caller: there may be one, there may be none, or the phone might ring off the hook with would-be suitors; all that matters is that every time there is a caller, if there is a caller, that caller, and not someone else, should be put off.

On logical grounds, then, variables are not the same thing as the more familiar "referential" pronouns that trigger number agreement (*he* meaning some particular guy, *they* meaning some particular bunch of guys). Some languages are considerate and offer their speakers different words for referential pronouns and for variables. But English is stingy; a referential pronoun must be drafted into service to lend its name when a speaker needs to use a variable. Since these are not real referential pronouns but only homonyms of them, there is no reason that the vernacular decision to borrow *they, their, them* for the task is any worse than the prescriptivists' recommendation of *he, him, his.* Indeed, *they* has the advantage of embracing both sexes and feeling right in a wider variety of sentences.

Through the ages, language mavens have deplored the way English speakers convert nouns into verbs. The following verbs have all been denounced in this century:

| | | |
|---|---|---|
| to caveat | to input | to host |
| to nuance | to access | to chair |
| to dialogue | to showcase | to progress |
| to parent | to intrigue | to contact |
| | to impact | |

As you can see, they range from varying degrees of awkwardness to the completely unexceptionable. In fact, easy conversion of nouns to verbs has been part of English grammar for centuries; it is one of the processes that make English English. I have estimated that about a fifth of all English verbs were originally nouns. Considering just the human body, you can *head a committee, scalp the missionary, eye a babe, nose around the office, mouth the lyrics, gum the biscuit, begin teething, tongue each note on the flute, jaw at the referee, neck in the back seat, back a candidate, arm the militia, shoulder the burden, elbow your way in, hand him a toy, finger the culprit, knuckle under, thumb a ride, wrist it into the net, belly up to the bar, stomach someone's com-*

*plaints, rib your drinking buddies, knee the goalie, leg it across town, heel on command, foot the bill, toe the line,* and several others that I cannot print in a family language book.

What's the problem? The concern seems to be that fuzzy-minded speakers are slowly eroding the distinction between nouns and verbs. But once again, the person in the street is not getting any respect. Remember a phenomenon we encountered in Chapter 5: the past tense of the baseball term *to fly out* is *flied,* not *flew*; similarly, we say *ringed the city,* not *rang,* and *grandstanded,* not *grandstood.* These are verbs that came from nouns (*a pop fly, a ring around the city, a grandstand*). Speakers are tacitly sensitive to this derivation. The reason they avoid irregular forms like *flew out* is that their mental dictionary entry for the baseball verb *to fly* is different from their mental dictionary entry for the ordinary verb *to fly* (what birds do). One is represented as a verb based on a noun root; the other, as a verb with a verb root. Only the verb root is allowed to have the irregular past-tense form *flew,* because only for verb roots does it make sense to have *any* past-tense form. The phenomenon shows that when people use a noun as a verb, they are making their mental dictionaries more sophisticated, not less so—it's not that words are losing their identities as verbs versus nouns; rather, there are verbs, there are nouns, and there are verbs based on nouns, and people store each one with a different mental tag.

The most remarkable aspect of the special status of verbs-from-nouns is that everyone unconsciously respects it. Remember from Chapter 5 that if you make up a new verb based on a noun, like someone's name, it is always regular, even if the new verb sounds the same as an existing verb that is irregular. (For example, Mae Jemison, the beautiful black female astronaut, *out-Sally-Rided Sally Ride,* not *out-Sally-Rode Sally Ride.*) My research team has tried this test, using about twenty-five new verbs made out of nouns, on hundreds of people—college students, respondents to an ad we placed in a tabloid newspaper asking for volunteers without college education, school-age children, even four-year-olds. They all behave like good intuitive grammarians: they inflect verbs that come from nouns differently from plain old verbs.

So is there anyone, anywhere, who does not grasp the principle? Yes—the language mavens. Look up *broadcasted* in Theodore Bernstein's *The Careful Writer,* and here is what you will find:

> If you think you have correctly forecasted the immediate future of English and have casted your lot with the permissivists, you may be receptive to *broadcasted,* at least in radio usage, as are some dictionaries. The rest of us, however, will decide that no matter how desirable it may be to convert all irregular verbs into regular ones, this cannot be done by ukase, nor can it be accomplished overnight. We shall continue to use *broadcast* as the past tense and participle, feeling that there is no reason for *broadcasted* other than one of analogy or consistency or logic, which the permissivists themselves so often scorn. Nor is this position inconsistent with our position on *flied,* the baseball term, which has a real reason for being. The fact—the inescapable fact—is that there are some irregular verbs.

Bernstein's "real reason" for *flied* is that it has a specialized meaning in baseball, but that is the wrong reason; *see a bet, cut a deal,* and *take the count* all have specialized meanings, but they get to keep their irregular pasts *saw, cut,* and *took,* rather than switching to *seed, cutted, taked.* No, the real reason is that *to fly out* means *to hit a fly,* and *a fly* is a noun. And the reason that people say *broadcasted* is the same: not that they want to convert all irregular verbs into regular ones overnight, but that they mentally analyze the verb *to broadcast* as "to make a broadcast," that is, as coming from the much more common noun *a broadcast.* (The original meaning of the verb, "to disperse seeds," is now obscure except among gardeners.) As a verb based on a noun, *to broadcast* is not eligible to have its own idiosyncratic past-tense form, so nonmavens sensibly apply the "add *-ed*" rule.

I am obliged to discuss one more example: the much-vilified *hopefully.* A sentence like *Hopefully, the treaty will pass* is said to be a grave error. The adverb *hopefully* comes from the adjective *hopeful,* meaning "in a manner of hope." Therefore, the mavens say, it should be used only when the sentence refers to a person who is doing something in a hopeful manner. If it is the writer or reader who is hopeful, one should say *It is hoped that the treaty will pass,* or *If hopes are realized, the treaty will pass,* or *I hope that the treaty will pass.*

Now consider the following:

1. It is simply not true that an English adverb must indicate the manner in which the actor performs the action. Adverbs come in two kinds: "verb phrase" adverbs like *carefully,* which do refer to the actor, and "sentence" adverbs like *frankly,* which indicate the attitude of the speaker toward the content of the sentence. Other examples of sentence adverbs include:

| | | |
|---|---|---|
| accordingly | curiously | oddly |
| admittedly | generally | parenthetically |
| alarmingly | happily | predictably |
| amazingly | honestly | roughly |
| basically | ideally | seriously |
| bluntly | incidentally | strikingly |
| candidly | intriguingly | supposedly |
| confidentially | mercifully | understandably |

Note that many of these fine sentence adverbs, like *happily, honestly,* and *mercifully,* come from verb phrase adverbs, and they are virtually never ambiguous in context. The use of *hopefully* as a sentence adverb, which has been around in writing at least since the 1930s (according to the *Oxford English Dictionary*) and in speech well before then, is a perfectly sensible application of this derivational process.

2. The suggested alternatives *It is hoped that* and *If hopes are realized* display four famous sins of bad writing: passive voice, needless words, vagueness, pomposity.

3. The suggested alternatives do not mean the same thing as *hopefully,* so the ban would leave certain thoughts unexpressible. *Hopefully* makes a hopeful prediction, whereas *I hope that* and *It is hoped that* merely describe certain people's mental states. Thus you

can say *I hope that the treaty will pass, but it isn't likely*, but it would be odd to say *Hopefully, the treaty will pass, but it isn't likely*.

4. We are supposed to use *hopefully* only as a verb phrase adverb, as in the following:

> Hopefully, Larry hurled the ball toward the basket with one second left in the game.
> Hopefully, Melvin turned the record over and sat back down on the couch eleven centimeters closer to Ellen.

Call me uncouth, call me ignorant, but these sentences do not belong to any language that I speak.

Imagine that one day someone announced that everyone has been making a grievous error. The correct name for the city in Ohio that people call Cleveland is really Cincinnati, and the correct name for the city that people call Cincinnati is really Cleveland. The expert gives no reasons, but insists that that is what is correct, and that anyone who cares about the language must immediately change the way that he (yes, *he*, not *they*) refers to the cities, regardless of the confusion and expense. You would surely think that this person is insane. But when a columnist or editor makes a similar pronouncement about *hopefully*, he is called an upholder of literacy and high standards.

I have debunked nine myths of the generic language maven, and now I would like to examine the mavens themselves. People who set themselves up as language experts differ in their goals, expertise, and common sense, and it is only fair to discuss them as individuals.

The most common kind of maven is the wordwatcher (a term invented by the biologist and wordwatcher Lewis Thomas). Unlike linguists, wordwatchers train their binoculars on the especially capricious, eccentric, and poorly documented words and idioms that get sighted from time to time. Sometimes a wordwatcher is a scholar in some other field, like Thomas or Quine, who indulges a lifelong hobby by writing a charming book on word origins. Sometimes it is a journalist assigned to the Question & Answer column of a newspaper. Here is a recent example from *Ask the Globe*:

Q. When we want to irritate someone, why do we say we want "to get his goat"? *J.E., Boston*

A. Slang experts aren't entirely sure, but some claim the expression comes from an old race track tradition of putting a goat in the same stall as a high-strung racing thoroughbred to keep the horse calm. Nineteenth century gamblers sometimes stole the goat to unnerve the horse and throw the race. Hence, the expression "get your goat."

This kind of explanation is satirized in Woody Allen's "Slang Origins":

How many of you have wondered where certain slang expressions come from? Like "She's the cat's pajamas," or to "take it on the lam." Neither have I. And yet for those who are interested in this sort of thing I have provided a brief guide to a few of the more interesting origins.

. . . "Take it on the lam" is English in origin. Years ago, in England, "lamming" was a game played with dice and a large tube of ointment. Each player in turn threw dice and then skipped around the room until he hemorrhaged. If a person threw seven or under he would say the word "quintz" and proceed to turn in a frenzy. If he threw over seven, he was forced to give every player a portion of his feathers and was given a good "lamming." Three "lammings" and a player was "kwirled" or declared a moral bankrupt. Gradually any game with feathers was called "lamming" and feathers became "lams." To "take it on the lam" meant to put on feathers and later, to escape, although the transition is unclear.

This passage captures my reaction to the wordwatchers. I don't think they do any harm, but (a) I never completely believe their explanations, and (b) in most cases I don't really care. Years ago a columnist recounted the origin of the word *pumpernickel*. During one of his campaigns in central Europe Napoleon stopped at an inn and was served a loaf of coarse, dark, sour bread. Accustomed to the delicate white baguettes of Paris, he sneered, "C'est pain pour Nicole," Nicole being his horse. When the columnist was challenged (the dictionaries say the word comes from colloquial German, meaning "farting goblin"), he confessed that he and some buddies had made up the story in a bar the night before. For me, wordwatching for its own sake has all the intellectual excitement of stamp collecting, with the added twist that an undetermined number of your stamps are counterfeit.

At the opposite end of the temperamental spectrum one finds the Jeremiahs, expressing their bitter laments and righteous prophecies of doom. An eminent dictionary editor, language columnist, and usage expert once wrote, quoting a poet:

> As a poet, there is only one political duty and that is to defend
> one's language from corruption. And that is particularly serious
> now. It is being corrupted. When it is corrupted, people lose
> faith in what they hear, and that leads to violence.

The linguist Dwight Bolinger, gently urging this man to get a grip, had to point out that "the same number of muggers would leap out of the dark if everyone conformed overnight to every prescriptive rule ever written."

In recent years the loudest Jeremiah has been the critic John Simon, whose venomous film and theater reviews are distinguished by their lengthy denunciations of actresses' faces. Here is a representative opening to one of his language columns:

> The English language is being treated nowadays exactly as slave
> traders once handled the merchandise in their slave ships, or as
> the inmates of concentration camps were dealt with by their
> Nazi jailers.

The grammatical error that inspired his tasteless comparison, incidentally, was Tip O'Neill's redundantly referring to his "fellow colleagues," which Simon refers to as "the rock bottom of linguistic ineptitude." Speaking of Black English Vernacular, Simon writes:

> Why should we consider some, usually poorly educated, sub-
> culture's notion of the relationship between sound and mean-

ing? And how could a grammar—any grammar—possibly describe that relationship?

As for "I be," "you be," "he be," etc., which should give us all the heebie-jeebies, these may indeed be comprehensible, but they go against all accepted classical and modern grammars and are the product not of a language with roots in history but of ignorance of how language works.

There is no point in refuting this malicious know-nothing, for he is not participating in any sincere discussion. Simon has simply discovered the trick used with great effectiveness by certain comedians, talk-show hosts, and punk-rock musicians: people of modest talent can attract the attention of the media, at least for a while, by being unrelentingly offensive.

The third kind of language maven is the entertainer, who shows off his collection of palindromes, puns, anagrams, rebuses, malapropisms, Goldwynisms, eponyms, sesquipedalia, howlers, and bloopers. Entertainers like Willard Espy, Dimitri Borgman, Gyles Brandreth, and Richard Lederer write books with titles like *Words at Play, Language on Vacation, The Joy of Lex,* and *Anguished English.* These rollicking exhibitions of linguistic zaniness are all in good fun, but when reading them I occasionally feel like Jacques Cousteau at a dolphin show, longing that these magnificent creatures be allowed to shake off their hula skirts and display their far more interesting natural talents in a dignified setting. Here is a typical example from Lederer:

> When we take the time to explore the paradoxes and vagaries of English, we find that hot dogs can be cold, darkrooms can be lit, homework can be done in school, nightmares can take place in broad daylight while morning sickness and daydreaming can take place at night. . . .
>
> Sometimes you have to believe that all English speakers should be committed to an asylum for the verbally insane. In what other language do people drive in a parkway and park in a driveway? In what other language do people recite a play and play at a recital? . . . How can a slim chance and a fat chance be

the same, while a *wise man* and a *wise guy* are opposites? . . . **Doughnut holes:** Aren't these little treats *doughnut balls*? The holes are what's left in the original doughnut. . . . **They're head over heels in love.** That's nice, but all of us do almost everything *head over heels*. If we are trying to create an image of people doing cartwheels and somersaults, why don't we say, *They're heels over head in love*?

Objection! (1) Everyone senses the difference between a compound, which can have a conventional meaning of its own, like any other word, and a phrase, whose meaning is determined by the meanings of its parts and the rules that put them together. A compound is pronounced with one stress pattern *(dárkroom)* and a phrase is pronounced with another *(dark róom)*. The supposedly "crazy" expressions, like *hot dog* and *morning sickness*, are obviously compounds, not phrases, so cold hot dogs and nighttime morning sickness do not violate grammatical logic in the least. (2) Isn't it obvious that *fat chance* and *wise guy* are sarcastic? (3) *Donut holes*, the trade name of a product of Dunkin' Donuts, is intentially whimsical—did someone not get the joke? (4) The preposition *over* has several meanings, including a static arrangement, as in *Bridge over troubled water*, and the path of a moving object, as in *The quick brown fox jumped over the lazy dog*. *Head over heels* involves the second meaning, describing the motion, not the position, of the inamorato's head.

I must also say something in defense of the college students, welfare applicants, and Joe Sixpacks whose language is so often held up to ridicule by the entertainers. Cartoonists and dialogue writers know that you can make anyone look like a bumpkin by rendering his speech quasi-phonetically instead of with conventional spelling ("sez," "cum," "wimmin," "hafta," "crooshul," and so on). Lederer occasionally resorts to this cheap trick in "Howta Reckanize American Slurivan," which deplores unremarkable examples of English phonological processes like "coulda" and "could of" *(could have)*, "forced" *(forest)*, "granite" *(granted)*, "neck store" *(next door)*, and "then" *(than)*. As we saw in Chapter 6, everyone but a science fiction robot slurs their speech (yes, *their* speech, dammit) in systematic ways.

Lederer also reproduces lists of "howlers" from student term papers, automobile insurance claim forms, and welfare applications, familiar to many people as faded mimeos tacked on the bulletin boards of university and government offices:

> In accordance with your instructions I have given birth to twins in the enclosed envelope.
>
> My husband got his project cut off two weeks ago and I haven't had any relief since.
>
> An invisible car came out of nowhere, struck my car, and vanished.
>
> The pedestrian had no idea which direction to go, so I ran over him.
>
> Artificial insemination is when the farmer does it to the cow instead of the bull.
>
> The girl tumbled down the stairs and lay prostitute on the bottom.
>
> Moses went up on Mount Cyanide to get the ten commandments. He died before he ever reached Canada.

These lists are good for a few laughs, but there is something you should know before you conclude that the teeming masses are comically inept at writing. Most of the howlers are probably fabrications.

The folklorist Jan Brunvand has documented hundreds of "urban legends," intriguing stories that everyone swears happened to a friend of a friend ("FOAF" is the technical term), and that circulate for years in nearly identical form in city after city, but that can never be documented as real events. The Hippie Baby Sitter, Alligators in the Sewers, the Kentucky Fried Rat, and Halloween Sadists (the ones who put razor blades in apples) are some of the more famous tales. The howlers, it turns out, are examples of a subgenre called xeroxlore. The employee who posts one of these lists admits that he did not compile the items himself but took them from a list someone gave him, which were taken from another list, which excerpted letters that someone in some office somewhere *really did receive*. Nearly identical lists have been circulating since World War I, and have been independently credited to offices in New England, Alabama, Salt Lake City, and so on. As Brunvand notes, the chances seem slim that the same amusing double entrendres are made in so many separate locations over so many years. The advent of electronic mail has quickened the creation and dissemination of these lists, and I receive one every now and again. But I smell intentional facetiousness (whether it is from the student or the professor is not clear), not accidentally hilarious incompetence, in howlers like "adamant: pertaining to original sin" and "gubernatorial: having to do with peanuts."

The final kind of maven is the sage, typified by the late Theodore Bernstein, a *New York Times* editor and the author of the delightful handbook *The Careful Writer,* and William Safire. They are known for taking a moderate, common-sense approach to matters of usage, and they tease their victims with wit rather than savaging them with invective. I enjoy reading the sages, and have nothing but awe for a pen like Safire's that can summarize the content of an anti-pornography statute as "It isn't the teat, it's the tumidity." But the sad fact is that even a sage like Safire, the closest thing we have to an enlightened language pundit, misjudges the linguistic sophistication of the common speaker and as a result misses the target in many of his commentaries. To prove this charge, I will walk you through a single column of his, from *The New York Times Magazine* of October 4, 1992.

The column had three stories, discussing six examples of questionable usage. The first story was a nonpartisan analysis of supposed pronoun case errors made by the two candidates in the 1992 U.S. presidential election. George Bush had recently adopted the slogan "Who do you trust?," alienating schoolteachers across the nation who noted that *who* is a "subject pronoun" (nominative or subjective case) and the question is asking about the object of *trust* (accusative or objective case). One would say *You do trust him,* not *You do trust he,* and so the question word should be *whom,* not *who.*

This, of course, is one of the standard prescriptivist complaints about common speech. In reply, one might point out that the *who/ whom* distinction is a relic of the English case system, abandoned by nouns centuries ago and found today only among pronouns in distinctions like *he/him*. Even among pronouns, the old distinction between subject *ye* and object *you* has vanished, leaving *you* to play both roles and *ye* as sounding completely archaic. *Whom* has outlived *ye* but is clearly moribund; it now sounds pretentious in most spoken contexts. No one demands of Bush that he say *Whom do ye trust?* If the langage can bear the loss of *ye,* using *you* for both subjects and objects, why insist on clinging to *whom,* when everyone uses *who* for both subjects and objects?

Safire, with his enlightened attitude toward usage, recognizes the problem and proposes

> Safire's Law of Who/Whom, which forever solves the problem troubling writers and speakers caught between the pedantic and the incorrect: "When *whom* is correct, recast the sentence." Thus, instead of changing his slogan to "Whom do you trust?"—making him sound like a hypereducated Yalie stiff—Mr. Bush would win back the purist vote with "Which candidate do you trust?"

But Safire's recommendation is Solomonic in the sense of being an unacceptable pseudo-compromise. Telling people to avoid a problematic construction sounds like common sense, but in the case of object questions with *who,* it demands an intolerable sacrifice. People ask questions about the objects of verbs and prepositions *a lot.* Here are just a few examples I culled from transcripts of conversations between parents and their children:

> I know, but who did we see at the other store?
> Who did we see on the way home?
> Who did you play with outside tonight?
> Abe, who did you play with today at school?
> Who did you sound like?

(Imagine replacing any of these with *whom!*) Safire's advice is to change such questions to *Which person* or *Which child.* But the advice would have people violate the most important maxim of good prose:

Omit needless words. It also would force them to overuse the word *which,* described by one stylist as "the ugliest word in the English language." Finally, it subverts the supposed goal of rules of usage, which is to allow people to express their thoughts as clearly and precisely as possible. A question like *Who did we see on the way home?* can embrace one person, many people, or any combination or number of adults, babies, children, and familiar dogs. Any specific substitution like *Which person?* forecloses some of these possibilities, contrary to the question-asker's intent. And how in the world would you apply Safire's Law to the famous refrain

Who're you gonna call? GHOSTBUSTERS!

Extremism in defense of liberty is no vice. Safire should have taken his observation about the pedantic sound of *whom* to its logical conclusion and advised the president that there is no reason to change the slogan, at least no grammatical reason.

Turning to the Democrats, Safire gets on Bill Clinton's case, as he puts it, for asking voters to "give Al Gore and I a chance to bring America back." No one would say *give I a break,* because the indirect object of *give* must have accusative case. So it should be *give Al Gore and me a chance.*

Probably no "grammatical error" has received as much scorn as "misuse" of pronoun case inside conjunctions (phrases containing two elements joined by *and* or *or*). What teenager has not been corrected for saying *Me and Jennifer are going to the mall?* A colleague of mine recalls that when she was twelve, her mother would not allow her to have her ears pierced until she stopped saying it. The standard story is that the accusative pronoun *me* does not belong in subject position—no one would say *Me is going to the mall*—so it should be *Jennifer and I.* People tend to misremember the advice as "When in doubt, 'say so-and-so and I,' not 'so-and-so and me,'" so they unthinkingly overapply it—a process linguists call hypercorrection—resulting in "mistakes" like *Al Gore and I a chance* and the even more despised *between you and I.*

But if the person on the street is so good at avoiding *Me is going* and *Give I a break,* and if even Ivy League professors and former Rhodes Scholars can't seem to avoid *Me and Jennifer are going* and *Give Al and I a chance,* might it not be the mavens that misunderstand English grammar, not the speakers? The mavens' case about case rests on one assumption: if an entire conjunction phrase has a grammatical feature like subject case, every word inside that phrase has to have that grammatical feature, too. But that is just false.

*Jennifer* is singular; you say *Jennifer is,* not *Jennifer are.* The pronoun *She* is singular; you say *She is,* not *She are.* But the conjunction *She and Jennifer* is not singular, it's plural; you say *She and Jennifer are,* not *She and Jennifer is.* So if a conjunction can have a different grammatical *number* from the pronouns inside it (She and Jennifer *are*), why must it have the same grammatical *case* as the pronouns inside it (Give Al Gore and *I* a chance)? The answer is that it need not. A conjunction is an example of a "headless' construction. Recall that the head of a phrase is the word that stands for the whole phrase. In the phrase *the tall blond man with one black shoe,* the head is the word *man,* because the entire phrase gets its properties from *man*— the phrase refers to a kind of man, and is third person singular, because that's what *man* is. But a conjunction has no head; it is not the same as any of its parts. If John and Marsha met, it does not mean that John met and that Marsha met. If voters give Clinton and Gore a chance, they are not giving Gore his own chance, added on to the chance they are giving Clinton; they are giving the entire ticket a chance. So just because *Me and Jennifer* is a subject that requires subject case, it does not mean that *Me* is a subject that requires subject case, and just because *Al Gore and I* is an object that requires object case, it does not mean that *I* is an object that requires object case. On grammatical grounds, the pronoun is free to have any case it wants. The linguist Joseph Emonds has analyzed the *Me and Jennifer/ Between you and I* phenomenon in great technical detail. He concludes that the language that the mavens want us to speak is not only not English, it is not a possible human language!

In the second story of his column, Safire replies to a diplomat who received a government warning about "crimes against tourists (primarily robberies, muggings, and pick-pocketings)." The diplomat writes,

> Note the State Department's choice of *pick-pocketings*. Is the doer of such deeds a *pickpocket* or a *pocket-picker*?

Safire replies, "The sentence should read 'robberies, muggings and pocket-pickings.' One picks pockets; no one pockets picks."

Significantly, Safire did not answer the question. If the perpetrator were called a *pocket-picker*, which is the most common kind of compound in English, then indeed the crime would be *pocket-picking*. But the name for the perpetrator is not really up for grabs; we all agree that he is called a *pickpocket*. And if he is called a pickpocket, not a pocket-picker, then what he does can perfectly well be called pick-pocketing, not pocket-picking, thanks to the ever-present English noun-to-verb conversion process, just as a cook cooks, a chair chairs, and a host hosts. The fact that no one pockets picks is a red herring—who said anything about a *pick-pocketer*?

The thing that is confusing Safire is that *pickpocket* is a special kind of compound, because it is headless—it is not a kind of pocket, as one would expect, but a kind of person. And though it is exceptional, it is not unique; there is a whole family of such exceptions. One of the delights of English is its colorful cast of characters denoted by headless compounds, compounds that describe a person by what he *does* or *has* rather than by what he *is*:

| | | |
|---|---|---|
| bird-brain | four-eyes | lazy-bones |
| blockhead | goof-off | loudmouth |
| boot-black | hard-hat | low-life |
| butterfingers | heart-throb | ne'er-do-well |
| cut-throat | heavyweight | pip-squeak |
| dead-eye | high-brow | redneck |
| egghead | hunchback | scarecrow |
| fathead | killjoy | scofflaw |
| flatfoot | know-nothing | wetback |

This list (sounding vaguely like a dramatis personae from Damon Runyon) shows that virtually everything in language falls into systematic patterns, even the seeming exceptions, if only you bother to look for them.

The third story deconstructs a breathless quote from Barbra Streisand, describing tennis star Andre Agassi:

> He's very, very intelligent; very, very, sensitive, very evolved; more than his linear years. . . . He plays like a Zen master. It's very in the moment.

Safire first speculates on the origin of Streisand's use of *evolved:* "It's change from the active to passive voice—from 'he *evolved from* the Missing Link' to 'He *is evolved*'—was probably influenced by the adoption of *involved* as a compliment."

These kinds of derivations have been studied intensively in linguistics, but Safire shows here that he does not understand how they work. He seems to think that people change words by being vaguely reminded of rhyming ones—*evolved* from *involved,* a kind of malapropism. But in fact people are not that sloppy and literal-minded. The lexical creations we have looked at—*Let me caveat that; They deteriorated the health care system; Boggs flied out to center field*—are based not on rhymes but on abstract rules that change a word's part-of-speech category and its cast of role-players, in the same precise ways across dozens or hundreds of words. For example, the transitive *to deteriorate the health care system* comes from the intransitive *the health care system deteriorated* in the same way that the transitive *to break the glass* comes from the intransitive *the glass broke.* Let's see, then, where *evolved* might have come from.

Safire's suggestion that it is an active-to-passive switch based on *involved* does not work at all. For *involved,* we can perhaps imagine a derivation from the active voice:

> Raising the child involved John. (active) →
> John was involved in raising his child. (passive) →
> John is very involved.

But for *evolved,* the parallel derivation would require a passive sentence, and before that an active sentence, that do not exist (I have marked them with asterisks):

  *Many experiences evolved John. →

  *John was evolved by many experiences. (or) *John was
   evolved in many experiences. →

  John is very evolved.

Also, if you're involved, it means that something involves you (you're
the object), whereas if you're evolved, it means that you have been
doing some evolving (you're the subject).

  The problem is that the conversion of *evolved from* to *very evolved*
is not a switch from the active voice of a verb to the passive voice, as
in *Andre beat Boris* → *Boris was beaten by Andre*. The source Safire
mentions, *evolved from,* is intransitive in modern English, with no
direct object. To passivize a verb in English you convert the direct
object into a subject, so *is evolved* could only have been passivized
from *Something evolved Andre,* which does not exist. Safire's explanation is like saying you can take *Bill bicycled from Lexington* and change
it to *Bill is bicycled* and then to *Bill is very bicycled.*

  This breakdown is a good illustration of one of the main scandals
of the language mavens: they show lapses in the most elementary
problems of grammatical analysis, like figuring out the part-of-speech
category of a word. Safire refers to the active and passive voice, two
forms of a verb. But is Barbra using *evolved* as a verb? One of the
major discoveries of modern generative grammar is that the part of
speech of a word—noun, verb, adjective—is not a label assigned by
convenience but an actual mental category that can be verified by
experimental assays, just as a chemist can verify whether a gem is a
diamond or zirconium. These tests are a standard homework problem
in the introductory course that linguists everywhere call Baby Syntax.
The method is to find as many constructions as you can in which
words that are clear-cut examples of a category, and no other kind of
word, can appear. Then when you are faced with a word whose category you do not know, you can see whether it can appear in that set
of constructions with some natural interpretation. By these tests we
can determine, for example, that the language maven Jacques Barzun
earned an "F" when he called a possessive noun like *Wellington's* an

adjective (as before, I have placed asterisks beside the phrases that sound wrong):

|  | REAL ADJECTIVE | IMPOSTER |
|---|---|---|
| 1. *very* X: | very intelligent | *very Wellington's |
| 2. *seems* X: | He seems intelligent | *This seems Wellington's |
| 3. *How* X: | How intelligent is he? | *How Wellington's is this ring? |
| 4. *more* X *than:* | more intelligent than | *more Wellington's than |
| 5. *a* Adj X Adj N: | A funny, intelligent old friend | *a funny, Wellington's old friend |
| 6. *un*-X: | unintelligent | *un-Wellington's |

Now let's apply this kind of test to Barbra's *evolved,* comparing it to a clear-cut verb in the passive voice like *was kissed by a passionate lover* (odd-sounding constructions are marked with an asterisk):

1. very evolved / *very kissed
2. He seems evolved / *He seems kissed
3. How evolved is he? / *How kissed is he?
4. He is more evolved now than he was last year / *He is more kissed now than he was yesterday
5. A thoughtful, evolved, sweet friend / *a tall, kissed, thoughtful man
6. He was unevolved / *He was unkissed by a passionate lover

Obviously, *evolved* does not behave like the passive voice of a verb; it behaves like an adjective. Safire was misled because adjectives can look like verbs in the passive voice and are clearly related to them, but they are not the same thing. This is the source of the running joke in the Bob Dylan song "Rainy Day Women #12 & 35":

> They'll stone you when you're riding in your car.
> They'll stone you when you're playing your guitar.
> But I would not feel so all alone.
> Everybody must get stoned.

This discovery steers us toward the real source of *evolved*. Since it is an adjective, not a verb in the passive voice, we no longer have to worry about the absence of the corresponding active voice sentence. To trace its roots, we must find a rule in English that creates adjectives from intransitive verbs. There is such a rule. It applies to the principle form of a certain class of intransitive verbs that refer to a change of state (what linguists call "unaccusative" verbs), and creates a corresponding adjective:

time that has elapsed → elapsed time
a leaf that has fallen → a fallen leaf
a man who has traveled widely → a widely traveled man
a testicle that has not descended into the scrotum → an undescended testicle
a Christ that has risen from the dead → a risen Christ
a window that has stuck → a stuck window
the snow which has drifted → the drifted snow
a Catholic who has lapsed → a lapsed Catholic
a lung that has collapsed → a collapsed lung
a writer who has failed → a failed writer

Take this rule and apply it to *a tennis player who has evolved,* and you get *an evolved player*. This solution also allows us to make sense of Streisand's meaning. When a verb is converted from the active to the passive voice, the verb's meaning is conserved. *Dog bites man = Man is bitten by dog.* But when a verb is converted to an adjective, the adjective can acquire idiosyncratic nuances. Not every woman who has fallen is a fallen woman, and if someone stones you you are not necessarily stoned. We all evolved from a missing link, but not all of us are evolved in the sense of being more spiritually sophisticated than our contemporaries.

Safire then rebukes Steisand for *more than his linear years*. He says.

> *Linear* means "direct, uninterrupted"; it has gained a pejorative vogue sense of "unimaginative," as in *linear thinking*, in contrast to insightful, inspired leaps of genius. I think what Ms. Streisand had in mind was "beyond his chronological years," which is better expressed as simply "beyond his years." You can see what she was getting at—the years lined up in an orderly fashion—but even in the anything-goes world of show-biz lingo, not everything goes. Strike the set on *linear*.

Like many language mavens, Safire underestimates the precision and aptness of slang, especially slang borrowed from technical fields. Streisand obviously is not using the sense of *linear* from Euclidean geometry, meaning "the shortest route between two points," and the associated image of years lined up in an orderly fashion. She is using the sense taken from analytic geometry, meaning "proportional" or "additive." If you take a piece of graph paper and plot the distance traveled at constant speed against the time that has elapsed, you get a straight line. This is called a linear relationship; for every hour that passes, you've traveled another 55 miles. In contrast, if you plot the amount of money in your compound-interest account, you get a nonlinear curve that swerves upward; as you leave your money in longer, the amount of interest you accrue in a year gets larger and larger. Streisand is implying that Agassi's level of evolvedness is not proportional to his age: whereas most people fall on a straight line that assigns them X spiritual units of evolvedness for every year they have lived, this young man's evolvedness has been compounding, and he floats above the line, with more units than his age would ordinarily entitle him to. Now, I cannot be sure that this is what Streisand had in mind (at the time of this writing, she has not replied to my inquiry), but this sense of *linear* is common in contemporary techno-pop cant (like *feedback, systems, holism, interface,* and *synergistic*), and it is unlikely that she blundered into a perfectly apt usage by accident, as Safire's analysis would imply.

Finally, Safire comments on *very in the moment:*

> This *very* calls attention to the use of a preposition or a noun
> as a modifier, as in "it's very *in*," or "It's very *New York*," or
> the ultimate fashion compliment, "It's very *you*." To be very
> *in the moment* (perhaps a variation of *of the moment* or *up to
> the minute*) appears to be a loose translation of the French
> *au courant,* variously translated as "up to date, fashionable,
> with-it."

Once again, by patronizing Streisand's language, Safire has misana-
lyzed both its form and its meaning. He has not noticed that: (1) The
word *very* is not connected to the preposition *in;* it's connected to the
entire prepositional phrase *in the moment.* (2) Streisand is not using
the intransitive *in,* with its special sense of "fashionable"; she is using
the conventional transitive *in* with a noun phrase object, *the moment.*
(3) Her use of a prepositional phrase as if it was an adjective to
describe some mental or emotional state follows a common pattern in
English: *under the weather, out of character, off the wall, in the dumps,
out to lunch, on the ball, in good spirits, on top of the world, out of his
mind,* and *in love.* (4) It's unlikely that Streisand was trying to say that
Agassi is *au courant* or fashionable; that would be a put-down implying shallowness, not a compliment. Her reference to Zen makes her
meaning entirely clear: that Agassi is very good at shutting out distrac-
tions and concentrating on the game or person he is involved with at
that moment.

So these are the language mavens. Their foibles can be blamed
on two blind spots. One is a gross underestimation of the linguistic
wherewithal of the common person. I am not saying that everything
that comes out of a person's mouth or pen is perfectly rule-governed
(remember Dan Quayle). But the language mavens would have a
much better chance of not embarrassing themselves if they saved the
verdict of linguistic incompetence for the last resort rather than jumping to it as a first conclusion. People come out with laughable verbiage
when they feel they are in a forum demanding an elevated, formal
style and know that their choice of words could have momentous con-

sequences for them. That is why the fertile sources of howlers tend to be politicians' speeches, welfare application letters, and student term papers (assuming there is some grain of truth in the reports). In less self-conscious settings, common people, no matter how poorly educated, obey sophisticated grammatical laws, and can express themselves with a vigor and grace that captivates those who listen seriously—linguists, journalists, oral historians, novelists with an ear for dialogue.

The other blind spot of the language mavens is their complete ignorance of the modern science of language—and I don't mean just the formal apparatus of Chomskyan theory, but basic knowledge of what kinds of constructions and idioms are found in English, and how people use them and pronounce them. In all fairness, much of the blame falls on members of my own profession for being so reluctant to apply our knowledge to the practical problems of style and usage and to everyone's natural curiosity about why people talk the way they do. With a few exceptions like Joseph Emonds, Dwight Bolinger, Robin Lakoff, James McCawley, and Geoffrey Nunberg, mainstream American linguists have left the field entirely to the mavens—or, as Bolinger calls them, the shamans. He has summed up the situation:

> In language there are no licensed practitioners, but the woods are full of midwives, herbalists, colonic irrigationists, bonesetters, and general-purpose witch doctors, some abysmally ignorant, others with a rich fund of practical knowledge—whom we shall lump together and call *shamans*. They require our attention not only because the fill a lack but because they are almost the only people who make the news when language begins to cause trouble and someone must answer the cry for help. Sometimes their advice is sound. Sometimes it is worthless, but still it is sought because no one knows where else to turn. We are living in an African village and Albert Schweitzer has not arrived yet.

So what should be done about usage? Unlike some academics in the 1960s, I am not saying that instruction in standard English grammar and composition is a tool to perpetuate an oppressive white patriarchal capitalist status quo and that The People should be liberated to write however they please. Some aspects of how people express themselves in some settings *are* worth trying to change. What I am calling for is innocuous: a more thoughtful discussion of language and how people use it, replacing *bubbe-maises* (old wives' tales) with the best scientific knowledge available. It is especially important that we not underestimate the sophistication of the actual cause of any instance of language use: the human mind.

It is ironic that the jeremiads wailing about how sloppy language leads to sloppy thought are themselves hairballs of loosely associated factoids and tangled non sequiturs. All the examples of verbal behavior that the complainer takes exception to for any reason are packed together in one unappealing mass and coughed up as proof of The Decline of the Language: teenage slang, sophistry, regional variations in pronunciation and diction, bureaucratic bafflegab, poor spelling and punctuation, pseudo-errors like *hopefully*, badly crafted prose, government euphemism, nonstandard grammar like *ain't*, misleading advertising, and so on (not to mention deliberate witticisms that go over the complainer's head).

I hope to have convinced you of two things. Many prescriptive rules of grammar are just plain dumb and should be deleted from the usage handbooks. And most of standard English is just that, standard, in the same sense that certain units of currency or household voltages are said to be standard. It is just common sense that people should be given every encouragement and opportunity to learn the dialect that has become the standard one in their society and to employ it in many formal settings. But there is no need to use terms like "bad grammar," "fractured syntax," and "incorrect usage" when referring to rural and black dialects. Though I am no fan of "politically correct" euphemism (in which, according to the satire, *white woman* should be replaced by *melanin-impoverished person of gender*), using terms like "bad grammar" for "nonstandard" is both insulting and scientifically inaccurate.

As for slang, I'm all for it! Some people worry that slang will somehow "corrupt" the language. We should be so lucky. Most slang lexicons are preciously guarded by their subcultures as membership badges. When given a glimpse into one of these lexicons, no true language-lover can fail to be dazzled by the brilliant wordplay and wit: from medical students (*Zorro-belly, crispy critter, prune*), rappers (*jaw-jacking, dissing*), college students (*studmuffin, veg out, blow off*), surfers (*gnarlacious, geeklified*), and hackers (*to flame, core-dump, crufty*). When the more passé terms get cast off and handed down to the mainstream, they often fill expressive gaps in the language beautifully. I don't know how I ever did without *to flame* (protest self-righteously), *to dis* (express disrespect for), and *to blow off* (dismiss an obligation), and there are thousands of now-unexceptionable English words like *clever, fun, sham, banter, mob, stingy, bully, junkie,* and *jazz* that began life as slang. It is especially hypocritical to oppose linguistic innovations reflexively and at the same time to decry the loss of distinctions like *lie* versus *lay* on the pretext of preserving expressive power. Vehicles for expressing thought are being created far more quickly than they are being lost.

There is probably a good explanation for the cult of inarticulateness, where speech is punctuated with *you know, like, sort of, I mean,* and so on. Everyone maintains a number of ways of speaking that are appropriate to different contexts defined by the status and solidarity they feel with respect to their interlocutor. It seems that younger Americans try to maintain lower levels of social distance than older generations are used to. I know many gifted prose stylists my age whose one-on-one speech is peppered with *sort of* and *you know,* their attempt to avoid affecting the stance of the expert who feels entitled to lecture the conversational partner with confident pronouncements. Some people find it grating, but most speakers can turn it off at will, and I find it no worse than the other extreme, certain older academics who hold court during social gatherings, pontificating eloquently to their trapped junior audiences.

The aspect of language use that is most worth changing is the clarity and style of written prose. Expository writing requires language to express far more complex trains of thought than it was biologically designed to do. Inconsistencies caused by limitations of short-term memory and planning, unnoticed in conversation, are not as tolerable when preserved on a page that is to be perused more leisurely. Also, unlike a conversational partner, a reader will rarely share enough background assumptions to interpolate all the missing premises that make language comprehensible. Overcoming one's natural egocentrism and trying to anticipate the knowledge state of a generic reader at every stage of the exposition is one of the most important tasks in writing well. All this makes writing a difficult craft that must be mastered through practice, instruction, feedback, and—probably most important—intensive exposure to good examples. There are excellent manuals of composition that discuss these and other skills with great wisdom, like Strunk and White's *The Elements of Style* and Williams's *Style: Toward Clarity and Grace*. What is most relevant to my point is how removed their practical advice is from the trivia of split infinitives and slang. For example, a banal but universally acknowledged key to good writing is to revise extensively. Good writers go through anywhere from two to twenty drafts before releasing a paper. Anyone who does not appreciate this necessity is going to be a bad writer. Imagine a Jeremiah exclaiming, "Our language today is threatened by an insidious enemy: the youth are not revising their drafts enough times." Kind of takes the fun out, doesn't it? It's not something that can be blamed on television, rock music, shopping mall culture, overpaid athletes, or any of the other signs of the decay of civilization. But if it's clear writing that we want, this is the kind of homely remedy that is called for.

Finally, a confession. When I hear someone use *disinterested* to mean "apathetic," I am apt to go into a rage. *Disinterested* (I suppose I must explain that it means "unbiased") is such a lovely word: it is ever-so-subtly different from *impartial* or *unbiased* in implying that the person has no stake in the matter, not that he is merely committed to being even-handed out of personal principle. It gets this fine meaning from its delicate structure: *interest* means "stake," as in *conflict of interest* and *financial interest;* adding *-ed* to a noun can make it pertain to someone that owns the referent of that noun, as in *moneyed, one-eyed,* or *hook-nosed; dis-* negates the combination. The grammatical logic reveals itself in the similarly structured *disadvantaged, disaf-fected, disillusioned, disjointed,* and *dispossessed.* Since we already have the word *uninterested,* there can be no reason to rob discerning lan-guage-lovers of *disinterested* by merging their meanings, except as a tacky attempt to sound more high-falutin'. And don't get me started on *fortuitous* and *parameter . . .*

Chill out, Professor. The original, eighteenth-century meaning of *disinterested* turns out to be—yes, "uninterested." And that, too, makes grammatical sense. The adjective *interested* meaning "engaged" (related to the participle of the verb *to interest*) is far more common than the noun *interest* meaning "stake," so *dis-* can be ana-lyzed as simply negating that adjective, as in *discourteous, dishonest, disloyal, disreputable,* and the parallel *dissatisfied* and *distrusted.* But these rationalizations are beside the point. Every component of a lan-guage changes over time, and at any moment a language is enduring many losses. But since the human mind does not change over time, the richness of a language is always being replenished. Whenever any of us gets grumpy about some change in usage, we would do well to read the words of Samuel Johnson in the preface to his 1755 *diction-ary,* a reaction to the Jeremiahs of his day:

> Those who have been persuaded to think well of my design, require that it should fix our language, and put a stop to those alterations which time and chance have hitherto been suffered to make in it without opposition. With this consequence I will confess that I have flattered myself for a while; but now begin to fear that I have indulged expectations which neither reason nor experience can justify. When we see men grow old and die at a certain time one after another, from century to century, we laugh at the elixir that promises to prolong life to a thousand years; and with equal justice may the lexicographer be derided, who being able to produce no example of a nation that has preserved their words and phrases from mutability, shall imag-ine that his dictionary can embalm his language, and secure it from corruption and decay, that it is in his power to change

sublunary nature, and clear the world at once from folly, vanity, and affectation. With this hope, however, academies have been instituted, to guard the avenues of their languages, to retain fugitives, and to repulse intruders; but their vigilance and activity have hitherto been vain; sounds are too volatile and subtle for legal restraints; to enchain syllables, and to lash the wind, are equally the undertakings of pride, unwilling to measure its desires by its strength.

# 13

❧

# Mind Design

*Early in this book I asked why you should believe that there is a language* instinct. Now that I have done my best to convince you that there is one, it is time to ask why you should care. Having a language, of course, is part of what it means to be human, so it is natural to be curious. But having hands that are not occupied in locomotion is even more important to being human, and chances are you would never have made it to the last chapter of a book about the human hand. People are more than curious about language; they are passionate. The reason is obvious. Language is the most accessible part of the mind. People want to know about language because they hope this knowledge will lead to insight about human nature.

This tie-in animates linguistic research, raising the stakes in arcane technical disagreements and attracting the attention of scholars from far-flung disciplines. Jerry Fodor, the philosopher and experimental psycholinguist, studies whether sentence parsing is an encapsulated mental module or blends in with general intelligence, and he is more honest than most in discussing his interest in the controversy:

> "But look," you might ask, "why do you care about
> modules so much? You've got tenure; why don't you take off

and go sailing?'' This is a perfectly reasonable question and one that I often ask myself. . . . Roughly, the idea that cognition saturates perception belongs with (and is, indeed, historically connected with) the idea in the philosophy of science that one's observations are comprehensively determined by one's theories; with the idea in anthropology that one's values are comprehensively determined by one's culture; with the idea in sociology that one's epistemic commitments, including especially one's science, are comprehensively determined by one's class affiliations; and with the idea in linguistics that one's metaphysics is comprehensively determined by one's syntax [i.e., the Whorfian hypothesis—SP]. All these ideas imply a kind of relativistic holism: because perception is saturated by cognition, observation by theory, values by culture, science by class, and metaphysics by language, rational criticism of scientific theories, ethical values, metaphysical world-views, or whatever can take place only *within* the framework of assumptions that—as a matter of geographical, historical, or sociological accident—the interlocutors happen to share. What you can't do is rationally criticize the framework.

The thing is: I *hate* relativism. I hate relativism more than I hate anything else, excepting, maybe, fiberglass powerboats. More to the point, I think that relativism is very probably false. What it overlooks, to put it briefly and crudely, is the fixed structure of human nature. (That is not, of course, a novel insight; on the contrary, the *malleability* of human nature is a doctrine that relativists are invariably much inclined to stress; see, for example, John Dewey. . . .) Well, in cognitive psychology the claim that there is a fixed structure of human nature traditionally takes the form of an insistence on the heterogeneity of cognitive mechanisms and the rigidity of the cognitive architecture that effects their encapsulation. If there are faculties and modules, then not everything affects everything else; not everything is plastic. Whatever the All is, at least there is more than One of it.

For Fodor, a sentence perception module that delivers the speaker's message verbatim, undistorted by the listener's biases and expectations, is emblematic of a universally structured human mind, the same in all places and times, that would allow people to agree on what is just and true as a matter of objective reality rather than of taste, custom, and self-interest. It is a bit of a stretch, but no one can deny that there is a connection. Modern intellectual life is suffused with a relativism that denies that there is such a thing as a universal human nature, and the existence of a language instinct in any form challenges that denial.

The doctrine underlying that relativism, the Standard Social Science Model (SSSM), began to dominate intellectual life in the 1920s. It was a fusion of an idea from anthropology and an idea from psychology.

1. Whereas animals are rigidly controlled by their biology, human behavior is determined by culture, an autonomous system of symbols and values. Free from biological constraints, cultures can vary from one another arbitrarily and without limit.

2. Human infants are born with nothing more than a few reflexes and an ability to learn. Learning is a general-purpose process, used in all domains of knowledge. Children learn their culture through indoctrination, reward and punishment, and role models.

The SSSM has not only been the foundation of the study of humankind within the academy, but serves as the secular ideology of our age, the position on human nature that any decent person should hold. The alternative, sometimes called "biological determinism," is said to assign people to fixed slots in the socio-political-economic hierarchy, and to be the cause of many of the horrors of recent centuries: slavery, colonialism, racial and ethnic discrimination, economic and social castes, forced sterilization, sexism, genocide. Two of the most famous founders of the SSSM, the anthropologist Margaret

Mead and the psychologist John Watson, clearly had these social implications in mind:

> We are forced to conclude that human nature is almost unbelievably malleable, responding accurately and contrastingly to contrasting cultural conditions. . . . The members of either or both sexes may, with more or less success in the case of different individuals, be educated to approximate [any temperament]. . . . If we are to achieve a richer culture, rich in contrasting values, we must recognize the whole gamut of human potentialities, and so weave a less arbitrary social fabric, one in which each diverse human gift will find a fitting place. [Mead, 1935]

> Give me a dozen healthy infants, well-formed, and my own specified world to bring them up in and I'll guarantee to take any one at random and train him to become any type of specialist I might select—doctor, lawyer, artist, merchant-chief, and yes, even beggarman and thief, regardless of his talents, penchants, tendencies, abilities, vocations, and race of his ancestors. [Watson, 1925]

At least in the rhetoric of the educated, the SSSM has attained total victory. In polite intellectual conversations and respectable journalism, any generalization about human behavior is carefully prefaced with SSSM shibboleths that distance the speaker from history's distasteful hereditarians, from medieval kings to Archie Bunker. "Our society," the discussions begin, even if no other society has been examined. "Socializes us," they continue, even if the experiences of the child are never considered. "To the role . . ." they conclude, regardless of the aptness of the metaphor of "role," a character or part arbitrarily assigned to be played by a performer.

Very recently, the newsmagazines tell us that "the pendulum is swinging back." As they describe the appalled pacifist feminist parents of a three-year-old gun nut son and a four-year-old Barbie-doll-obsessed daughter, they remind the reader that hereditary factors cannot be ignored and that all behavior is an interaction between nature and nurture, whose contributions are as inseparable as the length and width of a rectangle in determining its area.

I would be depressed if what we have learned about the language instinct were folded into the mindless dichotomies of heredity–environment (a.k.a. nature–nurture, nativism–empiricism, innate–acquired, biology–culture), the unhelpful bromides about inextricably intertwined interactions, or the cynical image of a swaying pendulum of scientific fashion. I think that our understanding of language offers a more satisfying way of studying the human mind and human nature.

To begin with, we can discard the pre-scientific, magical model in which the issues are usually framed:

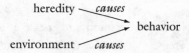

The "controversy" over whether heredity, environment, or some interaction between the two causes behavior is just incoherent. The organism has vanished; there is an environment without someone to perceive it, behavior without a behaver, learning without a learner. As Alice thought to herself when the Cheshire Cat vanished quite slowly, ending with the grin which remained some time after the rest of it had gone: "Well! I've often seen a cat without a grin, but a grin without a cat! It's the most curious thing I ever saw in all my life!"

The following model is also simplistic, but it is a much better beginning:

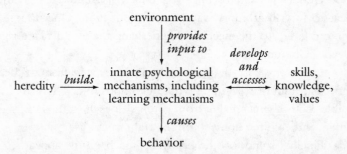

For we can now do justice to the complexity of the human brain, the immediate cause of all perception, learning, and behavior. Learning is not an alternative to innateness; without an innate mechanism to do the learning, it could not happen at all. The insights we have gained about the language instinct make this clear.

First, to reassure the nervous: yes, there are important roles for both heredity and environment. A child brought up in Japan ends up speaking Japanese; the same child, if brought up in the United States, would end up speaking English. So we know that the environment plays a role. If a child is inseparable from a pet hamster when growing up, the child ends up speaking a language, but the hamster, exposed to the same environment, does not. So we know that heredity plays a role. But there is much more to say.

- Since people can understand and speak an infinite number of novel sentences, it makes no sense to try to characterize their "behavior" directly—no two people's language behavior is the same, and a person's potential behavior cannot even be listed. But an infinite number of sentences can be generated by a finite rule system, a grammar, and it does make sense to study the mental grammar and other psychological mechanisms underlying language behavior.

- Language comes so naturally to us that we tend to be blasé about it, like urban children who think that milk just comes from a truck. But a close-up examination of what it takes to put words together into ordinary sentences reveals that mental language mechanisms must have a complex design, with many interacting parts.

- Under this microscope, the babel of languages no longer appear to vary in arbitrary ways and without limit. One now sees a common design to the machinery underlying the world's language, a Universal Grammar.

- Unless this basic design is built in to the mechanism that learns a particular grammar, learning would be impossible. There are many possible ways of generalizing from parents' speech to the language as a whole, and children home in on the right ones, fast.

- Finally, some of the learning mechanisms appear to be

designed for language itself, not for culture and symbolic behavior in general. We have seen Stone Age people with high-tech grammars, helpless toddlers who are competent grammarians, and linguistic idiot savants. We have seen a logic of grammar that cuts across the logic of common sense: the *it* of *It is raining* that behaves like the *John* of *John is running,* the *mice-eaters* who eat *mice* differing from the *rat-eaters* who eat *rats.*

The lessons of language have not been lost on the sciences of the rest of the mind. An alternative to the Standard Social Science Model has emerged, with roots in Darwin and William James and with inspiration from the research on language by Chomsky and the psychologists and linguists in his wake. It has been applied to visual perception by the computational neuroscientist David Marr and the psychologist Roger Shepard, and has been elaborated by the anthropologists Dan Sperber, Donald Symons, and John Tooby, the linguist Ray Jackendoff, the neuroscientist Michael Gazzaniga, and the psychologists Leda Cosmides, Randy Gallistel, Frank Keil, and Paul Rozin. Tooby and Cosmides, in their important recent essay "The Psychological Foundations of Culture," call it the Integrated Causal Model, because it seeks to explain how evolution caused the emergence of a brain, which causes psychological processes like knowing and learning, which cause the acquisition of the values and knowledge that make up a person's culture. It thus integrates psychology and anthropology into the rest of the natural sciences, especially neuroscience and evolutionary biology. Because of this last connection, they also call it Evolutionary Psychology.

Evolutionary psychology takes many of the lessons of human language and applies them to the rest of the psyche:

• Just as language is an improbable feat requiring intricate mental software, the other accomplishments of mental life that we take for granted, like perceiving, reasoning, and acting, require their own well-engineered mental software. Just as there is a universal design to the computations of grammar, there is a universal design to the rest of the human mind—an assumption that is not just a hopeful wish for human unity and brotherhood, but an actual discovery about the human species that is well motivated by evolutionary biology and genetics.

• Evolutionary psychology does not disrespect learning but seeks to explain it. In Molière's play *Le Malade Imaginaire,* the learned doctor is asked to explain how opium puts people to sleep, and cites its "sleep-producing power." Leibniz similarly ridiculed thinkers who invoke

> expressly occult qualities or faculties which they imagined to be like little demons or goblins capable of producing unceremoniously that which is demanded, just as if watches marked the hours by a certain horodeictic faculty without having need of wheels, or as if mills crushed grains by a fractive faculty without needing anything resembling millstones.

In the Standard Social Science Model, "learning" has been invoked in just these ways; in evolutionary psychology, there is no learning without some innate mechanism that makes the learning happen.

• Learning mechanisms for different spheres of human experience—language, morals, food, social relations, the physical world, and so on—are often found to work at cross-purposes. A mechanism designed to learn the right thing in one of these domains learns exactly the wrong thing in the others. This suggests that learning is accomplished not by some single general-purpose device but by different modules, each keyed to the peculiar logic and laws of one domain. People are flexible, not because the environment pounds or sculpts their minds into arbitrary shapes, but because their minds contain so many different modules, each with provisions to learn in its own way.

• Since biological systems with signs of complex engineering are unlikely to have arisen from accidents or coincidences, their organization must come from natural selection, and hence should have functions useful for survival and reproduction in the environments in which humans evolved. (This does not mean, however, that all aspects of mind are adaptations, or that the mind's adaptations are necessarily beneficial in evolutionary novel environments like twentieth-century cities.)

• Finally, culture is given its due, but not as some disembodied ghostly process or fundamental force of nature. "Culture" refers to the process whereby particular kinds of learning contagiously spread from person to person in a community and minds become coordinated into shared patterns, just as "a language" or "a dialect" refers to the process whereby the different speakers in a community acquire highly similar mental grammars.

A good place to begin discussing this new view of mind design is the place we began in discussing the language instinct: universality. Language, I noted early on, is universal among human societies, and as far as we know has been throughout the history of our species. Though languages are mutually unintelligible, beneath this superficial variation lies the single computational design of Universal Grammar, with its nouns and verbs, phrase structures and word structures, cases and auxiliaries, and so on.

At first glance, the ethnographic record seems to offer a stark contrast. Anthropology in this century has taken us through a mind-broadening fairground of human diversity. But might this carnival of taboos, kinship systems, shamanry, and all the rest be as superficial as the difference between *dog* and *hund,* hiding a universal human nature?

The culture of anthropologists themselves makes one apprehensive about their leitmotif that anything goes. One of America's most prominent, Clifford Geertz, has exhorted his colleagues to be "merchants of astonishment" who "hawk the anomalous, peddle the strange." "If we wanted only home truths," he adds, "we should have stayed at home." But this is an attitude that guarantees that anthropologists will miss any universal pattern in human ways. In fact, it can lead to outright error as the commonplace is cloaked as the anomalous, as in the Great Eskimo Vocabulary Hoax. As one young anthropologist wrote to me:

428 ⭆ The Language Instinct

The Eskimo vocabulary story will get its own section in a project of mine—a book whose working title is *One Hundred Years of Anthropological Malpractice*. I have been collecting instances of gross professional incompetence for years now: all of the anthropological chestnuts that turn out not to be true, but maintain their presence in textbooks anyway as the intellectual commonplaces of the field. Samoan free sex and the resultant lack of crime and frustration, the sex-reversed cultures like the "gentle" Arapesh (the men are head-hunters), the "stone-age" pristine Tasaday (a fabrication of the corrupt Philippine Minister of Culture—nearby villagers, dressed down as matriarchal "primitives"), the ancient matriarchies during the dawn of civilization, the fundamentally different Hopi concept of time, the cultures that everyone knows are out there where everything is the reverse of here, etc., etc.

One of the unifying threads will be that complete cultural relativism makes anthropologists far more credulous of almost any absurdity (Casteñeda's Don Juan novels—which I really enjoyed by the way—are in many textbooks as sober fact) than almost any ordinary person would be, equipped only with common sense. In other words, their professional "expertise" has made them complete and total gulls. Just as fundamentalism disposes you to accept accounts of miracles, being of the trained anthropologist faith disposes you to believe in any exotic account from Elsewhere. In fact, a lot of this nonsense is part of the standard intellectual equipment of every educated social scientist, providing a permanent obstacle to balanced reasoning about various psychological and social phenomena. I figure it will make me permanently unemployable, so I am not aiming to finish it any time soon.

The allusion to Samoan free sex pertains to Derek Freeman's 1983 bombshell showing how Margaret Mead got the facts wrong in her classic book *Coming of Age in Samoa*. (Among other things, her bored teenage informants enjoyed pulling her leg.) The other accusations are carefully documented in a recent review, *Human Universals,* written by another anthropologist, Donald E. Brown, who was trained in the standard ethnographic tradition. Brown has noted that behind anthropologists' accounts of the strange behavior of foreign peoples there are clear but abstract universals of human experience, such as rank, politeness, and humor. Indeed, anthropologists could not understand or live within other human groups unless they shared a rich set of common assumptions with them, what Dan Sperber calls a metaculture. Tooby and Cosmides note:

> Like fish unaware of the existence of water, anthropologists swim from culture to culture interpreting through universal human metaculture. Metaculture informs their every thought, but they have not yet noticed its existence. . . . When anthropologists go to other cultures, the experience of variation awakens them to things they had previously taken for granted in their own culture. Similarly, biologists and artificial intelligence researchers are "anthropologists" who travel to places where minds are far stranger than anywhere any ethnographer has ever gone.

Inspired by Chomsky's Universal Grammar (UG), Brown has tried to characterize the Universal People (UP). He has scrutinized archives of ethnography for universal patterns underlying the behavior of all documented human cultures, keeping a skeptical eye out both for claims of the exotic belied by the ethnographers' own reports, and for claims of the universal based on flimsy evidence. The outcome is stunning. Far from finding arbitrary variation, Brown was able to characterize the Universal People in gloriously rich detail. His findings contain something to startle almost anyone, and so I will reproduce the substance of them here. According to Brown, the Universal People have the following:

Value placed on articulateness. Gossip. Lying. Misleading. Verbal humor. Humorous insults. Poetic and rhetorical speech forms. Narrative and storytelling. Metaphor. Poetry with repetition of linguistic elements and three-second lines separated by pauses. Words for days, months, seasons, years, past, present, future, body parts, inner states (emotions, sensations, thoughts), behavioral propensities, flora, fauna, weather, tools, space, motion, speed, location, spatial dimensions, physical properties, giving, lending, affecting things and people, numbers (at the very least "one," "two," and "more than two"), proper names, possession. Distinctions between mother and father. Kinship categories, defined in terms of mother, father, son, daughter, and age sequence. Binary distinctions, including male and female, black and white, natural and cultural, good and bad. Measures. Logical relations including "not," "and," "same," "equivalent," "opposite," general versus particular, part versus whole. Conjectural reasoning (inferring the presence of absent and invisible entities from their perceptible traces).

Nonlinguistic vocal communication such as cries and squeals. Interpreting intention from behavior. Recognized facial expressions of happiness, sadness, anger, fear, surprise, disgust, and contempt. Use of smiles as a friendly greeting. Crying. Coy flirtation with the eyes. Masking, modifying, and mimicking facial expressions. Displays of affection.

Sense of self versus other, responsibility, voluntary versus involuntary behavior, intention, private inner life, normal versus abnormal mental states. Empathy. Sexual attraction. Powerful sexual jealousy. Childhood fears, especially of loud noises, and, at the end of the first year, strangers. Fear of snakes. "Oedipal" feelings (possessiveness of mother, coolness toward her consort). Face recognition. Adornment of bodies and arrangement of hair. Sexual attractiveness, based in part on signs of health and, in women, youth. Hygiene. Dance. Music. Play, including play fighting.

Manufacture of, and dependence upon, many kinds of tools, many of them permanent, made according to culturally transmitted motifs, including cutters, pounders, containers, string, levers, spears. Use of fire to cook food and for other purposes. Drugs, both medicinal and recreational. Shelter. Decoration of artifacts.

A standard pattern and time for weaning. Living in groups, which claim a territory and have a sense of being a distinct people.

Families built around a mother and children, usually the biological mother, and one or more men. Institutionalized marriage, in the sense of publicly recognized right of sexual access to a woman eligible for childbearing. Socialization of children (including toilet training) by senior kin. Children copying their elders. Distinguishing of close kin from distant kin, and favoring of close kin. Avoidance of incest between mothers and sons. Great interest in the topic of sex.

Status and prestige, both assigned (by kinship, age, sex) and achieved. Some degree of economic inequality. Division of labor by sex and age. More child care by women. More aggression and violence by men. Acknowledgment of differences between male and female natures. Domination by men in the public political sphere. Exchange of labor, goods, and services. Reciprocity, including retaliation. Gifts. Social reasoning. Coalitions. Government, in the sense of binding collective decisions about public affairs. Leaders, almost always nondictatorial, perhaps ephemeral. Laws, rights, and obligations, including laws against violence, rape, and murder. Punishment. Conflict, which is deplored. Rape. Seeking of redress for wrongs. Mediation. Ingroup/out-group conflicts. Property. Inheritance of property. Sense of right and wrong. Envy.

Etiquette. Hospitality. Feasting. Diurnality. Standards of sexual modesty. Sex generally in private. Fondness for sweets. Food taboos. Discreetness in elimination of body wastes. Supernatural beliefs. Magic to sustain and increase life, and to attract the opposite sex. Theories of fortune and misfortune. Explanations of disease and death. Medicine. Rituals, including rites of passage. Mourning the dead. Dreaming, interpreting dreams.

Obviously, this is not a list of instincts or innate psychological propensities; it is a list of complex interactions between a universal human nature and the conditions of living in a human body on this planet. Nor, I hasten to add, is it a characterization of the inevitable, a demarcation of the possible, or a prescription of the desirable. A list of human universals a century ago could have included the absence of ice cream, oral contraceptives, movies, rock and roll, women's suffrage, and books about the language instinct, but that would not have stood in the way of these innovations.

Like the identical twins reared apart who dipped buttered toast in their coffee, Brown's Universal People jolts our preconceptions about human nature. And just as the discoveries about twins do not call for a buttered-toast-in-coffee gene, the discoveries about universals do not implicate a universal toilet-training instinct. A theory of the universal mind is doubtless going to be as abstractly related to the Universal People as X-bar theory is related to a list of universals of word order. But it seems certain that any such theory will have to put more in the human head than a generalized tendency to learn or to copy an arbitrary role model.

With the assumption of an infinitely variable human nature from anthropology out of the way, let's look at the assumption of an infinitely acquisitive learning ability from psychology. How might we make sense of the concept of a general, multipurpose learning device?

Explicit pedagogy—learning by being told—is one kind of general-purpose learning, but most would agree it is the least important. Few people have been convinced by arguments like "No one ever teaches children how Universal Grammar works, but they respect it anyway; therefore it must be innate." Most learning, everyone agrees, takes place outside of classroom lessons, by generalizing from examples. Children generalize from role models, or from their own behaviors that are rewarded or not rewarded. The power comes from the generalization according to similarity. A child who echoed back a parent's sentences verbatim would be called autistic, not a powerful learner; children generalize to sentences that are *similar* to their parents', not to those sentences exactly. Likewise, a child who observes that barking German shepherds bite should generalize to barking Doberman pinschers and other similar dogs.

Similarity is thus the mainspring of a hypothetical general multipurpose learning device, and there is the rub. In the words of the logician Nelson Goodman, similarity is "a pretender, an imposter, a quack." The problem is that similarity is in the mind of the beholder—just what we are trying to explain—not in the world. Goodman writes:

> Consider baggage at an airport check-in station. The spectator may notice shape, size, color, material, and even make of luggage; the pilot is more concerned with weight, and the passenger with destination and ownership. Which pieces of baggage are more alike than others depends not only upon what properties they share, but upon who makes the comparison, and when. Or suppose we have three glasses, the first two filled with colorless liquid, the third with a bright red liquid. I might be likely to say the first two are more like each other than either is like the third. But it happens that the first glass is filled with water and the third with water colored by a drop of vegetable dye, while the second is filled with hydrochloric acid—and I am thirsty.

The unavoidable implication is that a sense of "similarity" must be innate. This much is not controversial; it is simple logic. In behaviorist psychology, when a pigeon is rewarded for pecking a key in the presence of a red circle, it pecks more to a red ellipse, or to a pink circle, than it does to a blue square. This "stimulus generalization" happens automatically, without extra training, and it entails an innate "similarity space"; otherwise the animal would generalize to everything or to nothing. These subjective spacings of stimuli are necessary for learning, so they cannot all be learned themselves. Thus even the behaviorist is "cheerfully up to his neck" in innate similarity-determining mechanisms, as the logician W. V. O. Quine pointed out (and his colleague B. F. Skinner did not demur).

For language acquisition, what is the innate similarity space that allows children to generalize from sentences in their parents' speech to the "similar" sentences that define the rest of English? Obviously, "Red is more similar to pink than to blue," or "Circle is more similar to ellipse than to triangle," is of no help. It must be some kind of mental computation that makes *John likes fish* similar to *Mary eats apples,* but not similar to *John might fish;* otherwise the child would say *John might apples*. It must make *The dog seems sleepy* similar to *The men seem happy*, but not similar to *The dog seems sleeping*, so that the child will avoid that false leap. That is, the "similarity" guiding the child's generalization has to be an analysis of speech into nouns and verbs and phrases, computed by the Universal Grammar built into the learning mechanisms. Without such innate computation defining which sentence is similar to which other ones, the child would have no way of correctly generalizing—any sentence is "similar," in one sense, to nothing but a verbatim repetition of itself, and also "similar," in another sense, to any random rearrangement of those words, and "similar," in still other senses, to all kinds of other inappropriate word strings. This is why it is no paradox to say that flexibility in learned *behavior* require innate constraints on the *mind*. The chapter on language acquisition (see p. 291) offers a good example: the ability of children to generalize to an infinite number of potential sentences depends on their analyzing parental speech using a fixed set of mental categories.

So learning a grammar from examples requires a special similarity space (defined by Universal Grammar). So does learning the meanings of words from examples, as we saw in Quine's *gavagai* problem, in which a word-learner has no logical basis for knowing whether *gavagai* means "rabbit," "hopping rabbit," or "undetached rabbit parts." What does this say about learning everything else? Here is how Quine reports, and defuses, what he calls the "scandal of induction":

> It makes one wonder the more about other inductions, where what is sought is a generalization not about our neighbor's verbal behavior but about the harsh impersonal world. It is reasonable that our [mental] quality space should match our neighbor's, we being birds of a feather; and so the general trustworthiness of induction in the . . . learning of words was a put-up job. To trust induction as a way of access to the truths of nature, on the other hand, is to suppose, more nearly, that our quality space matches that of the cosmos. . . . [But] why does our innate subjective spacing of qualities accord so well

with the functionally relevant groupings in nature as to make our inductions tend to come out right? Why should our subjective spacing of qualities have a special purchase on nature and a lien on the future?

There is some encouragement in Darwin. If people's innate spacing of qualities is a gene-linked trait, then the spacing that has made for the most successful inductions will have tended to predominate through natural selection. Creatures inveterately wrong in their inductions have a pathetic but praiseworthy tendency to die before reproducing their kind.

Quite right, though the cosmos is heterogeneous, and thus the computations of similarity that allow our generalizations to harmonize with it must be heterogeneous, too. Qualities that make two utterances equivalent in terms of learning the grammar, such as being composed of the same sequence of nouns and verbs, should not make them equivalent in terms of scaring away animals, such as being a certain loudness. Qualities that should make bits of vegetation equivalent in terms of causing or curing an illness, such as being different parts of a kind of plant, are not the qualities that should make them equivalent for nutrition, like sweetness; equivalent for feeding a fire, like dryness; equivalent for insulating a shelter, like bulk; or equivalent for giving as a gift, like beauty. The qualities that should classify people as potential allies, such as showing signs of affection, should not necessarily classify them as potential mates, such as showing signs of fertility and not being close blood relatives. There must be many similarity spaces, defined by different instincts or modules, allowing those modules to generalize intelligently in some domain of knowledge such as the physical world, the biological world, or the social world.

Since innate similarity spaces are inherent to the logic of learning, it is not surprising that human-engineered learning systems in artificial intelligence are always innately designed to exploit the constraints in some domain of knowledge. A computer program intended to learn the rules of baseball is pre-programmed with the assumptions underlying competitive sports, so that it will not interpret players' motions as a choreographed dance or a religious ritual. A program designed to learn the past tense of English verbs is given only the verb's sound as its input; a program designed to learn a verb's dictionary entry is given only its meaning. This requirement is apparent in what the designers do, though not always in what they say. Working within the assumptions of the Standard Social Science Model, the computer scientists often hype their programs as mere demos of powerful general-purpose learning systems. But because no one would be so foolhardy as to try to model the entire human mind, the researchers can take advantage of this allegedly practical limitation. They are free to hand-tailor their demo program to the kind of problem it is charged with solving, and they can be a deus ex machina funneling just the right inputs to the program at just the right time. Which is not a criticism; that's the way learning systems have to work!

So what are the modules of the human mind? A common academic parody of Chomsky has him proposing innate modules for bicycling, matching ties with shirts, rebuilding carburetors, and so on. But the slope from language to carburetor repair is not that slippery. We can avoid the skid with some obvious footholds. Using engineering analyses, we can examine what a system would need, in principle, to do the right kind of generalizing for the problem it is solving (for example, in studying how humans perceive shapes, we can ask whether a system that learns to recognize different kinds of furniture can also recognize different faces, or whether it needs special shape analyzers for faces). Using biological anthropology, we can look for evidence that a problem is one that our ancestors had to solve in the environments in which they evolved—so language and face recognition are at least candidates for innate modules, but reading and driving are not. Using data from psychology and ethnography, we can test the following prediction: when children solve problems for which they have mental modules, they should look like geniuses, knowing things they have not been taught; when they solve problems that their minds are not equipped for, it should be a long hard slog. Finally, if a module for some problem is real, neuroscience should discover that the brain tissue computing the problem has some kind of physiological cohesiveness, such as constituting a circuit or subsystem.

Being a bit foolhardy myself, I will venture a guess as to what kinds of modules, or families of instincts, might eventually pass these tests, aside from language and perception (for justification, I refer you to a recent compendium called *The Adapted Mind*):

1. Intuitive mechanics: knowledge of the motions, forces, and deformations that objects undergo.
2. Intuitive biology: understanding of how plants and animals work.
3. Number.
4. Mental maps for large territories.
5. Habitat selection: seeking of safe, information-rich, productive environments, generally savannah-like.
6. Danger, including the emotions of fear and caution, phobias for stimuli such as heights, confinement, risky social encounters, and venomous and predatory animals, and a motive to learn the circumstances in which each is harmless.
7. Food: what is good to eat.
8. Contamination, including the emotion of disgust, reactions to certain things that seem inherently disgusting, and intuitions about contagion and disease.
9. Monitoring of current well-being, including the emotions of happiness and sadness, and moods of contentment and restlessness.
10. Intuitive psychology: predicting other people's behavior from their beliefs and desires.
11. A mental Rolodex: a database of individuals, with blanks for kinship, status or rank, history of exchange of favors, and inherent skills and strengths, plus criteria that valuate each trait.
12. Self-concept: gathering and organizing information about one's value to other people, and packaging it for others.

13. Justice: sense of rights, obligations, and deserts, including the emotions of anger and revenge.
14. Kinship, including nepotism and allocations of parenting effort.
15. Mating, including feelings of sexual attraction, love, and intentions of fidelity and desertion.

To see how far standard psychology is from this conception, just turn to the table of contents of any textbook. The chapters will be: Physiological, Learning, Memory, Attention, Thinking, Decision-Making, Intelligence, Motivation, Emotion, Social, Development, Personality, Abnormal. I believe that with the exception of Perception and, of course, Language, not a single curriculum unit in psychology corresponds to a cohesive chunk of the mind. Perhaps this explains the syllabus-shock experienced by Introductory Psychology students. It is like explaining how a car works by first discussing the steel parts, then the aluminum parts, then the red parts, and so on, instead of the electrical system, the transmission, the fuel system, and so on. (Interestingly, textbooks on the brain are more likely to be organized around what I think of as real modules. Mental maps, fear, rage, feeding, maternal behavior, language, and sex are all common sections in neuroscience texts.)

For some readers, the preceding list will be the final proof that I have lost my mind. An innate module for doing biology? Biology is a recently invented academic discipline. Students struggle through it. The person in the street, and tribes around the world, are fonts of superstition and misinformation. The idea seems only slightly less mad than the innate carburetor repair instinct.

But recent evidence suggests otherwise; there may be an innate "folk biology" that gives people different basic intuitions about plants and animals than they have about other objects, like man-made artifacts. The study of folk biology is young compared with the study of language, and the idea might be wrong. (Maybe we reason about living things using two modules, one for plants and one for animals.

Maybe we use a bigger module, one that embraces other natural kinds like rocks and mountains. Or maybe we use an inappropriate module, like folk psychology.) But the evidence so far is suggestive enough that I can present folk biology as an example of a possible cognitive module other than language, giving you an idea of the kinds of things an instinct-populated mind might contain.

To begin with, as hard as it may be for a supermarket-jaded city dweller to believe, "stone age" hunter-gatherers are erudite botanists and zoologists. They typically have names for hundreds of wild plant and animal species, and copious knowledge of those species' life cycles, ecology, and behavior, allowing them to make subtle and sophisticated inferences. They might observe the shape, freshness, and direction of an animal's tracks, the time of day and year, and the details of the local terrain to predict what kind of animal it is, where it has gone, and how old, hungry, tired, and scared it is likely to be. A flowering plant in the spring might be remembered through the summer and returned to in the fall for its underground tuber. The use of medicinal drugs, recall, is part of the lifestyle of the Universal People.

What kind of psychology underlies this talent? How does our mental similarity space accord with this part of the cosmos? Plants and animals are special kinds of objects. For a mind to reason intelligently about them, it should treat them differently from rocks, islands, clouds, tools, machines, and money, among other things. Here are four of the basic differences. First, organisms (at least, sexual organisms) belong to populations of interbreeding individuals adapted to an ecological niche; this makes them fall into species with a relatively unified structure and behavior. For example, all robins are more or less alike, but they are different from sparrows. Second, related species descended from a common ancestor by splitting off from a lineage; this makes them fall into non-overlapping, hierarchically included classes. For example, sparrows and robins are alike in being birds, birds and mammals are alike in being vertebrates, vertebrates and insects are alike in being animals. Third, because an organism is a complex, self-preserving system, it is governed by dynamic physiological processes that are lawful even when hidden. For example, the bio-chemical organization of an organism enables it to grow and move, and is lost when it dies. Fourth, because organisms have separate genotypes and phenotypes, they have a hidden "essence" that is con-served as they grow, change form, and reproduce. For example, a cat-erpillar, chrysalis, and butterfly are in a crucial sense the same animal.

Remarkably, people's unschooled intuition about living things seems to mesh with these core biological facts, including the intu-itions of young children who cannot read and have not set foot in a biology lab.

The anthropologists Brent Berlin and Scott Atran have studied folk taxonomies of flora and fauna. They have found that, universally, people group local plants and animals into kinds that correspond to the *genus* level in the Linnaean classification system of profes-sional biology (species-genus-family-order-class-phylum-kingdom). Since most locales contain a single species from any genus, these folk categories usually correspond to species as well. People also classify kinds into higher-level life-forms, like tree, grass, moss, quadruped, bird, fish, and insect. Most of the life-form categories of animals coin-cide with the biologist's level of *class*. Folk classifications, like profes-sional biologist's classifications, are strictly hierarchical: every plant or animal belongs to one and only one genus; every genus belongs to only one life-form; every life-form is either a plant or an animal; plants and animals are living things, and every object is either a living thing or not. All this gives people's intuitive biological concepts a logical structure that is different from the one that organizes their other con-cepts, such as human-made artifacts. Whereas people everywhere say that an animal cannot be both fish and fowl, they are perfectly happy with saying, for example, that a wheelchair can be both furniture and vehicle, or that a piano can be both musical instrument and furniture. And this in turn makes reasoning about natural kinds different from reasoning about artifacts. People can deduce that if a trout is a kind of fish and a fish is a kind of animal, then a trout is a kind of animal. But they do not infer that if a car seat is a kind of chair and a chair is a kind of furniture, then a car seat is a kind of furniture.

Special intuitions about living things begin early in life. Recall that the human infant is far from being a bag of reflexes, mewling and puking in the nurse's arms. Three- to six-month infants, well before they can move about or even see very well, know about objects and their possible motions, how they causally impinge on one another, their properties like compressibility, and their number and how it changes with addition and subtraction. The distinction between living and nonliving things is appreciated early, perhaps before the first birthday. The cut initially takes the form of a difference between inanimate objects that move around according to the laws of billiard-ball physics and objects like people and animals that are self-propelled. For example, in an experiment by the psychologist Elizabeth Spelke, a baby is shown a ball rolling behind a screen and another ball emerging from the other side, over and over again to the point of boredom. If the screen is removed and the infant sees the expected hidden event, one ball hitting the other and launching it on its way, the baby's interest is only momentarily revived; presumably this is what the baby had been imagining all along. But if the screen is removed and the baby sees the magical event of one object stopping dead in its tracks without reaching the second ball, and the second ball taking off mysteriously on its own, the baby stares for much longer. Crucially, infants expect inanimate balls and animate people to move according to different laws. In another scenario, people, not balls, disappeared and appeared from behind the screen. After the screen was removed, the infants showed little surprise when they saw one person stop short and the other up and move; they were more surprised by a collision.

By the time children are of nursery school and kindergarten age, they display a subtle understanding that living things fall into kinds with hidden essences. The psychologist Frank Keil has challenged children with pixilated questions like these:

Doctors took a raccoon [shows picture of a raccoon] and shaved away some of its fur. They dyed what was left all black. Then they bleached a single stripe all white down the center of its back. Then, with surgery, they put in its body a sac of super smelly yucky stuff, just like a skunk has. When they were all done, the animal looked like this [shows picture of skunk]. After the operation, was this a skunk or a raccoon?

Doctors took a coffeepot that looked like this [shows picture of a coffeepot]. They sawed off the handle, sealed the top, took off the top knob, closed the spout, and sawed it off. They also sawed off the base and attached a flat piece of metal. They attached a little stick, cut a window in it, and filled the metal container with birdfood. When they were done, it looked like this [shows picture of a birdfeeder]. After the operation, was this a coffeepot or a birdfeeder?

Doctors took this toy [shows picture of a wind-up bird]. You wind it up with a key, and its mouth opens and a little machine inside plays music. The doctors did an operation on it. They put on real feathers to make it nice and soft and they gave it a better beak. Then they took off the wind-up key and put in a new machine so that it flapped its wings and flew, and chirped [shows picture of a bird]. After the operation, was it a real bird or a toy bird?

For artifacts like a coffeepot turning into a bird feeder (or a deck of cards turning into toilet paper), the children accepted the changes at face value: a birdfeeder is anything that is meant to feed birds, so that thing is a birdfeeder. But for natural kinds like a raccoon turning into a skunk (or a grapefruit turning into an orange), they were more resistant; there was some invisible raccoonhood lingering in the skunk's clothing, and they were less likely to say that the new creature was a skunk. And for violations of the boundary between artifacts and natural kinds, like a toy turning into a bird (or a porcupine turning into a hairbrush), they were adamant: a bird is a bird and a toy is a toy. Keil also showed that children are uncomfortable with the idea of a horse that has cow insides and cow parents and cow babies, even though they have no problem with a key that is made of melted-down pennies and is then melted down to make pennies again.

And of course adults from other cultures have the same sorts of intuitions. Illiterate rural Nigerians were given the following kind of question:

> Some students took a pawpaw [shows picture of a pawpaw] and stuck some green, pointed leaves on the top. Then they put small, prickly patches all over it. Now it looks like this [shows picture of a pineapple]—is it a pawpaw or a pineapple?

A typical response was, "It's a pawpaw, because a pawpaw has its own structure from heaven and a pineapple its own origin. One cannot turn into the other."

Little children also sense that animal kinds fall into larger categories, and their generalizations follow the similarity defined by category membership, not mere similarity of appearance. Susan Gelman and Ellen Markman showed three-year-old children a picture of a flamingo, a picture of a bat, and a picture of a blackbird, which looked a lot more like the bat than like the flamingo. They told the kids that a flamingo feeds its babies mashed-up food but a bat feeds its babies milk, and asked them what the blackbird feeds its babies. With no further information, children went by appearances and predicted milk. But all it took was a mention that flamingos and blackbirds were birds, and the children lumped them together and predicted mashed-up food.

And if you really doubt that we have botany instincts, consider one of the oddest of human motives: looking at flowers. A huge industry specializes in breeding and growing flowers for people to use in decorating dwellings and parks. Some research shows that bringing flowers to hospital patients is more than a warm gesture; it may actually improve the patient's mood and recovery rate. Since people rarely eat flowers, this diversion of effort and resources seems inexplicably frivolous. But if we evolved as intuitive botanists, it makes some sense. A flower is a microfiche of botanical information. When plants are not in bloom, they blend into a sea of green. A flower is often the only way to identify a plant species, even for a professional taxonomist. Flowers also signal seasons and terrains of expected bounty and the exact locations of future fruits and seeds. A motive to pay attention to flowers, and to be where they are, would obviously have been useful in environments where there were no year-round salad bars.

Intuitive biology is, of course, very different from what professors of biology do in their laboratories. But professional biology may have intuitive biology at its foundation. Folk taxonomy was obviously the predecessor to Linnaean taxonomy, and even today, professional taxonomists rarely contradict indigenous tribes when they classify the local species. The intuitive conviction that living things have a hidden essence and are governed by hidden processes is clearly what impelled the first professional biologists to try to understand the nature of plants and animals by bringing them into the laboratory and putting bits of them under a microscope. Anyone who announced he was trying to understand the nature of chairs by bringing them into a laboratory and putting bits of them under a microscope would surely be dismissed as mad, not given a grant. Indeed, probably all of science and mathematics is driven by intuitions coming from innate modules like number, mechanics, mental maps, even law. Physical analogies (heat is a fluid, electrons are particles), visual metaphors (linear function, rectangular matrix), and social and legal terminology (attraction, obeying laws) are used throughout science. And if you will allow me to sneak in one more offhand remark that really deserves a book of its own, I would guess that most other human "cultural" practices (competitive sports, narrative literature, landscape design, ballet), no matter how much they seem like arbitrary outcomes of a Borgesian lottery, are clever technologies we have invented to exercise and stimulate mental modules that were originally designed for specific adaptive functions.

So the language instinct suggests a mind of adapted computational modules rather than the blank slate, lump of wax, or general-purpose computer of the Standard Social Science Model. But what does this view say about the secular ideology of equality and opportunity that the model has provided us? If we abandon the SSSM, are we forced to repugnant doctrines like "biological determinism"?

Let me begin with what I hope are obvious points. First, the human brain works however it works. Wishing for it to work in some way as a shortcut to justifying some ethical principle undermines both the science and the ethics (for what happens to the principle if the scientific facts turn out to go the other way?). Second, there is no foreseeable discovery in psychology that could bear on the self-evident truth that ethically and politically, all people are created equal, that they are endowed with certain inalienable rights, and that among these are life, liberty, and the pursuit of happiness. Finally, radical empiricism is not necessarily a progressive, humanitarian doctrine. A blank slate is a dictator's dream. Some psychology textbooks mention the "fact" that Spartan and samurai mothers smiled upon hearing that their sons fell in battle. Since history is written by generals, not mothers, we can dismiss this incredible claim, but it is clear what purposes it must have served.

With those points out of the way, I do want to point out some implications of the theory of cognitive instincts for heredity and humankind, for they are the opposite of what many people expect. It is a shame that the following two claims are so often confused:

Differences between people are innate.
Commonalities among all people are innate.

The two claims could not be more different. Take number of legs. The reason that some people have fewer legs than others is 100% due to the environment. The reason that all uninjured people have exactly two legs (rather than eight, or six, or none) is 100% due to heredity. But claims that a universal human nature is innate are often run together with claims that differences between individuals, sexes, or races are innate. One can see the misguided motive for running them together: if *nothing* in the mind is innate, then differences between people's minds cannot be innate; thus it would be good if the mind had no structure because then decent egalitarians would have nothing to worry about. But the logical inverse is false. Everyone could be born with identical, richly structured minds, and all differences among them could be bits of acquired knowledge and minor perturbations that accumulate through people's history of life experiences. So even for people who, inadvisably in my view, like to conflate science and ethics, there is no need for alarm at the search for innate mental structure, whatever the truth turns out to be.

One reason innate commonalities and innate differences are so easy to confuse is that behavior geneticists (the scientists who study inherited deficits, identical and fraternal twins, adopted and biological children, and so on) have usurped the word "heritable" as a technical term referring to the proportion of *variation* in some trait that correlates with genetic differences within a species. This sense is different from the everyday term "inherited" (or genetic), which refers to traits whose inherent structure or organization comes from information in the genes. Something can be ordinarily inherited but show zero heritability, like number of legs at birth or the basic structure of the mind. Conversely, something can be not inherited but have 100% heritability. Imagine a society where all and only the red-haired people were made priests. Priesthood would be highly "heritable," though of course not inherited in any biologically meaningful sense. For this reason, people are bound to be confused by claims like "Intelligence is 70% heritable," especially when the newsmagazines report them in the same breath (as they inevitably do, alas) with research in cognitive science on the basic workings of the mind.

All claims about a language instinct and other mental modules are claims about the commonalities among all normal people. They have virtually nothing to do with possible genetic differences between people. One reason is that, to a scientist interested in how complex biological systems work, differences between individuals are so *boring!* Imagine what a dreary science of language we would have if instead of trying to figure out how people put words together to express their thoughts, researchers have begun by developing a Language Quotient (LQ) scale, and busied themselves by measuring thousands of people's relative language skills. It would be like asking how lungs work and being told that some people have better lungs than others, or asking how compact disks reproduce sound and being given a consumer magazine that ranked them instead of an explanation of digital sampling and lasers.

But emphasizing commonalities is not just a matter of scientific taste. The design of any adaptive biological system—the explanation of how it works—is almost certain to be uniform across individuals in a sexually reproducing species, because sexual recombination would fatally scramble the blueprints for qualitatively different designs. There is, to be sure, a great deal of genetic diversity among individuals; each person is biochemically unique. But natural selection is a process that feeds on that variation, and (aside from functionally equivalent varieties of molecules) when natural selection creates adaptive designs, it does so by using the variation up: the variant genes that specify more poorly designed organs disappear when their owners starve, get eaten, or die mateless. To the extent that mental modules are complex products of natural selection, genetic variation will be limited to quantitative variations, not differences in basic design. Genetic differences among people, no matter how fascinating they are to us in love, biography, personnel, gossip, and politics, are of minor interest to us when we appreciate what makes minds intelligent at all.

Similarly, an interest in mind design puts possible innate differences between sexes (as a psycholinguist I refuse to call them "genders") and races in a new light. With the exception of the maleness-determining gene on the Y-chromosome, every functioning gene in a man's body is also found in a woman's and vice versa. The maleness gene is a developmental switch that can activate some suites of genes and deactivate others, but the same blueprints are in both kinds of bodies, and the default condition is identity of design. There is some evidence that the sexes depart from this default in the case of the psychology of reproduction and the adaptive problems directly and indirectly related to it, which is not surprising; it seems unlikely that peripherals as different as the male and female reproductive systems would come with the same software. But the sexes face essentially similar demands for most of the rest of cognition, including language, and I would be surprised if there were differences in design between them.

Race and ethnicity are the most minor differences of all. The human geneticists Walter Bodmer and Luca Cavalli-Sforza have noted a paradox about race. Among laypeople, race is lamentably salient, but for biologists it is virtually invisible. Eighty-five percent of human genetic variation consists of the differences between one person and another within the same ethnic group, tribe, or nation. Another eight percent is between ethnic groups, and a mere seven percent is between "races." In other words, the genetic difference between, say, two randomly picked Swedes is about twelves times as large as the genetic difference between the average of Swedes and the average of Apaches or Warlpiris. Bodmer and Cavalli-Sforza suggests that the illusion is the result of an unfortunate coincidence. Many of the systematic differences among races are adaptations to climate: melanin protects skin against the tropical sun, eyelid folds insulate eyes from dry cold and snow. But the skin, the part of the body seen by the weather, is also the part of the body seen by other people. Race is, quite literally, skin-deep, but to the extent that perceivers generalize from external to internal differences, nature has duped them into thinking that race is important. The X-ray vision of the molecular geneticist reveals the unity of our species.

And so does the X-ray vision of the cognitive scientist. "Not speaking the same language" is a virtual synonym for incommensurability, but to a psycholinguist, it is a superficial difference. Knowing about the ubiquity of complex language across individuals and cultures and the single mental design underlying them all, no speech seems foreign to me, even when I cannot understand a word. The banter among New Guinean highlanders in the film of their first contact with the rest of the world, the motions of a sign language interpreter, the prattle of little girls in a Tokyo playground—I imagine seeing through the rhythms to the structures underneath, and sense that we all have the same minds.

# Notes

## 1. An Instinct to Acquire an Art

1. Amorous octopuses: adapted from Wallace, 1980. Cherry stains: *Parade* magazine, April 5, 1992, p. 16. *All My Children:* adapted from *Soap Opera Digest,* March 30, 1993.

3. Horse graveyard: Lambert & The Diagram Group, 1987. Megafauna extinctions: Martin & Klein, 1984.

3. Cognitive science: Gardner, 1985; Posner, 1989; Osherson & Lasnik, 1990; Osherson, Kosslyn, & Hollerbach, 1990; Osherson & Smith, 1990.

6. Instinct to acquire an art: Darwin, 1874, pp. 101–102.

7. The *why* of instinctive acts: James, 1892/1920, p. 394.

8. Chomsky: Chomsky, 1959, 1965, 1975, 1980a, 1988, 1991; Kasher, 1991.

9. Chomsky on mental organs: Chomsky, 1975, pp. 9–11.

10. Top ten list: from *Arts and Humanities Citation Index;* Kim Vandiver, Chairman of the Faculty, MIT, citation for Noam Chomsky's Killian Faculty Achievement Award, MIT, March 1992.

11. Standard Social Science Model: Brown, 1991; Tooby & Cosmides, 1992; Degler, 1991. Challenging Chomsky: Harman, 1974; Searle,

1971; Piatelli-Palmarini, 1980; commentators in Chomsky, 1980b; Modgil & Modgil, 1987; Botha, 1989; Harris, 1993. Putnam on Chomsky: Piatelli-Palmarini, 1980, p. 287.

## 2. Chatterboxes

13. First contact: Connolly & Anderson, 1987.
14. Language is universal: Murdoch, 1975; Brown, 1991.
14. No primitive languages: Sapir, 1921; Voegelin & Voegelin, 1977. Plato and swineherds: Sapir, 1921, p. 219.
14. Bantu syntax: Bresnan & Moshi, 1988; Bresnan, 1990. Cherokee pronouns: Holmes & Smith, 1977.
16. Logic of nonstandard English: Labov, 1969.
19. Putnam on general multipurpose learning strategies: Piatelli-Palmarini, 1980; Putnam, 1971; see also Bates, Thal, & Marchman, 1991.
21. Creoles: Holm, 1988; Bickerton, 1981, 1984.
24. Sign language: Klima & Bellugi, 1979; Wilbur, 1979.
24. Lenguaje de Signos Nicaragüense and Idioma de Signos Nicaragüense: Kegl & Lopez, 1990; Kegal & Iwata, 1989.
25. Children acquiring ASL: Petitto, 1988. Adults acquiring language (signed and spoken): Newport, 1990.
26. Simon: Singleton & Newport, 1993. Sign languages as creoles: Woodward, 1978; Fischer, 1978. Unlearnability of artificial sign systems: Supalla, 1986.
29. Aunt Mae: Heath, 1983, p. 84.
29. Structure dependence: Chomsky, 1975.
30. Children, Chomsky, and Jabba: Crain & Nakayama, 1986.
31. Universal auxiliaries: Steele et al., 1981. Language universals: Greenberg, 1963; Comrie, 1981; Shopen, 1985. Fluent backwards talkers: Cowan, Braine, & Leavitt, 1985.
32. Language development: Brown, 1973; Pinker, 1989; Ingram, 1989.
33. Sarah masters agreement: Brown, 1973. Examples are from a com-

puter search of Sarah's transcripts in the Child Language Data Exchange System; MacWhinney, 1991.

34. Children's creative errors (*be's, gots, do's*): Marcus, Pinker, Ullman, Hollander, Rosen, & Xu, 1992.

35. Recovered aphasic: Gardner, 1974, p. 402. Permanent aphasic: Gardner, 1974, pp. 60–61.

37. Language mutants: Gopnik, 1990a, b; Gopnik & Crago, 1991; Gopnik, 1993.

39. Blatherers: Cromer, 1991.

41. More blatherers: Curtiss, 1989.

41. Williams syndrome: Bellugi et al., 1991, 1992.

## 3. *Mentalese*

44. Newspeak: Orwell, 1949, pp. 246–247, 255.

46. Language and animal rights: Singer, 1992. General Semantics: Korzybski, 1933; Hayakawa, 1964; Murphy, 1992.

46. Sapir: Sapir, 1921. Whorf: Carroll, 1956.

48. Sapir: Sapir, 1921. Boas school: Degler, 1991; Brown, 1991. Whorf: Carroll, 1956.

50. Early Whorf critics: Lenneberg, 1953; Brown, 1958.

51. Die Schrecken der Deutschen Sprache: quoted in Brown, 1958, p. 232; see also Espy, 1989, p. 100.

51. Color lexicons: Crystal, 1987, p. 106.

52. Color vision: Hubel, 1988.

52. Color universals: Berlin & Kay, 1969. New Guineans learn red: Heider, 1972.

52. Timeless Hopi: Carroll, 1956, p. 57. Also pp. 55, 64, 140, 146, 153, 216–17.

53. Hopi prayer hour: Malotki, 1983, p. 1.

53. Hopi time: Brown, 1991; Malotki, 1983.

53. The Great Eskimo Vocabulary Hoax: Martin, 1986; Pullum, 1991.

54. Pullum on Eskimos: Pullum, 1991, pp. 162, 165–166. "Polysynthetic perversity" is an in-joke, from the linguist's classification of

Eskimo languages as "polysynthetic"; compare Freud's "polymorphous perversity."

56. Whorf in the lab: Cromer, 1991b; Kay & Kempton, 1984.
56. Subjunctives and the Chinese mind: Bloom, 1981, 1984; Au, 1983, 1984; Liu, 1985; Takano, 1989.
58. A man without words: Schaller, 1991.
59. Baby thoughts: Spelke et al., 1992. Baby arithmetic: Wynn, 1992.
59. Animal thinking: Gallistel, 1992. Monkey friends and relations: Cheney & Seyfarth, 1992.
61. Visual thinkers: Shepard, 1978; Shepard & Cooper, 1982. Einstein: Kosslyn, 1983.
62. Mind's eye: Shepard & Cooper, 1982; Kosslyn, 1983; Pinker, 1985.
68. Representational theory of mind: in Haugeland, 1981, articles by Haugeland, Newell & Simon, Pylyshyn, Dennett, Marr, Searle, Putnam, and Fodor; in Pinker and Mehler, 1988, articles by Fodor & Pylyshyn and Pinker & Prince; Jackendoff, 1987.
69. English versus mentalese: Fodor, 1975; McDermott, 1981.
69. Headlines: Columbia Journalism Review, 1980.
72. An example of mentalese: Jackendoff, 1987; Pinker, 1989.

## 4. *How Language Works*

75. Arbitrary sound-meaning relation: Saussure, 1916/1959.
75. Infinite use of finite media: Humboldt, 1836/1972.
75. Discrete combinatorial systems: Chomsky, 1991; Abler, 1989; Studdert-Kennedy, 1990.
76. Discrete inheritance and evolution: Dawkins, 1986.
77. 110-word Shavian sentence: example from Jacques Barzun; cited in Bolinger, 1980.
78. Faulker example (with modifications): Espy, 1989.
79. Sentences commenting on their own ungrammaticality: David Moser, cited in Hofstadter, 1985.
80. Nineteenth-century nonsense: Hofstadter, 1985.

80. Sleeping esophagus: Twain, "Double-Barreled Detective Story." Example from Lederer, 1990.
80. Pobbles: Edward Lear, "The Pobble Who Has No Toes." Jabberwocky: Carroll, 1871/1981. Colorless green ideas: Chomsky, 1957.
81. Automated news story: Frayn, 1965. Example from Miller, 1967.
83. Gobbledygook generators: Brandreth, 1980; Bolinger, 1980; *Spy* magazine, January 1993.
84. Approximations to English: Miller & Selfridge, 1950.
85. Finite-state devices and their problems: Chomsky, 1957; Miller & Chomsky, 1963; Miller, 1967. *TV Guide* example from Gleitman, 1981.
94. Cook with round bottom: Columbia Journalism Review, 1980; Lederer, 1987.
96. Impenetrable Chomsky: Chomsky, 1986, p. 79. Textbooks on modern grammatical theory: Friedin, 1992; Radford, 1988; Riemsdijk & Williams, 1986.
101. Sex between parked cars: Columbia Journalism Review, 1980.
103. X-bar syntax: Jackendoff, 1977; Kornai & Pullum, 1990.
103. Word-order correlations: Greenberg, 1963; Dryer, 1992.
105. Verbs' demands: Grimshaw, 1990; Pinker, 1989.
113. Blinkenlights: Raymond, 1991.
113. Deep structure: Chomsky, 1965, 1988. Chomsky on doing without d-structure: Chomsky, 1991. Chomsky still believes that there are several phrase structures underlying a sentence; he simply wants to eliminate the idea that there is a special one called d-structure, a single framework defined for the entire sentence into which the verbs are then plugged. The suggested replacement is to have each verb come with a chunk of phrase structure preinstalled; the sentence is assembled by snapping together the various chunks.

## 5. *Words, Words, Words*

119. Grammatical Man: Campbell, 1982. Chomsky in *Rolling Stone*: issue 631, May 28, 1992, p. 42. The Whore of Mensa: Allen, 1983.

120. Bantu verbs: Bresnan & Moshi, 1988; Wald, 1990.

122. Part-Vulcans and other novel forms: Sproat, 1992.

123. Word-building machinery: Aronoff, 1976; Chomsky & Halle, 1968/1991; Di Sciullo & Williams, 1987; Kiparsky, 1982; Selkirk, 1982; Sproat, 1992; Williams, 1981. The *anti-missile missile* example is from Yehoshua Bar-Hillel.

124. Inflectional rules as linguistic fruit flies: Pinker & Prince, 1988, 1992; Pinker, 1991.

125. People versus artificial neural networks: Prasada & Pinker, 1993; Sproat, 1992; McClelland & Rumelhart, 1986.

127. Man sold as pet fish: Columbia Journalism Review, 1980.

128. Heads of words: Williams, 1981; Selkirk, 1982.

131. Hackitude: Raymond, 1991.

132. Irregular verbs: Chomsky & Halle, 1968/1991; Kiparsky, 1982; Pinker & Prince, 1988, 1992; Pinker, 1991; Mencken, 1936. Irregular doggerel: author unknown, from Espy, 1975.

134. Dizzy Dean: Staten, 1992; Espy, 1975.

135. Irregularity and young minds: Yourcenar, 1961; quotation from Michael Maratsos.

136. Flying out: Kiparsky, 1982; Kim, Pinker, Prince, & Prasada, 1991; Kim, Marcus, Pinker, Hollander, & Coppola, in press; Pinker & Prince, 1992; Marcus, Clahsen, Brinkmann, Wiese, Woest, and Pinker, 1993.

136. *Walkmans* versus *Walkmen: Newsweek,* August 7, 1989, p. 68.

140. Mice-eaters: Kiparsky, 1982; Gordon, 1986.

141. Morphological products, syntactic atoms, and listemes: Di Sciullo and Williams, 1987.

143. Shakespeare's vocabulary: Bryson, 1990; Kučera, 1992. Shakespeare used about 30,000 different word forms, but many of these were inflected variants of a single word, like *angel* and *angels* or *laugh* and *laughed.* Applying statistics from contemporary English, one would get an estimate of about 18,000 word types, but this must be adjusted downward to about 15,000 because Shakespeare used more inflections than we do; for example, he used both *-eth* and *-s.*

143. Counting words: Miller, 1977, 1991; Carey, 1978; Lorge & Chall, 1963.

144. Typical vocabulary size: Miller, 1991.

145. Word as arbitrary symbol: Saussure, 1916/1959; Hurford, 1989.

147. "You" and "me" in ASL: Petitto, 1988.

147. "Gavagai!": Quine, 1960.

149. Categories: Rosch, 1978; Anderson, 1990.

151. Babies and objects: Spelke et al., 1992; Baillargeon, in press.

151. Children learning words: Markman, 1989.

152. Children, words, and kinds: Markman, 1989; Keil, 1989; Clark, 1993; Pinker, 1989, 1994. Sibbing: Brown, 1957; Gleitman, 1990.

## 6. *The Sounds of Silence*

153. Sine-wave speech: Remez et al., 1981.

154. "Duplex" perception of speech components: Liberman & Mattingly, 1989.

155. McGurk effect: McGurk & MacDonald, 1976.

155. Speech segmentation: Cole & Jakimik, 1980.

155. Oronyms: Brandreth, 1980.

156. Pullet surprises: Lederer, 1987; Brandreth, 1980; LINGUIST electronic bulletin board, 1992.

157. Smeared phonemes: Liberman et al., 1967.

157. Rate of speech perception: Miller, 1967; Liberman et al., 1967; Cole & Jakimik, 1980.

157. DragonDictate: Bamberg & Mandel, 1991.

159. Vocal tract: Crystal, 1987; Lieberman, 1984; Denes & Pinson, 1973; Miller, 1991; Green, 1976; Halle, 1990.

163. Phonetic symbolism: Brown, 1958.

163. *Fiddle-faddle, flim-flam:* Cooper & Ross, 1975; Pinker & Birdsong, 1979.

166. *Razzle-dazzle, rub-a-dub-dub:* Cooper & Ross, 1975; Pinker & Birdsong, 1979.

166. Speech gestures and distinctive features: Halle, 1983, 1990.

167. Speech sounds across the world: Halle, 1990; Crystal, 1987.

168. Speaking in tongues: Thomason, 1984; Samarin, 1972.

169. "Giacche Enne Binnestaucche": Espy, 1975.

170. Syllables and feet: Kaye, 1989; Jackendoff, 1987.

171. Phonological rules: Kenstowicz & Kisseberth, 1979; Kaye, 1989; Halle, 1990; Chomsky & Halle, 1968/1991.

175. Phonology with tiers: Kaye, 1989.

176. Shaw: Preface to *Pygmalion*. Slurvian: Lederer, 1987.

176. American pronunciation: Cassidy, 1985. Teachers with accents: *Boston Globe*, July 10, 1992.

177. Speaker versus hearer: Bolinger, 1980; Liberman & Mattingly, 1989; Pinker & Bloom, 1990.

178. Quine on redundancy: Quine, 1987.

178. Graceful motion: Jordan & Rosenbaum, 1989.

179. Why speech recognition is hard: Liberman et al., 1967; Mattingly & Studdert-Kennedy, 1991; Lieberman, 1984; Bamberg & Mandel, 1991; Cole & Jakimik, 1980.

179. Nonsense in noise: Miller, 1967. Phonemic restoration effect: Warren, 1970.

182. Problems with top-down perception: Fodor, 1983.

183. Mondegreens: LINGUIST electronic bulletin board, 1992.

184. HEARSAY system: Lesser et al., 1975.

184. DragonDictate: Bamberg & Mandel, 1991.

185. Spelling poem: quoted in C. Chomsky, 1970.

185. Shaw: from Crystal, 1987, p. 216.

186. Written versus spoken language: Liberman et al., 1967; Miller, 1991.

186. Writing systems: Crystal, 1987; Miller, 1991; Logan, 1986.

187. Two tragedies in life: from *Man and Superman*.

187. Rationality of English orthography: Chomsky & Halle, 1968/1991; C. Chomsky, 1970.

188. Twain on foreigners: from *The Innocents Abroad*.

## 7. *Talking Heads*

190. Artificial Intelligence: Winston, 1992; Wallich, 1991; *The Economist*, 1992.

191. Turing Test of whether machines can think: Turing, 1950.

193. ELIZA: Weizenbaum, 1976.

193. Loebner Prize competition: Shieber, in press.

194. Fast comprehension: Garrett, 1990; Marslen-Wilson, 1975.

194. Style: Williams, 1990.

196 Parsing: Smith, 1991; Ford, Bresnan, & Kaplan, 1982; Wanner & Maratsos, 1978; Yngve, 1960; Kaplan, 1972; Berwick et al., 1991; Wanner, 1988; Joshi, 1991; Gibson, in press.

200. Magical number seven: Miller, 1956.

200. Dangling sentences: Yngve, 1960; Bever, 1970; Williams, 1990.

201. Memory and grammatical load: Bever, 1970; Kuno, 1974; Hawkins, 1988.

202. Right-, left-, and center-embedding: Yngve, 1960; Miller & Chomsky, 1963; Miller, 1967; Kuno, 1974; Chomsky, 1965.

204. Number of rules for child to learn: Pinker, 1984.

209. Breadth-first dictionary lookup: Swinney, 1979; Seidenberg et al., 1982.

210. Killer sentenced to die twice: Columbia Journalism Review, 1980; Lederer, 1987.

211. Garden path sentences: Bever, 1970; Ford, Bresnan, & Kaplan, 1982; Wanner, 1988; Gibson, in press.

212. Multiple trees in memory: MacDonald, Just, and Carpenter, 1992; Gibson, in press.

213. Modularity of mind: Fodor, 1983. Modularity debate: Fodor, 1985; Garfield, 1987; Marslen-Wilson, 1989.

214. General smarts and understanding sentences: Trueswell, Tanenhaus, and Garnsey, in press.

214. Verbs help parsing, pro and con: Trueswell, Tanenhaus, & Kello, in press; Ford et al., 1982; Frazier, 1989; Ferreira & Henderson, 1990.

215. Computer parsers: Joshi, 1991.

215. Late closure and minimal attachment, pro and con: Frazier & Fodor, 1978; Ford et al., 1982; Wanner, 1988; Garfield, 1987.

216. The language of judges: Solan, 1993. Language and law: Tiersma, 1993.

218 Fillers and gaps: Wanner & Maratsos, 1978; Bever & McElree, 1988; MacDonald, 1989; Nicol & Swinney, 1989; Garnsey, Tanenhaus, & Chapman, 1989; Kluender & Kutas, 1993; J. D. Fodor, 1989.

220. Shortening filler-gap distances: Bever, 1970; Yngve, 1960; Williams, 1990. Bounding phrase movement to help parsing: Berwick & Weinberg, 1984.

221. Watergate transcripts: Committee on the Judiciary, U.S. House of Representatives, 1974; *New York Times* Staff, 1974.

224. *Masson* v. *The New Yorker Magazine: Time,* July 1, 1991, p. 68; *Newsweek,* July 1, 1991, p. 67.

226. Discourse, pragmatics, and inference: Grice, 1975; Levinson, 1983; Sperber & Wilson, 1986; Leech, 1983; Clark & Clark, 1977.

227. Scripts and stereotypes: Schanck & Riesbeck, 1981. Programming common sense: Freedman, 1990; Wallich, 1991; Lenat & Guha, 1990.

228. Logic of conversation: Grice, 1975; Sperber & Wilson, 1986.

229. Letter of recommendation: Grice, 1975; Norman & Rumelhart, 1975.

230. Politeness: Brown & Levinson, 1987.

230. Conduit metaphor: Lakoff & Johnson, 1980.

## 8. *The Tower of Babel*

232. Variation without limit: Joos, 1957, p. 96. One Earthly language: Chomsky, 1991.

232. Language differences: Crystal, 1987; Comrie, 1990; Department of Linguistics, Ohio State University.

234. Language universals: Greenberg, 1963; Greenberg, Ferguson, & Moravscik, 1978; Comrie, 1981; Hawkins, 1988; Shopen, 1985; Keenan, 1976; Bybee, 1985.

234. History versus typology: Kiparsky, 1976; Wang, 1976; Aronoff, 1987.

235. SOV, SVO, and center-embedding: Kuno, 1974.

237. Crosslinguistic meaning of "subject": Keenan, 1976; Pinker, 1984, 1987.

238. Human versus animal communication: Hockett, 1960.

239. Evolution disfavoring change for change's sake: Williams, 1966.

241. Babel speeds evolution: Dyson, 1979; Babel provides women: Crystal, 1987, p. 42.

242. Languages and species: Darwin, 1874, p. 106.

243. Evolution of innateness and learning: Williams, 1966; Lewontin, 1966; Hinton & Nowlan, 1987.

244. Why there is language learning: Pinker & Bloom, 1990.

245. Linguistic innovation as contagious disease: Cavalli-Sforza & Feldman, 1981.

246. Reanalysis and language change: Aitchison, 1991; Samuels, 1972; Kiparsky, 1976; Pyles & Algeo, 1982; Department of Linguistics, Ohio State University, 1991.

248. American English: Cassidy, 1985; Bryson, 1990.

249. History of English: Jespersen, 1938/1982; Pyles & Algeo, 1982; Aitchison, 1991; Samuels, 1972; Bryson, 1990; Department of Linguistics, Ohio State University, 1991.

251. Apprehending adolescents and catching kids: Williams, 1991.

253. The Great Vowel Shift as dudespeak: Burling, 1992.

253. Germanic and Indo-European: Pyles & Algeo, 1982; Renfrew, 1987; Crystal, 1987.

255. First European farmers: Renfrew, 1987; Ammerman & Cavalli-Sforza, 1984; Sokal, Oden, & Wilson, 1991; Roberts, 1992.

256. Language families: Comrie, 1990; Crystal, 1987; Ruhlen, 1987; Katzner, 1977.

257. Language of the Americas: Greenberg, 1987; Cavalli-Sforza et al., 1988; Diamond, 1990.

258. Language lumpers: Wright, 1991; Ross, 1991; Shevoroshkin & Markey, 1986.

260. Correlations between genes and language families: Cavalli-Sforza et al., 1988; Cavalli-Sforza, 1991. African Eve: Stringer & Andrews, 1988; Stringer, 1990; Gibbons, 1993.

261. Genes and languages in Europe: Harding & Sokal, 1988. Lack of

correlation between language families and genetic groups: Guy, 1992.

262. Proto-World: Shevoroshkin, 1990; Wright, 1991; Ross, 1991.

262. Language extinctions: Hale et al., 1992.

263. Another perspective on language extinctions: Ladefoged, 1992.

## 9. *Baby Born Talking—Describes Heaven*

266. Infant speech perception: Eimas et al., 1971; Werker, 1991.

268. Learning French in utero: Mehler et al., 1988.

268. Infants learn phonemes: Kuhl et al., 1992.

269. Babbling: Locke, 1992; Petitto & Marentette, 1991.

269. Babbling robots: Jordon & Rosenbaum, 1989.

269. First words: Clark, 1993; Ingram, 1989.

270. Finding word boundaries: Peters, 1983. Children's examples are from Peters, family memories, *Life* magazine, and MIT librarian Pat Claffey. The *Hill Street Blues* example is from Mark Aronoff.

271. First word combinations: Braine, 1976; Brown, 1973; Pinker, 1984; Ingram, 1989.

272. Infant comprehension: Hirsh-Pasek & Golinkoff, 1991.

272. Speech bottleneck in children: Brown, 1973, p. 205.

273. Language blasts off: Ingram, 1989, p. 235; Brown, 1973; Limber, 1973; Pinker, 1984; Bickerton, 1992.

273. Adam and Eve: Brown, 1973; MacWhinney, 1991.

275. Children avoid tempting errors: Stromswold, 1990.

277. Language acquisition across the globe: Slobin, 1985, 1992.

278. Alligator goed kerplunk: Marcus, Pinker, Ullman, Hollander, Rosen, & Xu, 1992.

279. Don't giggle me: Bowerman, 1982; Pinker, 1989.

281. Wild children: Tartter, 1986; Curtiss, 1989; Rymer, 1993.

282. Thurber & White: from "Is Sex Necessary?" Example from Donald Symons.

282. Language from television: Ervin-Tripp, 1973. Understanding Motherese from content words: Slobin, 1977. Children as mind-readers: Pinker, 1979, 1984.

283. Motherese: Newport, et al., 1977; Fernald, 1992.

284. Mute child: Stromswold, 1994.

284. No parental feedback: Brown & Hanlon, 1970; Braine, 1971; Morgan & Travis, 1989; Marcus, 1993.

286. Learning language without feedback: Pinker, 1979, 1984, 1989; Wexler & Culicover, 1980; Osherson, Stob, & Weinstein, 1985; Berwick, 1985; Marcus et al., 1992.

287. Language acquisition close up: Pinker, 1979, 1984; Wexler & Culicover, 1980.

293. Human versus other primate gestation periods: Corballis, 1991.

293. Brain growth & language development: Bates, Thal, & Janowsky, 1992; Locke, 1992; Huttenlocher, 1990.

294. Children's language in evolution: Williams, 1966.

295. Linguistic development and motor development: Lenneberg, 1967.

295. Foreign language learning: Hakuta, 1986; Grosjean, 1982; Bley-Vroman, 1990; Birdsong, 1989.

296. Critical ages for second language acquisition: Lieberman, 1984; Bley-Vroman, 1990; Newport, 1990; Long, 1990.

296. Critical periods for first language acquisition: Deaf: Newport, 1990. Genie: Curtiss, 1989; Rymer, 1992. Isabelle: Tartter, 1986. Chelsea: Curtiss, 1989.

298. Recovery from brain injury: Curtiss, 1989; Lenneberg, 1967.

298. Biology of the life cycle: Williams, 1966.

300. Evolution of the critical period: Hurford, 1991.

301. Senescence: Williams, 1957; Medawar, 1957.

## 10. *Language Organs and Grammar Genes*

302. Associated Press story: February 11, 1992. Kilpatrick: Universal Press Syndicate, February 28, 1992. Bombeck: March 5, 1992.

304. Broca: Caplan, 1987. Language on the left: Caplan, 1987, 1992; Corballis, 1991; Geschwind, 1979; Geschwind & Galaburda, 1987; Gazzaniga, 1983.

305. Left-hemisphere language and the Psalms: example from Michael Corballis.

306. Language affects scalp electrodes: Neville et al., 1991; Kluender & Kutas, 1993.

306. Language lights up brains: Wallesch et al., 1985; Peterson et al., 1988, 1990; Mazoyer et al., 1992; Zatorre et al., 1992; Poeppel, 1993.

307. Language, not language-like stimuli and responses, in the left: Gardner, 1974; Etcoff, 1986. Sign language in the left, gesturing the right: Poizner, Klima, & Bellugi, 1990; Corina, Vaid, & Bellugi, 1992.

308. Bilateral symmetry: Corballis, 1991. Symmetry is sexy: Cronin, 1992.

309. Twisted chordates: Kinsbourne, 1978. Snail anatomy: Buchsbaum, 1948.

310. Lopsided animals: Corballis, 1991.

311. Lopsided brains: Corballis, 1991; Kosslyn, 1987; Gazzaniga, 1978, 1989.

312. Southpaws: Corballis, 1991; Coren, 1992. Parsing by relatives of southpaws: Bever et al., 1989.

313. Perisylvian cortex as the language organ: Caplan, 1987; Gazzaniga, 1989.

313. Peter Hogan's aphasia: Goodglass, 1973.

314. Broca's aphasia: Caplan, 1987, 1992; Gardner, 1974; Zurif, 1989.

315. ERP and PET pick up language in left anterior perisylvian: Kluender & Kutas, 1993; Neville et al., 1991; Mazoyer et al., 1992; Wallesch et al., 1985; Stromswold, Caplan, & Alpert, 1993.

316. Anatomy of Broca's aphasia: Caplan, 1987; Dronkers et al., 1992. Parkinson's and language: Lieberman et al., 1992. Broca's aphasics detect ungrammaticality: Linebarger, Schwartz, & Saffran, 1983; Cornell, Fromkin, & Mauner, 1993.

316. Wernicke's aphasic: Gardner, 1974.

318. Wernicke's and related aphasias: Gardner, 1974; Geschwind, 1979; Caplan, 1987, 1992.

318. Anomia: Gardner, 1974; Caplan, 1987. The man with no nouns: Baynes & Iven, 1991.

319. Words and EEG's: Neville et al., 1991. Words and PET: Peterson et al., 1990; Poeppel, 1993.

319. Different aphasias in different people: Caplan, 1987, 1992; Miceli et al., 1989. Losing derivational morphology while keeping inflectional morphology: Miceli & Caramazza, 1988.

321. Banananomia: Warrington & McCarthy, 1987; Hillis & Caramazza, 1991; Hart, Berndt, & Caramazza, 1985; Farah, 1990.

321. Anomalies and variation in language localization: Caplan, 1987; Basso et al., 1985; Bates, Thal, & Janowsky, 1992.

321. Visual areas: Hubel, 1988. Neuroscience: Gazzaniga, 1992; see also the special issue of *Scientific American* on "Mind and Brain," September 1992.

322. Stimulation of circumscribed but variable language spots: Ojemann & Whitaker, 1978; Ojemann, 1991.

322. Words as hubs: Damasio and Damasio, 1992.

323. Moving language around in baby brains: Curtiss, 1989; Caplan, 1987; Bates, Thal, & Janowsky, 1992; Basso et al., 1985.

324. Functional MRI: Belliveau et al., 1991; MEG: Gallen, 1994.

324. Computing in neural networks: McCulloch & Pitts, 1943; Rumelhart & McClelland, 1986.

325. Computing language in neural networks: McClelland & Rumelhart, 1986; Pinker & Prince, 1988; Pinker & Mehler, 1988.

327. Neural development: Rakic, 1988; Shatz, 1992; Dodd & Jessell, 1988; von der Malsburg & Singer, 1988.

330. Transgenic pig: Brian Duffy, North America Syndicate.

330. Genetics of stuttering and dyslexia: Ludlow & Cooper, 1983. Genetics of SLI: Gopnik & Crago, 1991; Gopnik, 1993; Stromswold, 1994. Pronunciation errors in twins: Locke & Mather, 1989. Grammar in twins: Mather & Black, 1984; Munsinger & Douglas, 1976; Fahey, Kamitomo, & Cornell, 1978; Bishop, North, & Donlan, 1993; Adopted babies' language development: Hardy-Brown, Plomin, & DeFries, 1981.

331. Three generations of SLI: Gopnik, 1990a, 1990b, 1993; Gopnik & Crago, 1991.

333. Universal human nature and individual uniqueness: Tooby & Cosmides, 1990a.

335. Separated at birth: Holden, 1987; Lykken et al., 1992.

336. Behavior genetics: Bouchard et al., 1990; Lykken et al., 1992; Plomin, 1990.

337. Bushspeak: The Editors of *The New Republic,* 1992. Quaylespeak: Goldsman, 1992.

338. Linguistic geniuses: Yogi Berra, from Safire, 1991; Lederer, 1987. Dr. Seuss (Theodore Geisel), from *On Beyond Zebra,* 1955. Nabokov, from *Lolita,* 1958. King, from the march on Washington, 1963. Shakespeare, from *Hamlet,* Act 2, Scene 2.

## 11. *The Big Bang*

340. Elephants: Williams, 1989; Carrington, 1958.

342. Darwinian explanations of the language instinct: Pinker & Bloom, 1990; Pinker, in press; Hurford, 1989, 1991; Newmeyer, 1991; Brandon & Hornstein, 1986; Corballis, 1991.

333. Animal communication: Wilson, 1972; Gould and Marler, 1987.

343. Nonlinguistic communication and the brain: Deacon, 1988, 1989; Caplan, 1987; Myers, 1976; Robinson, 1976.

343. Gua and Viki: Tartter, 1986.

344. Sarah: Premack & Premack, 1972; Premack, 1985. Kanzi: Savage-Rumbaugh, 1991; Greenfield & Savage-Rumbaugh, 1991. Washoe: Gardner & Gardner, 1969, 1974. Koko: Patterson, 1978. See Wallman, 1992, for a review.

344. Nice guys in the animal kingdom: Sagan & Druyan, 1992. Quotation from excerpt in *Parade* magazine, September 20, 1992.

346. Nim: Terrace, 1979; Terrace et al., 1979. Ape language debunkers: Terrace et al., 1979; Seidenberg & Petitto, 1979; Petitto & Seidenberg, 1979; Seidenberg, 1986; Seidenberg & Petitto, 1987; Petitto, 1988; see Wallman, 1992, for a review. Threatened lawsuit: Wallman, 1992, p. 5.

346. Deaf signer observing chimps: Neisser, 1983, pp. 214–216.

347. The misbehavior of organisms: Breland & Breland, 1961.

351. Bates on Big Bangs: Bates, Thal, & Marchman, 1991, pp. 30, 35.

352. Chains, ladders, and bushes in evolution: Mayr, 1982; Dawkins, 1986; Gould, 1985.

357. Featherless biped: example from Wallman, 1992.

359. Logical impossibility of the liver: Lieberman, 1990, pp. 741–742.

359. New modules in evolution: Mayr, 1982.

360. Broca's area in monkeys: Deacon, 1988, 1989; Galaburda & Pandya, 1982.

361. Chimp and human DNA: King & Wilson, 1975; Miyamoto, Slightom, & Goodman, 1987.

362. Bow-wow, ding-dong, gestural, and other theories of transitional language: Harnad, Steklis, & Lancaster, 1976.

363. Dating language origins: Pinker, 1992, in press; Bickerton, 1990. Evolution of modern humans: Stringer & Andrews, 1988; Stringer, 1990; Gibbons, 1993.

364. Descent of larynx and Neanderthal speech: Lieberman, 1984. Neanderthal fans: Gibbons, 1992. Heimlich maneuver: *Parade*, June 28, 1992.

365. Chomsky denigrates natural selection: Chomsky, 1972, pp. 97–98; Chomsky, 1988, p. 167.

366. Logic of natural selection: Darwin, 1859/1964; Williams, 1966, 1992; Mayr, 1983; Dawkins, 1986; Tooby & Cosmides, 1990b; Maynard Smith, 1984, 1986; Dennett, 1983.

369. Just-so stories: Gould & Lewontin, 1979; Piatelli-Palmarini, 1989. It's just not so: Dawkins, 1986; Mayr, 1983; Maynard Smith, 1988; Tooby & Cosmides, 1990a, b; Pinker & Bloom, 1990; Dennett, 1983.

373. Natural language and natural selection: Pinker & Bloom, 1990.

373. Chomsky on the physics of brains: in Piatelli-Palmarini, 1980.

374. Language in dwarfs: Lenneberg, 1967. Language in normal hydrocephalics: Lewin, 1980. Normal brains and analytic processing in SLI: Gopnik, 1990b.

375. The throwing madonna: Calvin, 1991.

376. Demystifying language evolution: Pinker & Bloom, 1990.

377. Bates on three quarters of a rule: Bates, Thal, & Marchman, 1991, p. 31.

377. Bickerton on protolanguage and the Big Bang: Bickerton, 1990; Pinker, 1992.

378. Premack on mastodon-hunters: Premack, 1985, pp. 281–282.

378. Advantages of complex language: Burling, 1986. Cognitive arms race: Cosmides & Tooby, 1992. Gossip: Barkow, 1992. Some of the passages in this section are based on Pinker & Bloom, 1990.

381. Descent versus modification: Tooby & Cosmides, 1989.

## 12. The Language Mavens

385. On language mavens: Bolinger, 1980; Bryson, 1990; Lakoff, 1990.

386. History of prescriptive grammar: Bryson, 1990; Crystal, 1987; Lakoff, 1990; McCrum, Cran, & MacNeil, 1986; Nunberg, 1992.

388. *Write, wrote; bite, bote*: Lederer, 1990, p. 117.

391. Everyone and their brother: LINGUIST electronic bulletin board, Oct. 9, 1991.

392. A fifth of English verbs were nouns: Prasada & Pinker, 1993.

393. Flying out and Sally Ride: Kim, Pinker, Prince, & Prasada, 1991; Kim, Marcus, Pinker, Hollander, & Coppola, in press.

394. Bernstein on *broadcasted*: Bernstein, 1977, p. 81.

396. Wordwatchers: Quine, 1987; Thomas, 1990.

396. The *Boston Globe* on *get your goat*: December 23, 1992.

397. Taking it on the lam: Allen, 1983.

398. Bad grammar leading to violence: Bolinger, 1980, pp. 4–6.

398. Shock-grammarian: Simon, 1980, pp. 97, 165–166.

399. Crazy English: Lederer, 1990, pp. 15–21.

400. Slurvian: Lederer, 1987, pp. 114–117.

401. Howlers: Lederer, 1987; Brunvand, 1989.

401. Urban legends and xeroxlore: Brunvand, 1989.

402. Language sages: Bernstein, 1977; Safire, 1991.

403. Child language transcripts: MacWhinney, 1991.

404. *Me and Jennifer/Between you and I*: Emonds, 1986.

406. Low-lifes, cut-throats, ne'er-do-wells, and other disreputable compounds: Quirk et al., 1985.

408. Barzun on parts of speech: quoted in Bolinger, 1980, p. 169.

410. Adjectives from participles: Bresnan, 1982.

## 13. *Mind Design*

419. Language as a window into human nature: Rymer, 1993.

420. Sentence understanding, relativism, and fiberglass powerboats: Fodor, 1985, p. 5.

421. Standard Social Science Model: Tooby & Cosmides, 1992; Degler, 1991; Brown, 1991.

421. "Biological determinism": Gould, 1981; Lewontin, Rose, & Kamin, 1984; Kitcher, 1985; Chorover, 1979; See Degler, 1991.

422. Educating either sex: Mead, 1935. Training a dozen infants: Watson, 1925.

425. Evolutionary psychology: Darwin, 1872, 1874; James, 1892/ 1920; Marr, 1982; Symons, 1979, 1992; Sperber, 1985, in press; Tooby & Cosmides, 1990a, b, 1992; Jackendoff, 1987, 1992; Gazzaniga, 1992; Keil, 1989; Gallistel, 1990; Cosmides & Tooby, 1987; Shepard, 1987; Rozin & Schull, 1988; See also Konner, 1982, and the contributions to Barkow, Cosmides, & Tooby, 1992, and Hirschfeld & Gelman, in press.

427. Merchants of astonishment: Geertz, 1984.

428. Mead in Samoa: Freeman, 1983.

429. Anthropologists swimming through metaculture: Brown, 1991; Sperber, 1982; Tooby & Cosmides, 1992, p. 92.

429. Universal People: Brown, 1991.

432. Strictures on similarity: Goodman, 1972, p. 445.

433. Innate similarity space: Quine, 1969.

435. Artificial learning systems: Pinker, 1979, 1989; Pinker & Prince, 1988; Prasada & Pinker, 1993.

436. Modules of mind: Chomsky, 1975, 1980b, 1988; Marr, 1982; Tooby & Cosmides, 1992; Jackendoff, 1992; Sperber, in press. For a different conception, see Fodor, 1983, 1985.

439. Biological erudition of hunter-gatherers: Konner, 1982; Kaplan, 1992.

440. Folk biological taxonomies: Berlin, Breedlove, & Raven, 1973; Atran, 1987, 1990.

441. The cerebral infant: Spelke et al., 1992; Wynn, 1992; Flavell, Miller, & Miller, 1993.

441. Skunks turning into raccoons: Keil, 1989.

443. Pawpaws and pineapples among the Yoruba: Jeyifous, 1986.

443. Flamingos, blackbirds, and bats: Gelman & Markman, 1987.

443. Flower power: Kaplan, 1992; see also Orians & Heerwagen, 1992.

444. Folk science turning into science: Carey, 1985; Keil, 1989; Atran, 1990. Analogy and metaphor in mathematics and physical science: Gentner & Jeziorski, 1989; Lakoff, 1987. Stimulating our mental modules: Tooby & Cosmides, 1990b; Barkow, 1992.

445. Innateness versus heritability: Tooby & Cosmides, 1990a, 1992.

447. Universal human nature and unique individuals: Tooby & Cosmides, 1990a, 1992.

447. Sex differences in the psychology of sex: Symons, 1979, 1980, 1992; Daly & Wilson, 1988; Wilson & Daly, 1992.

448. Race as illusion: Bodmer & Cavalli-Sforza, 1970; Gould, 1977; Lewontin, Rose, & Kamin, 1984; Lewontin, 1982; Tooby & Cosmides, 1990a.

# References

Abler, W. L. 1989. On the particulate principle of self-diversifying systems. *Journal of Social and Biological Structures, 12,* 1–13.

Aitchison, J. 1991. *Language change: Progress or decay?* 2nd ed. New York: Cambridge University Press.

Allen, W. 1983. *Without feathers.* New York: Ballantine.

Ammerman, A. J., & Cavalli-Sforza, L. L. 1984. *The neolithic transition and the genetics of populations in Europe.* Princeton, N.J.: Princeton University Press.

Anderson, J. R. 1990. *The adaptive character of thought.* Hillsdale, N.J.: Erlbaum.

Aronoff, M. 1976. *Word formation in generative grammar.* Cambridge, Mass.: MIT Press.

Aronoff, M. 1987. Review of J. L. Bybee's "Morphology: A study of the relation between meaning and form." *Language, 63,* 115–129.

Atran, S. 1987. Folkbiological universals as common sense. In Modgil & Modgil, 1987.

Atran, S. 1990. *The cognitive foundations of natural history.* New York: Cambridge University Press.

Au, T. K.-F. 1983. Chinese and English counterfactuals: the Sapir-Whorf hypothesis revisited. *Cognition, 15,* 155–187.

Au, T. K.-F. 1984. Counterfactuals: In reply to Alfred Bloom. *Cognition, 17,* 155–187.

Baillargeon, R. In press. The object concept revisited: New directions in the investigation of infants' physical knowledge. In C. Granrud (Ed.), *Visual perception and cognition in infancy.* Hillsdale, N.J.: Erlbaum.

Bamberg, P. G., & Mandel, M. A. 1991. Adaptable phoneme-based models for large-vocabulary speech recognition. *Speech Communication, 10,* 437–451.

Barkow, J. H. 1992. Beneath new culture is old psychology: Gossip and social stratification. In Barkow, Cosmides, & Tooby, 1992.

Barkow, J. H., Cosmides, L., & Tooby, J. (Eds.) 1992. *The adapted mind: Evolutionary psychology and the generation of culture.* New York: Oxford University Press.

Basso, A., Lecours, A. R., Moraschini, S., & Vanier, M. 1985. Anatomoclinical correlations of the aphasias as defined through computerized tomography: Exceptions. *Brain and Language, 26,* 201–229.

Bates, E., Thal, D., & Janowsky, J. S. 1992. Early language development and its neural correlates. In I. Rapin & S. Segalowitz (Eds.), *Handbook of neuropsychology, Vol. 6: Child neurology.* Amsterdam: Elsevier.

Bates, E., Thal, D., & Marchman, V. 1991. Symbols and syntax: A Darwinian approach to language development. In Krasnegor et al., 1991.

Baynes, K., & Iven, C. 1991. Access to the phonological lexicon in an aphasic patient. Paper presented to the annual meeting of the Academy of Aphasia.

Belliveau, J. W., Kennedy, D. N., McKinstry, R. C., Buchbinder, B. R., Weisskoff, R. M., Cohen, M. S., Vevea, J. M., Brady, T. J., & Rosen, B. R. 1991. Functional mapping of the human visual cortex by Magnetic Resonance Imaging. *Science, 254,* 716–719.

Bellugi, U., Bihrle, A., Jernigan, T., Trauner, D., & Doherty, S. 1991. Neuropsychological, neurological, and neuroanatomical profile of Williams Syndrome. *American Journal of Medical Genetics Supplement, 6,* 115–125.

Bellugi, U., Bihrle, A., Neville, H., Doherty, S., & Jernigan, T. 1992. Language, cognition, and brain organization in a neurodevelopmental disorder. In M. Gunnar & C. Nelson (Eds.), *Developmental behavioral neuroscience: The Minnesota Symposia on Child Psychology.* Hillsdale, N.J.: Erlbaum.

Berlin, B., Breedlove, D., & Raven, P. 1973. General principles of classification and nomenclature in folk biology. *American Anthropologist, 87,* 298–315.

Berlin, B., & Kay, P., 1969. *Basic color terms: Their universality and evolution.* Berkeley: University of California Press.

Bernstein, T. M. 1977. *The careful writer: A modern guide to English usage.* New York: Atheneum.

Berwick, R. C. 1985. *The acquisition of syntactic knowledge.* Cambridge, Mass.: MIT Press.

Berwick, R. C., Abney, S. P., & Tenny, C. (Eds.), 1991. *Principle-based parsing: Computation and psycholinguistics.* Dordrecht, Netherlands: Kluwer.

Berwick, R. C., & Weinberg, A. 1984. *The grammatical basis of linguistic performance.* Cambridge, Mass.: MIT Press.

Bever, T. G. 1970. The cognitive basis for linguistic structures. In J. R. Hayes (Ed.), *Cognition and the development of language.* New York: Wiley.

Bever, T. G., Carrithers, C., Cowart, W., & Townsend, D. J. 1989. Language processing and familial handedness. In A. M. Galaburda (Ed.), *From reading to neurons.* Cambridge, Mass.: MIT Press.

Bever, T. G., & McElree, B. 1988. Empty categories access their antecedents during comprehension. *Linguistic Inquiry, 19,* 35–45.

Bickerton, D. 1981. *Roots of language.* Ann Arbor, Mich.: Karoma.

Bickerton, D., & commentators. 1984. The language bioprogram hypothesis. *Behavioral and Brain Sciences, 7,* 173–221.

Bickerton, D. 1990. *Language and species.* Chicago: University of Chicago Press.

Bickerton, D. 1992. The pace of syntactic acquisition. In L. A. Sutton, C. Johnson, & R. Shields (Eds.), *Proceedings of the 17th Annual Meeting of the Berkeley Linguistics Society: General Session and Parasession on the Grammar of Event Structure.* Berkeley, Calif.: Berkeley Linguistics Society.

Birdsong, D. 1989. *Metalinguistic performance and interlinguistic competence*. New York: Springer-Verlag.

Bishop, D., V. M., North, T., & Conlan, D. 1993. Genetic basis for Specific Language Impairment: Evidence from a twin study. Unpublished manuscript, Medical Research Council Applied Psychology Unit, Cambridge, U.K.

Bley-Vroman, R. 1990. The logical problem of foreign language learning. *Linguistic Analysis, 20,* 3–49.

Bloom, A. H. 1981. *The linguistic shaping of thought: A study in the impact of language on thinking in China and the west*. Hillsdale, N.J.: Erlbaum.

Bloom, A. H. 1984. Caution—the words you use may affect what you say: A response to Au. *Cognition, 17,* 275–287.

Bodmer, W. F., & Cavalli-Sforza, L. L. 1970. Intelligence and race. *Scientific American,* October.

Bolinger, D. 1980. *Language: The loaded weapon*. New York: Longman.

Botha, R. P. 1989. *Challenging Chomsky*. Cambridge, Mass.: Blackwell.

Bouchard, T. J., Jr., Lykken, D. T., McGue, M., Segal, N. L., & Tellegen, A. 1990. Sources of human psychological differences: The Minnesota study of twins reared apart. *Science, 250,* 223–228.

Bowerman, M. 1982. Evaluating competing linguistic models with language acquisition data: Implications of developmental errors with causative verbs. *Quaderni di Semantica, 3,* 5–66.

Braine, M. D. S. 1971. On two types of models of the internalization of grammars. In D. I. Slobin (Ed.), *The ontogenesis of grammar: A theoretical symposium*. New York: Academic Press.

Braine, M. D. S. 1976. Children's first word combinations. *Monographs of the Society for Research in Child Development, 41.*

Brandon, R. N., & Hornstein, N. 1986. From icons to symbols: Some speculations on the origin of language. *Biology and Philosophy, 1,* 169–189.

Brandreth, G. 1980. *The joy of lex*. New York: Morrow.

Breland, K., & Breland, M. 1961. The misbehavior of organisms. *American Psychologist, 16,* 681–684.

Bresnan, J. 1982. *The mental representation of grammatical relations.* Cambridge, Mass.: MIT Press.

Bresnan, J. 1990. Levels of representation in locative inversion: A comparison of English and Chichewa. Unpublished manuscript, Department of Linguistics, Stanford University.

Bresnan, J., & Moshi, L. 1988. Applicatives in Kivunjo (Chaga): Implications for argument structure and syntax. Unpublished manuscript, Department of Linguistics, Stanford University.

Brown, D. E. 1991. *Human universals.* New York: McGraw-Hill.

Brown, P., & Levinson, S. C. 1987. *Politeness: Some universals in language usage.* New York: Cambridge University Press.

Brown, R. 1957. Linguistic determinism and parts of speech. *Journal of Abnormal and Social Psychology, 55,* 1–5.

Brown, R. 1958. *Words and things.* New York: Free Press.

Brown, R. 1973. *A first language: The early stages.* Cambridge, Mass.: Harvard University Press.

Brown, R., and Hanlon, C. 1970. Derivational complexity and order of acquisition in child speech. In J. R. Hayes (Ed.), *Cognition and the development of language.* New York: Wiley.

Brunvand, J. H. 1989. *Curses! Broiled again! The hottest urban legends going.* New York: Norton.

Bryson, B. 1990. *The mother tongue.* New York: Morrow.

Buchsbaum, R. 1948. *Animals without backbones* (2nd ed.). Chicago: University of Chicago Press.

Burling, R. 1986. The selective advantage of complex language. *Ethology and Sociobiology, 7,* 1–16.

Burling, R. 1992. *Patterns of language: Structure, variation, change.* New York: Academic Press.

Bybee, J. 1985. *Morphology: A study of the relation between meaning and form.* Philadelphia: Benjamins.

Calvin, W. H. 1983. *The throwing madonna: Essays on the brain.* New York: McGraw-Hill.

Campbell, J. 1982. *Grammatical man.* New York: Simon & Schuster.

Caplan, D. 1987. *Neurolinguistics and linguistic aphasiology.* New York: Cambridge University Press.

Caplan, D. 1992. *Language: Structure, processing, and disorders.* Cambridge, Mass.: MIT Press.

Carey, S. 1978. The child as word-learner. In M. Halle, J. Bresnan, & G. A. Miller (Eds.), *Linguistic theory and psychological reality.* Cambridge, Mass.: MIT Press.

Carey, S. 1985. *Conceptual change in childhood.* Cambridge, Mass.: MIT Press.

Carrington, R. 1958. *Elephants.* London: Chatto & Windus.

Carroll, J. B. (Ed.) 1956. *Language, thought, and reality: Selected writings of Benjamin Lee Whorf.* Cambridge, Mass.: MIT Press.

Carroll, L. 1871/1981. *Alice's adventures in Wonderland and Through the looking-glass.* New York: Bantam Books.

Cassidy, F. G. (Ed.). 1985. *Dictionary of American regional English.* Cambridge, Mass.: Harvard University Press.

Cavalli-Sforza, L. L. 1991. Genes, peoples, and languages. *Scientific American, 265,* 104–110.

Cavalli-Sforza, L. L., & Feldman, M. W. 1981. *Cultural transmission and evolution: A quantitative approach.* Princeton, N.J.: Princeton University Press.

Cavalli-Sforza, L. L., Piazza, A., Menozzi, P., & Mountain, J. 1988. Reconstruction of human evolution: Bringing together genetic, archaeological, and linguistic data. *Proceedings of the National Academy of Science, 85,* 6002–6006.

Cheney, D. L., & Seyfarth, R. M. 1992. The representation of social relations by monkeys. *Cognition, 37,* 167–196. Also in Gallistel, 1992.

Chomsky, C. 1970. Reading, writing, and phonology. *Harvard Educational Review, 40,* 287–309.

Chomsky, N. 1957. *Syntactic structures.* The Hague: Mouton.

Chomsky, N. 1959. A review of B. F. Skinner's "Verbal Behavior." *Language, 35,* 26–58.

Chomsky, N. 1965. *Aspects of the theory of syntax.* Cambridge, Mass.: MIT Press.

Chomsky, N. 1972. *Language and mind* (enl. ed.). New York: Harcourt Brace Jovanovich.

Chomsky, N. 1975. *Reflections on language.* Pantheon.

Chomsky, N. 1980a. *Rules and representations.* New York: Columbia University Press.

Chomsky, N., & commentators. 1980b. Rules and representations. *Behavioral and Brain Sciences, 3,* 1–61.

Chomsky, N. 1986. *Barriers.* Cambridge: MIT Press.

Chomsky, N. 1988. *Language and problems of knowledge: The Managua lectures.* Cambridge, Mass.: MIT Press.

Chomsky, N. 1991. Linguistics and cognitive science: Problems and mysteries. In Kasher, 1991.

Chomsky, N., & Halle, M. 1968/1991. *The sound pattern of English.* Cambridge, Mass.: MIT Press.

Chorover, S. 1979. *From genesis to genocide.* Cambridge, Mass.: MIT Press.

Clark, E. V. 1993. *The lexicon in acquisition.* New York: Cambridge University Press.

Clark, H. H., & Clark, E. V. 1977. *Psychology and language.* New York: Harcourt Brace Jovanovich.

Clemens, S. L. 1910. The horrors of the German language. In *Mark Twain's speeches.* New York: Harper.

Cole, R. A., & Jakimik, J. 1980. A model of speech perception. In R. A. Cole (Ed.), *Perception and production of fluent speech.* Hillsdale, N.J.: Erlbaum.

Columbia Journalism Review. (Ed.) 1980. *Squad helps dog bite victim.* New York: Doubleday.

Committee on the Judiciary, United States House of Representatives, 93rd Congress. 1974. *Transcripts of eight recorded presidential conversations.* Serial No. 34. Washington, D.C.: U.S. Government Printing Office.

Comrie, B. 1981. *Language universals and linguistic typology.* Chicago: University of Chicago Press.

Comrie, B. 1990. *The world's major languages.* New York: Oxford University Press.

Connolly, B., & Anderson, R. 1987. *First contact: New Guinea highlanders encounter the outside world.* New York: Viking Penguin.

Cooper, W. E., & Ross, J. R. 1975. World order. In R. E. Grossman, L. J.

San, & T. J Vance (Eds.), *Papers from the parasession on functionalism*. Chicago: Chicago Linguistics Society.

Corballis, M. 1991. *The lopsided ape*. New York: Oxford University Press.

Corén, S. 1992. *The left-hander syndrome: The causes and consequences of left-handedness*. New York: Free Press.

Corina, D. P., Vaid, J., & Bellugi, U. 1992. The linguistic basis of left hemisphere specialization. *Science, 255,* 1258–1260.

Cornell, T. L., Fromkin, V. A., & Mauner, G. 1993. The syntax-there-but-not-there paradox: A linguistic account. *Current Directions in Psychological Science, 2.*

Cosmides, L. & Tooby, J. 1987. From evolution to behavior: Evolutionary psychology as the missing link. In J. Dupré (Ed.), *The latest on the best: Essays on evolution and optimality*. Cambridge, Mass.: MIT Press.

Cosmides, L., & Tooby, J. 1992. Cognitive adaptations for social exchange. In Barkow, Cosmides, & Tooby, 1992.

Cowan, N., Braine, M. D. S., & Leavitt, L. A. 1985. The phonological and metaphonological representation of speech: Evidence from fluent backward talkers. *Journal of Memory and Language, 24,* 679–698.

Crain, S., & Nakayama, M. 1986. Structure dependence in children's language. *Language, 62,* 522–543.

Cromer, R. F. 1991. The cognition hypothesis of language acquisition? In R. F. Cromer, *Language and thought in normal and handicapped children*. Cambridge, Mass.: Blackwell.

Cronin, H. 1992. *The ant and the peacock*. New York: Cambridge University Press.

Crystal, D. 1987. *The Cambridge encyclopedia of language*. New York: Cambridge University Press.

Curtiss, S. 1989. The independence and task-specificity of language. In A. Bornstein & J. Bruner (Eds.), *Interaction in human development*. Hillsdale, N.J.: Erlbaum.

Daly, M., & Wilson, M. 1988. *Homicide*. Hawthorne, N.Y.: Aldine de Gruyter.

Damasio, A. R., & Damasio, H. 1992. Brain and language. *Scientific American, 267* (September), 88–95.

References  ❧  477

Darwin, C. R. 1859/1964. *On the origin of species.* Cambridge, Mass.: Harvard University Press.

Darwin, C. R. 1872. *The expression of emotion in man and animals.* London: Murray.

Darwin, C. R. 1874. *The descent of man and selection in relation to sex* (2nd ed.). New York: Hurst & Co.

Dawkins, R. 1986. *The blind watchmaker.* New York: Norton.

Deacon, T. W. 1988. Evolution of human language circuits. In H. Jerison & I. Jerison (Eds.), *Intelligence and evolutionary biology.* New York: Springer-Verlag.

Deacon, T. W. 1989. The neural circuitry underlying primate cells and human language. *Human Evolution, 4,* 367–401.

Degler, C. N. 1991. *In search of human nature: The decline and revival of Darwinism in American social thought.* New York: Oxford University Press.

Denes, P. B., & Pinson, E. N. 1973. *The speech chain: The physics and biology of spoken language.* Garden City, N.Y.: Anchor/Doubleday.

Dennett, D. C., & commentators. 1983. Intentional systems in cognitive ethology: The "Panglossian Paradigm" defended. *Behavioral and Brain Sciences, 6,* 343–390.

Department of Linguistics, Ohio State University. 1991. *Language files* (5th ed.). Columbus: Ohio State University.

DiSciullo, A. M., & Williams, E. 1987. *On the definition of word.* Cambridge, Mass.: MIT Press.

Diamond, J. M. 1990. The talk of the Americas. *Nature, 344,* 589–590.

Dodd, J., & Jessell, T. M. 1988. Axon guidance and the patterning of neuronal projections in vertebrates. *Science, 242,* 692–699.

Dronkers, N. F., Shapiro, J., Redfern, B., & Knight, R. 1992. The role of Broca's area in Broca's aphasia. *Journal of Clinical and Experimental Neuropsychology, 14,* 52–53.

Dryer, M. S. 1992. The Greenbergian word order correlations. *Language, 68,* 81–138.

Dyson, F. 1979. *Disturbing the universe.* New York: Harper.

*The Economist.* 1992. Minds in the making: A survey of Artificial Intelligence. March 14, 1992, 1–24.

The Editors of *The New Republic*. 1992. *Bushisms*. New York: Workman.

Elimas, P. D., Siqueland, E. R., Jusczyk, P., & Vigorito, J. 1971. Speech perception in infants. *Science, 171*, 303–306.

Emonds, J. 1986. Grammatically deviant prestige constructions. In *A festschrift for Sol Saporta*. Seattle: Noit Amrofer.

Ervin-Tripp, S. 1973. Some strategies for the first two years. In T. E. Moore (Ed.), *Cognitive development and the acquisition of language*. New York: Academic Press.

Espy, W. R. 1975. *An almanac of words at play*. New York: Clarkson Potter.

Espy, W. R. 1989. *The word's gotten out*. New York: Clarkson Potter.

Etcoff, N.L. 1986. The neuropsychology of emotional expression. In G. Goldstein & R. E. Tarter (Eds.), *Advances in Clinical Neuropsychology, Vol. 3*. New York: Plenum.

Fahey, V., Kamitomo, G. A., & Cornell, E. H. 1978. Heritability in syntactic development: a critique of Munsinger and Douglass. *Child Development, 49*, 253–257.

Farah, M. J. 1990. *Visual agnosia*. Cambridge, Mass.: MIT Press.

Fernald, A. 1992. Human maternal vocalizations to infants as biologically relevant signals: An evolutionary perspective. In Barkow, Cosmides, & Tooby, 1992.

Ferreira, F., & Henderson, J. M. 1990. The use of verb information in syntactic parsing: A comparison of evidence from eye movements and word-by-word self-paced reading. *Journal of Experimental Psychology: Learning, Memory and Cognition, 16*, 555–568.

Fischer, S. D. 1978. Sign language and creoles. In Siple, 1978.

Flavell, J. H., Miller, P. H., & Miller, S. A. 1993. *Cognitive development* (3rd ed.). Englewood Cliffs, N.J.: Prentice Hall.

Fodor, J. A. 1975. *The language of thought*. New York: Crowell.

Fodor, J. A. 1983. *The modularity of mind*. Cambridge, Mass.: MIT Press.

Fodor, J. A., & commentators. 1985. Précis and multiple book review of "The Modularity of Mind." *Behavioral and Brain Sciences, 8*, 1–42.

Fodor, J. D. 1989. Empty categories in sentence processing. *Language and Cognitive Processes, 4*, 155–209.

Ford, M., Bresnan, J., & Kaplan, R. M. 1982. A competence-based theory of syntactic closure. In Bresnan, 1982.

Frazier, L. 1989. Against lexical generation of syntax. In Marslen-Wilson, 1989.

Frazier, L., & Fodor, J. D. 1978. The sausage machine. A new two-stage parsing model. *Cognition, 6,* 291–328.

Freedman, D. H. 1990. Common sense and the computer. *Discover,* August, 65–71.

Freeman, D. 1983. *Margaret Mead and Samoa: The making and unmaking of an anthropological myth.* Cambridge, Mass.: Harvard University Press.

Friedin, R. 1992. *Foundations of generative syntax.* Cambridge, Mass.: MIT Press.

Galaburda, A.M., & Pandya, D. N. 1982. Role of architectonics and connections in the study of primate brain evolution. In E. Armstrong & D. Falk (Eds.), *Primate brain evolution.* New York: Plenum.

Gallen, C. 1994. Neuromagnetic assessment of human cortical function and dysfunction: Magnetic source imaging. In P. Tallal (Ed.), *Neural and cognitive mechanisms underlying speech, language, and reading.* Cambridge, Mass.: Harvard University Press.

Gallistel, C. R. 1990. *The organization of learning.* Cambridge, Mass.: MIT Press.

Gallistel, C. R. (Ed.) 1992. *Aminal cognition.* Cambridge, Mass.: MIT Press.

Gardner, B. T., & Gardner, R. A. 1974. Comparing the early utterances of child and chimpanzee. In A. Pick (Ed.), *Minnesota symposium on child psychology, Vol. 8.* Minneapolis: University of Minnesota Press.

Gardner, H. 1974. *The shattered mind.* New York: Vintage.

Gardner, H. 1985. *The mind's new science: A history of the cognitive revolution.* New York: Basic Books.

Gardner, R. A., & Gardner, B. T. 1969. Teaching sign language to a chimpanzee. *Science, 165,* 664–672.

Garfield, J. (Ed.) 1987. *Modularity in knowledge representation and natural-language understanding.* Cambridge, Mass.: MIT Press.

Garnsey, S. M., Tanenhaus, M. D., & Chapman, R. M. 1989. Evoked potentials and the study of sentence comprehension. *Journal of Psycholinguistic Research, 18,* 51–60.

Garrett, M. 1990. Sentence processing. In Osherson & Lasnik, 1990.

Gazzaniga, M. S. 1978. *The integrated mind.* New York: Plenum.

Gazzaniga, M. S. 1983. Right hemisphere language following brain bisection: A 20-year perspective. *American Psychologist, 38,* 528–549.

Gazzaniga, M. S. 1989. Organization of the human brain. *Science, 245,* 947–952.

Gazzaniga, M. S. 1992. *Nature's mind.* New York: Basic Books.

Geertz, C. 1984. Anti anti-relativism. *American Anthropologist, 86,* 263–278.

Geisel, T. S. 1955. *On beyond zebra, by Dr. Seuss.* New York: Random House.

Gelman, S. A., & Markman, E. 1987. Young children's inductions from natural kinds: The role of categories and appearances. *Child Development, 58,* 1532–1540.

Gentner, D., & Jeziorski, M. 1989. Historical shifts in the use of analogy in science. In B. Gholson, W. R. Shadish, Jr., R. A. Beimeyer, & A. Houts (Eds.), *The psychology of science: Contributions to metascience.* New York: Cambridge University Press.

Geschwind, N. 1979. Specializations of the human brain. *Scientific American,* September.

Geschwind, N., & Galaburda, A. 1987. *Cerebral lateralization: Biological mechanisms, associations, and pathology.* Cambridge, Mass.: MIT Press.

Gibbons, A. 1992. Neanderthal language debate: Tongues wag anew. *Science, 256,* 33–34.

Gibbons, A. 1993. Mitochondrial Eve refuses to die. *Science, 259,* 1249–1250.

Gibson, E. In press. *A computational theory of human linguistic processing: Memory limitations and processing breakdown.* Cambridge, Mass.: MIT Press.

Gleitman, L. R. 1981. Maturational determinants of language growth. *Cognition, 10,* 103–114.

Gleitman, L. R. 1990. The structural sources of verb meaning. *Language Acquisition, 1,* 3–55.

Goldsman, M. 1992. Quayle quotes. Various computer networks.

Goodglass, H. 1973. Studies on the grammar of aphasics. In H. Goodglass & S. E. Blumstein (Eds.), *Psycholinguistics and aphasia*. Baltimore: Johns Hopkins University Press.

Goodman, N. 1972. Seven strictures on similarity. In *Problems and projects*. Indianapolis: Bobbs-Merrill.

Gopnik, M. 1990a. Dysphasia in an extended family. *Nature, 344,* 715.

Gopnik, M. 1990b. Feature blindness: A case study, *Language Acquisition, 1,* 139–164.

Gopnik, M. 1993. The absence of obligatory tense in genetic language impairment. Unpublished manuscript, Department of Linguistics, McGill University.

Gopnik, M., & Crago, M. 1991. Familial aggregation of a developmental language disorder. *Cognition, 39,* 1–50.

Gordon, P. 1986. Level-ordering in lexical development. *Cognition, 21,* 73–93.

Gould, J. L., & Marler, P. 1987. Learning by instinct. *Scientific American,* January.

Gould, S. J. 1977. Why we should not name human races: A biological view. In S. J. Gould, *Ever since Darwin*. New York: Norton.

Gould, S. J. 1981. *The mismeasure of man*. New York: Norton.

Gould, S. J. 1985. *The flamingo's smile: Reflections in natural history*. New York: Norton.

Gould, S. J., & Lewontin, R. C. 1979. The spandrels of San Marco and the Panglossian paradigm: A critique of the adaptationist programme. *Proceedings of the Royal Society of London, 205,* 281–288.

Green, D. M. 1976. *An introduction to hearing*. Hillsdale, N.J.: Erlbaum.

Greenberg, J. H. (Ed.) 1963. *Universals of language*. Cambridge, Mass.: MIT Press.

Greenberg, J. H. 1987. *Language in the Americas*. Stanford, Calif.: Stanford University Press.

Greenberg, J. H., Ferguson, C. A., & Moravcsik, E. A. (Eds.) 1978. *Universals of human language* (4 vols.). Stanford, Calif.: Stanford University Press.

Greenfield, P. M., & Savage-Rumbaugh, E. S. 1991. Imitation, grammatical development, and the invention of protogrammar by an ape. In Krasnegor et al., 1991.

Grice, H. P. 1975. Logic and conversation. In P. Cole & J. L. Morgan (Eds.), *Syntax and Semantics 3: Speech acts*. New York: Academic Press.

Grimshaw, J. 1990. *Argument structure*. Cambridge, Mass.: MIT Press.

Grosjean, F., 1982. *Life with two languages: An introduction to bilingualism*. Cambridge, Mass.: Harvard University Press.

Guy, J. 1992. Genes, peoples, and languages? An examination of a hypothesis by Cavalli-Sforza. LINGUIST electronic bulletin board, January 27.

Hakuta, K. 1986. *Mirror of language: The debate on bilingualism*. New York: Basic Books.

Hale, K., Krauss, M., Watahomigie, L., Yamamoto, A., Craig, C., Jeanne, L. M., & England, N. 1992. Endangered languages. *Language, 68*, 1–42.

Halle, M. 1983. On distinctive features and their articulatory implementation. *Natural Language and Linguistic Theory, 1*, 91–105.

Halle, M. 1990. Phonology. In Osherson & Lasnik, 1990.

Harding, R. M., & Sokal, R. R. 1988. Classification of the European language families by genetic distance. *Proceedings of the National Academy of Science, 85*, 9370–9372.

Hardy-Brown, K., Plomin, R., & DeFries, J. C. 1981. Genetic and environmental influences on the rate of communicative development in the first year of life. *Developmental Psychology, 17*, 704–717.

Harman, G. (Ed.) 1974. *On Noam Chomsky: Critical essays*. New York: Doubleday.

Harnad, S. R., Steklis, H. S., & Lancaster, J. (Eds.) 1976. *Origin and evolution of language and speech* (special volume). *Annals of the New York Academy of Sciences, 280*.

Harris, R. A. 1993. *The linguistics wars*. New York: Oxford University Press.

Hart, J., Berndt, R. S., & Caramazza, A. 1985. Category-specific naming deficit following cerebral infarction. *Nature, 316*, 439–440.

Haugeland, J. (Ed.) 1981. *Mind design*. Cambridge, Mass.: MIT Press.

Hawkins, J. (Ed.) 1988. *Explaining language universals*. Basil Blackwell.

Hayakawa, S. I. 1964. *Language in thought and action* (2nd ed.). New York: Harcourt Brace.

Heath, S. B. 1983. *Ways with words: Language, life and work in communities and classrooms.* New York: Cambridge University Press.

Heider, E. R. 1972. Universals in color naming and memory. *Cognitive Psychology, 3,* 337–354.

Hillis, A. E., & Caramazza, A. 1991. Category-specific naming and comprehension impairment: A double dissociation. *Brain, 114,* 2081–2094.

Hinton, G. E., & Nowlan, S. J. 1987. How learning can guide evolution. *Complex Systems, 1,* 495–502.

Hirschfeld, L. A., & Gelman, S. A. (Eds.) In press. *Domain specificity in cognition and culture.* New York: Cambridge University Press.

Hirsh-Pasek, K., & Golinkoff, R. M. 1991. Language comprehension: A new look at some old themes. In Krasnegor et al., 1991.

Hockett, C. F. 1960. The origin of speech. *Scientific American, 203,* 88–111.

Hofstadter, D. R. 1985. *Metamagical themas.* New York: Basic Books.

Holden, C. 1987. The genetics of personality. *Science, 237,* 598–601.

Holm, J. 1988. *Pidgins and creoles* (2 vols.). New York: Cambridge University Press.

Holmes, R. B., & Smith, B. S. 1977. *Beginning Cherokee* (2nd ed.). Norman, Okla.: University of Oklahoma Press.

Hubel, D. 1988. *Eye, brain, and vision.* San Francisco: Freeman.

Humboldt, W. von. 1836/1972. *Linguistic variability and intellectual development* (G. C. Buck & F. Raven, Trans.). Philadelphia: University of Pennsylvania Press.

Hurford, J. R. 1989. Biological evolution of the Saussurean sign as a component of the language acquisition device. *Lingua, 77,* 187–222.

Hurford, J. R. 1991. The evolution of the critical period in language acquisition. *Cognition, 40,* 159–201.

Huttenlocher, P. R. 1990. Morphometric study of human cerebral cortex development. *Neuropsychologia, 28,* 517–527.

Ingram, D. 1989. *First language acquisition: Method, description, and explanation.* New York: Cambridge University Press.

Jackendoff, R. S. 1977. *X-bar syntax: A study of phrase structure.* Cambridge, Mass.: MIT Press.

Jackendoff, R. S. 1987. *Consciousness and the computational mind.* Cambridge, Mass.: MIT Press.

Jackendoff, R. S. 1992. *Languages of the mind.* Cambridge, Mass.: MIT Press.

James, W. 1892/1920. *Psychology: Briefer course.* New York: Henry Holt & Company.

Jespersen, O. 1938/1982. *Growth and structure of the English language.* Chicago: University of Chicago Press.

Jeyifous, S. 1986. Atimodemo: Semantic conceptual development among the Yoruba. Doctoral dissertation, Cornell University.

Johnson, S. 1755. Preface to the *Dictionary.* Reprinted in E. L. McAdam, Jr., and G. Milne (Eds.), 1964, *Samuel Johnson's Dictionary: A modern selection.* New York: Pantheon.

Joos, M. (Ed.) 1957. *Readings in linguistics: The development of descriptive linguistics in America since 1925.* Washington, D.C.: American Council of Learned Societies.

Jordan, M. I., & Rosenbaum, D. 1989. Action. In Posner, 1989.

Joshi, A. K. 1991. Natural language processing. *Science, 253,* 1242–1249.

Kaplan, R. 1972. Augmented transition networks as psychological models of sentence comprehension. *Artificial Intelligence, 3,* 77–100.

Kaplan, S. 1992. Environmental preference in a knowledge-seeking, knowledge-using organism. In Barkow, Cosmides, & Tooby, 1992.

Kasher, A. (Ed.) 1991. *The Chomskyan turn.* Cambridge, Mass.: Blackwell.

Katzner, K. 1977. *The languages of the world.* New York: Routledge & Kegan Paul.

Kay, P., & Kempton, W. 1984. What is the Sapir-Whorf hypothesis? *American Anthropologist, 86,* 65–79.

Kaye, J. 1989. *Phonology: A cognitive view.* Hillsdale, N.J.: Erlbaum.

Keenan, E. O. 1976. Towards a universal definition of "subject." In C. Li (Ed.), *Subject and Topic.* New York: Academic Press.

Kegl, J. & Iwata, G. A. 1989. Lenguage de Signos Nicaragüense: A pidgin sheds light on the "creole?" ASL. *Proceedings of the Fourth Annual*

*Meeting of the Pacific Linguistics Conference*. Eugene, Ore.: University of Oregon.

Kegl, J., & Lopez, A. 1990. The deaf community in Nicaragua and their sign language(s). Unpublished paper, Department of Molecular and Behavioral Neuroscience, Rutgers University, Newark, N.J. Originally presented at Encuentro Latinamericano y del Caribe de Educadores de Sordos: Il Encuentro Nacional de Especialistas en la Educacion del Sordo, November 12–17.

Keil, F. 1989. *Concepts, kinds, and conceptual development*. Cambridge, Mass.: MIT Press.

Kenstowicz, M., & Kisseberth, C. 1979. *Generative phonology*. New York: Academic Press.

Kim, J. J., Pinker, S., Prince, A., & Prasada, S. 1991. Why no mere mortal has ever flown out to center field. *Cognitive Science, 15,* 173–218.

Kim, J. J., Marcus, G. F., Pinker, S., Hollander, M., & Coppola, M. In press. Sensitivity of children's inflection to morphological structure. *Journal of Child Language*.

King, M., & Wilson, A. 1975. Evolution at two levels in humans and chimpanzees. *Science, 188,* 107–116.

Kinsbourne, M. 1978. Evolution of language in relation to lateral action. In M. Kinsbourne (Ed.), *Asymmetrical function of the brain*. New York: Cambridge University Press.

Kiparsky, P. 1976. Historical linguistics and the origin of language. In Harnad, Steklis, & Lancaster, 1976.

Kiparsky, P. 1982. Lexical phonology and morphology. In I. S. Yang (Ed.), *Linguistics in the morning calm*. Seoul: Hansin.

Kitcher, P. 1985. *Vaulting ambition: Sociobiology and the quest for human nature*. Cambridge, Mass.: MIT Press.

Klima, E., & Bellugi, U. 1979. *The signs of language*. Cambridge, Mass.: Harvard University Press.

Kluender, R., & Kutas, M. 1993. Bridging the gap: Evidence from ERPs on the processing of unbounded dependencies. *Journal of Cognitive Neuroscience, 4.*

Konner, M. 1982. *The tangled wing: Biological constraints on the human spirit*. Harper.

Kornai, A., & Pullum, G. K. 1990. The X-bar theory of phrase structure. *Language, 66,* 24–50.

Korzybski, A. 1933. *Science and sanity: An introduction to non-Aristotelian systems and General Semantics.* Lancaster, Penn.: International Non-Aristotelian Library.

Kosslyn, S. M. 1983. *Ghosts in the mind's machine: Creating and using images in the brain.* New York: Norton.

Kosslyn, S. M. 1987. Seeing and imagining in the cerebral hemispheres: A computational approach. *Psychological Review, 94,* 144–175.

Krasnegor, N. A., Rumbaugh, D. M., Schiefelbusch, R. L., & Studdert-Kennedy, M. (Eds.) 1991. *Biological and behavioral determinants of language development.* Hillsdale, N.J.: Erlbaum.

Kučera, H. 1992. The mathematics of language. In *The American Heritage Dictionary of the English language* (3rd ed.). Boston: Houghton Mifflin.

Kuhl, P., & Williams, K. A., Lacerda, F., Stevens, K. N., & Lindblom, B. 1992. Linguistic experience alters phonetic perception in infants by six months of age. *Science, 255,* 606–608.

Kuno, S. 1974. The position of relative clauses and conjunctions. *Linguistic Inquiry, 5,* 117–136.

Labov, W. 1969. The logic of nonstandard English. *Georgetown Monographs on Language and Linguistics, 22,* 1–31.

Ladefoged, P. 1992. Another view of endangered languages. *Language, 68,* 809–811.

Lakoff, G. 1987. *Women, fire, and dangerous things.* Chicago: University of Chicago Press.

Lakoff, G., & Johnson, M. 1980. *Metaphors we live by.* Chicago: University of Chicago Press.

Lakoff, R. 1990. *Talking power: The politics of language in our lives.* New York: Basic Books.

Lambert, D., & The Diagram Group. 1987. *The field guide to early man.* New York: Facts on File Publications.

Lederer, R. 1987. *Anguished English.* Charleston: Wyrick.

Lederer, R. 1990. *Crazy English.* New York: Pocket Books.

Leech, G. N. 1983. *Principles of pragmatics.* London: Longman.

Lenat, D. B., & Guha, D. V. 1990. *Building large knowledge-based systems*. Reading, Mass.: Addison-Wesley.

Lenneberg, E. H. 1953. Cognition and ethnolinguistics. *Language, 29,* 463–471.

Lenneberg, E. H. 1967. *Biological foundations of language*. New York: Wiley.

Lesser, V. R., Fennel, R. D., Erman, L. D., & Reddy, R. D. 1975. The Hearsay II speech understanding system. *IEEE Transactions on Acoustics, Speech, and Signal Processing, 23,* 11–24.

Levinson, S. C. 1983. *Pragmatics*. New York: Cambridge University Press.

Lewin, R. 1980. Is your brain really necessary? *Science, 210,* 1232–1234.

Lewontin, R. C. 1966. Review of G. C. Williams' "Adaptation and natural selection." *Science, 152,* 338–339.

Lewontin, R. C. 1982. *Human diversity*. San Francisco: Scientific American.

Lewontin, R. C., Rose, S., & Kamin, L. 1984. *Not in our genes*. New York: Pantheon.

Liberman, A. M., Cooper, F. S., Shankweiler, D. P., & Studdert-Kennedy, M. 1967. Perception of the speech code. *Psychological Review, 74,* 431–461.

Liberman, A. M., & Mattingly, I. G. 1989. A specialization for speech perception. *Science, 243,* 489–494.

Lieberman, P. 1984. *The biology and evolution of language*. Cambridge, Mass.: Harvard University Press.

Lieberman, P. 1990. Not invented here. In Pinker & Bloom, 1990.

Lieberman, P., Kako, E., Friedman, J., Tajchman, G., Feldman, L. S., & Jiminez, E. B. 1992. Speech production, syntax comprehension, and cognitive deficits in Parkinson's Disease. *Brain and Language, 43,* 169–189.

Limber, J. 1973. The genesis of complex sentences. In T. E. Moore (Ed.), *Cognitive development and the acquisition of language*. New York: Academic Press.

Linebarger, M., Schwartz, M. F., & Saffran, E. M. 1983. Sensitivity to grammatical structure in so-called agrammatic aphasics. *Cognition, 13,* 361–392.

Liu, L. G. 1985. Reasoning counterfactually in Chinese: Are there any obstacles? *Cognition, 21,* 239–270.

Locke, J. L. 1992. Structure and stimulation in the ontogeny of spoken language. *Developmental Psychobiology, 28,* 430–440.

Locke, J. L., & Mather, P. L. 1989. Genetic factors in the ontogeny of spoken language: Evidence from monozygotic and dizygotic twins. *Journal of Child Language, 16,* 553–559.

Logan, R. K. 1986. *The alphabet effect.* New York: St. Martin's Press.

Long, M. H. 1990. Maturational constraints on language development. *Studies in Second Language Acquisition, 12,* 251–285.

Lorge, I., & Chall, J. 1963. Estimating the size of vocabularies of children and adults: An analysis of methodological issues. *Journal of Experimental Education, 32,* 147–157.

Ludlow, C. L., & Cooper, J. A. (Eds.) 1983. *Genetic aspects of speech and language disorders.* New York: Academic Press.

Lykken, D. T., McGue, M., Tellegen, A., & Bouchard, T. J., Jr. 1992. Emergenesis: Genetic traits that may not run in families. *American Psychologist, 47,* 1565–1577.

MacDonald, M. C. 1989. Priming effects from gaps to antecedents. *Language and Cognitive Processes, 4,* 1–72.

MacDonald, M. C., Just, M. A., & Carpenter, P. A. 1992. Working memory constraints on the processing of syntactic ambiguity. *Cognitive Psychology, 24,* 56–98.

MacWhinney, B. 1991. *The CHILDES Project: Tools for Analyzing Talk.* Hillsdale, N.J.: Erlbaum.

Malotki, E. 1983. *Hopi time: A linguistic analysis of temporal concepts in the Hopi language.* Berlin: Mouton.

Marcus, G. F. 1993. Negative evidence in language acquisition. *Cogniton, 46,* 53–85.

Marcus, G. F., Brinkmann, U., Clahsen, H., Wiese, R., Woest, A., & Pinker, S. 1993. German inflection: The exception that proves the rule. MIT Center for Cognitive Science Occasional Paper #47.

Marcus, G. F., Pinker, S., Ullman, M., Hollander, M., Rosen, T. J., &

Xu, F. 1992. Overregularization in language acquisition. *Monographs of the Society for Research in Child Development, 57.*

Markman, E. 1989. *Categorization and naming in children: Problems of induction.* Cambridge, Mass.: MIT Press.

Marr, D. 1982. *Vision.* San Francisco: Freeman.

Marslen-Wilson, W. 1975. Sentence comprehension as an interactive, parallel process. *Science, 189,* 226–228.

Marslen-Wilson, W. (Ed.) 1989. *Lexical representation and process.* Cambridge, Mass.: MIT Press.

Martin, L. 1986. "Eskimo words for snow": A case study in the genesis and decay of an anthropological example. *American Anthropologist, 88,* 418–423.

Martin, P., & Klein, R. 1984. *Quaternary extinctions.* Tucson: University of Arizona Press.

Mather, P., & Black, K. 1984. Hereditary and environmental influences on preschool twins' language skills. *Developmental Psychology, 20,* 303–308.

Mattingly, I. G., & Studdert-Kennedy, M. (Eds.) 1991. *Modularity and the motor theory of speech perception.* Hillsdale, N.J.: Erlbaum.

Maynard Smith, J. 1984. Optimization theory in evolution. In E. Sober (Ed.), *Conceptual issues in evolutionary biology.* Cambridge, Mass.: MIT Press.

Maynard Smith, J. 1986. *The problems of biology.* Oxford: Oxford University Press.

Maynard Smith, J. 1988. *Games, sex, and evolution.* New York: Harvester Wheatsheaf.

Mayr, E. 1982. *The growth of biological thought.* Cambridge, Mass.: Harvard University Press.

Mayr, E. 1983. How to carry out the adaptationist program. *American Naturalist, 121,* 324–334.

Mazoyer, B. M., Dehaene, S., Tzourio, N., Murayama, N., Cohen, L., Levrier, O., Salamon, G., Syrota, A., & Mehler, J. 1992. The cortical representation of speech. Unpublished manuscript, Laboratoire de Sciences Cognitives et Psycholinguistique, Centre Nationale de la Recherche Scientifique, Paris.

McClelland, J. L., Rumelhart, D. E., & The PDP Research Group. 1986. *Parallel distributed processing: Explorations in the microstructure of cognition, Vol. 2: Psychological and biological models.* Cambridge, Mass.: MIT Press.

McCrum, R., Cran, W., & MacNeil, R. 1986. *The story of English.* New York: Viking.

McCulloch, W. S., & Pitts, W. 1943. A logical calculus of the ideas immanent in nervous activity. *Bulletin of Mathematical Biophysics, 5,* 115–133.

McDermott, D. 1981. Artificial intelligence meets natural stupidity. In Haugeland, 1981.

McGurk, H., & MacDonald, J. 1976. Hearing lips and seeing voices. *Nature, 264,* 746–748.

Mead, M. 1935. *Sex and temperament in three primitive societies.* New York: Morrow.

Medawar, P. B. 1957. An unsolved problem in biology. In P. B. Medawar, *The uniqueness of the individual.* London: Methuen.

Mehler, J., Jusczyk, P. W., Lambertz, G., Halsted, N., Bertoncini, J., & Amiel-Tison, C. 1988. A precursor to language acquisition in young infants. *Cognition, 29,* 143–178.

Mencken, H. 1936. *The American language.* New York: Knopf.

Miceli, G., & Caramazza A. 1988. Dissociation of inflectional and derivational morphology. *Brain and Language, 35,* 24–65.

Miceli, G., Silveri, M. C., Romani, C., & Caramazza, A. 1989. Variation in the pattern of omissions and substitutions of grammatical morphemes in the spontaneous speech of so-called agrammatic patients. *Brain and Language, 36,* 447–492.

Miller, G. A. 1956. The magical number seven, plus or minus two: Some limits on our capacity for processing information. *Psychological Review, 63,* 81–96.

Miller, G. A. 1967. *The psychology of communication.* London: Penguin Books.

Miller, G. A. 1977. *Spontaneous apprentices: Children and language.* New York: Seabury Press.

Miller, G. A. 1991. *The science of words.* New York: Freeman.

Miller, G. A., & Chomsky, N. 1963. Finitary models of language users. In R. D. Luce, R. Bush, and E. Galanter (Eds.), *Handbook of mathematical psychology, Vol. 2.* New York: Wiley.

Miller, G. A., & Selfridge, J. 1950. Verbal context and the recall of meaningful material. *American Journal of Psychology, 63,* 176–185.

Miyamoto, M. M., Slightom, J. L., & Goodman, M. 1987. Phylogenetic relations of humans and African apes from DNA sequences in the ψη-globin region. *Science, 238,* 369–373.

Modgil, S., & Modgil, C. (Eds.) 1987. *Noam Chomsky: Consensus and controversy.* New York: Falmer Press.

Morgan, J. L., & Travis, L. L. 1989. Limits on negative information in language learning. *Journal of Child Language, 16,* 531–552.

Munsinger, H., & Douglass, A. 1976. The syntactic abilities of identical twins, fraternal twins and their siblings. *Child Development, 47,* 40–50.

Murdock, G. P. 1975. *Outline of world's cultures* (5th ed.). New Haven, Conn.: Human Relations Area Files.

Murphy, K. 1992. "To be" in their bonnets. *Atlantic Monthly,* February.

Myers, R. E. 1976. Comparative neurology of vocalization and speech: Proof of a dichotomy. In Harnad, Steklis, & Lancaster, 1976.

Nabokov, V. 1958. *Lolita.* New York: Putnam.

Neisser, A. 1983. *The other side of silence.* New York: Knopf.

Neville, H., Nicol, J. L., Barss, A., Forster, K. I., & Garrett, M. F. 1991. Syntactically based sentence processing classes: Evidence from event-related brain potentials. *Journal of Cognitive Neuroscience, 3,* 151–165.

*New York Times* Staff. 1974. *The White House Transcripts.* New York: Bantam Books.

Newmeyer, F. 1991. Functional explanation in linguistics and the origin of language. *Language and Communication, 11,* 3–96.

Newport, E. 1990. Maturational constraints on language learning. *Cognitive Science, 14,* 11–28.

Newport, E., Gleitman, H., & Gleitman, E. 1977. Mother I'd rather do it myself: Some effects and non-effects of maternal speech style. In C. E. Snow and C. A. Ferguson (Eds.), *Talking to children: Language input and acquisition.* Cambridge: Cambridge University Press.

Nicol, J., & Swinney, D. A. 1989. Coreference processing during sentence comprehension. *Journal of Psycholinguistic Research, 18,* 5–19.

Norman, D., & Rumelhart, D. E. (Eds.) 1975. *Explorations in cognition.* San Francisco: Freeman.

Nunberg, G. 1992. Usage in The American Heritage Dictionary: The place of criticism. In *The American Heritage Dictionary of the English language* (3rd ed.). Boston: Houghton Mifflin.

Ojemann, G. A. 1991. Cortical organization of language. *Journal of Neuroscience, 11,* 2281–2287.

Ojemann, G. A., & Whitaker, H. A. 1978. Language localization and variability. *Brain and Language, 6,* 239–260.

Orians, G. H., & Heerwagen, J. H. 1992. Evolved responses to landscapes. In Barkow, Cosmides, & Tooby, 1992.

Osherson, D. N., Stob, M., and Weinstein, S. 1985. *Systems that learn.* Cambridge, Mass.: MIT Press.

Osherson, D. N., & Lasnik, H. (Eds.) 1990. *Language: An invitation to cognitive science, Vol. 1.* Cambridge, Mass.: MIT Press.

Osherson, D. N., Kosslyn, S. M., & Hollerbach, J. M. (Eds.). 1990. *Visual cognition and action: An invitation to cognitive science, Vol. 2.* Cambridge, Mass.: MIT Press.

Osherson, D. N., & Smith, E. E. (Eds.), 1990. *Thinking: An invitation to cognitive science, Vol. 3.* Cambridge, Mass.: MIT Press.

Patterson, F. G. 1978. The gestures of a gorilla: Language acquisition in another pongid. *Brain and Language, 5,* 56–71.

Peters, A. M. 1983. *The units of language acquisition.* New York: Cambridge University Press.

Peterson, S. E., Fox, P. T., Posner, M. I., Mintun, M., & Raichle, M. E. 1988. Positron emission tomographic studies of the cortical anatomy of single-word processing. *Nature, 331,* 585–589.

Peterson, S. E., Fox, P. T., Snyder, A. Z., & Raichle, M. E. 1990. Activation of extrastriate and frontal cortical areas by visual words and wordlike stimuli. *Science, 249,* 1041–1044.

Petitto, L. A. 1988. "Language" in the prelinguistic child. In F. Kessel (Ed.), *The development of language and of language researchers: Papers presented to Roger Brown.* Hillsdale, N.J.: Erlbaum.

Petitto, L. A., & Marentette, P. F. 1991. Babbling in the manual mode: Evidence for the ontogeny of language. *Science, 251,* 1493–1496.

Petitto, L. A., & Seidenberg, M. S. 1979. On the evidence for linguistic abilities in signing apes. *Brain and Language, 8,* 162–183.

Piattelli-Palmarini, M. (Ed.) 1980. *Language and learning: The debate between Jean Piaget and Noam Chomsky.* Cambridge, Mass.: Harvard University Press.

Piattelli-Palmarini, M. 1989. Evolution, selection, and cognition: From "learning" to parameter setting in biology and the study of language, *Cognition, 31,* 1–44.

Pinker, S. 1979. Formal models of language learning. *Cognition, 7,* 217–283.

Pinker, S. 1984. *Language learnability and language development.* Cambridge, Mass.: Harvard University Press.

Pinker, S. (Ed.) 1985. *Visual cognition.* Cambridge, Mass.: MIT Press.

Pinker, S. 1987. The bootstrapping problem in language acquisition. In B. MacWhinney (Ed.), *Mechanisms of language acquisition.* Hillsdale, N.J.: Erlbaum.

Pinker, S. 1989. *Learnability and cognition: The acquisition of argument structure.* Cambridge, Mass.: MIT Press.

Pinker, S. 1990. Language acquisition. In Osherson & Lasnik, 1990.

Pinker, S. 1991. Rules of language. *Science, 253,* 530–535.

Pinker, S. 1992. Review of Bickerton's "Language and Species." *Language, 68,* 375–382.

Pinker, S. 1994. How could a child use verb syntax to learn verb semantics? *Lingua, 92.*

Pinker, S. In press. Facts about human language relevant to its evolution. In J.-P. Changeux (Ed.), *Origins of the human brain.* New York: Oxford University Press.

Pinker, S., & Birdsong, D. 1979. Speakers' sensitivity to rules of frozen word order. *Journal of Verbal Learning and Verbal Behavior, 18,* 497–508.

Pinker, S., & Bloom, P., & commentators. 1990. Natural language and natural selection. *Behavioral and Brain Sciences, 13,* 707–784.

Pinker, S., & Mehler, J. (Eds.) 1988. *Connections and symbols*. Cambridge, Mass.: MIT Press.

Pinker, S., and Prince, A. 1988. On language and connectionism: Analysis of a Parallel-Distributed Processing model of language acquisition. *Cognition, 28,* 73–193.

Pinker, S., and Prince, A. 1992. Regular and irregular morphology and the psychological status of rules of grammar. In L. A. Sutton, C. Johnson, & R. Shields (Eds.), *Proceedings of the 17th Annual Meeting of the Berkeley Linguistics Society: General Session and Parasession on the Grammar of Event Structure*. Berkeley, Calif.: Berkeley Linguistics Society.

Plomin, R. 1990. The role of inheritance in behavior. *Science, 248,* 183–188.

Poeppel, D. 1993. PET studies of language: A critical review. Unpublished manuscript, Department of Brain and Cognitive Sciences, MIT.

Poizner, H., Klima, E. S., & Bellugi, U. 1990. *What the hands reveal about the brain*. Cambridge, Mass.: MIT Press.

Posner, M. I. (Ed.) 1989. *Foundations of cognitive science*. Cambridge, Mass.: MIT Press.

Prasada, S., & Pinker, S. 1993. Generalizations of regular and irregular morphology. *Language and Cognitive Processes, 8,* 1–56.

Premack, A. J., & Premack, D. 1972. Teaching language to an ape. *Scientific American,* October.

Premack, D. 1985. "Gavagai!" or the future history of the animal language controversy. *Cognition, 19,* 207–296.

Pullum, G. K. 1991. *The great Eskimo vocabulary hoax and other irreverent essays on the study of language*. Chicago: University of Chicago Press.

Putnam, H. 1971. The "innateness hypothesis" and explanatory models in linguistics. In J. Searle (Ed.), *The philosophy of language*. New York: Oxford University Press.

Pyles, T., & Algeo, J. 1982. *The origins and development of the English language* (3rd ed.). New York: Harcourt Brace Jovanovich.

Quine, W. V. O. 1960. *Word and object*. Cambridge, Mass.: MIT Press.

Quine, W. V. O. 1969. Natural kinds. In *Ontological relativity and other essays.* New York: Columbia University Press.

Quine, W. V. O. 1987. *Quiddities: An intermittently philosophical dictionary.* Cambridge, Mass.: Harvard University Press.

Quirk, R., Greenbaum, S., Leech, G., & Svartvik, J. 1985. *A comprehensive grammar of the English language.* New York: Longman.

Radford, A. 1988. *Transformational syntax: A first course* (2nd ed.). New York: Cambridge University Press.

Rakic, P. 1988. Specification of cerebral cortical areas. *Science, 241,* 170–176.

Raymond, E. S. (Ed.) 1991. *The new hacker's dictionary.* Cambridge, Mass.: MIT Press.

Remez, R. E., Rubin, P. E., Pisoni, D. B., & Carrell, T. D. 1981. Speech perception without traditional speech cues. *Science, 212,* 947–950.

Renfrew, C. 1987. *Archaeology and language: The puzzle of Indo-European origins.* New York: Cambridge University Press.

Riemsdijk, H. van, & Williams, E. 1986. *Introduction to the theory of grammar.* Cambridge, Mass.: MIT Press.

Roberts, L. 1992. Using genes to track down Indo-European migrations. *Science, 257,* 1346.

Robinson, B. W. 1976. Limbic influences on human speech. In Harnad, Steklis, & Lancaster, 1976.

Rosch, E. 1978. Principles of categorization. In E. Rosch & B. Lloyd (Eds.), *Cognition and categorization.* Hillsdale, N.J.: Erlbaum.

Ross, P. E. 1991. Hard words. *Scientific American,* April, 138–147.

Rozin, P., & Schull, J. 1988. The adaptive-evolutionary point of view in experimental psychology. In R. C. Atkinson, R. J. Herrnstein, G. Lindzey, & R. D. Luce (Eds.), *Stevens's handbook of experimental psychology.* New York: Wiley.

Ruhlen, M. 1987. *A guide to the world's languages, Vol. 1.* Stanford University Press.

Rumelhart, D. E., McClelland, J. L., & The PDP Research Group. 1986. *Parallel distributed processing: Explorations in the microstructure of cognition, Vol. 1: Foundations.* Cambridge, Mass.: MIT Press.

Rymer, R. 1993. *Genie: An abused child's flight from silence,* New York: HarperCollins.

Safire, W. 1991. *Coming to terms.* New York: Henry Holt.

Sagan, C., & Druyan, A. 1992. *Shadows of forgotten ancestors.* New York: Random House.

Samarin, W. J. 1972. *Tongues of men and angels: The religious language of Pentecostalism.* New York: Macmillan.

Samuels, M. L. 1972. *Linguistic evolution.* New York: Cambridge University Press.

Sapir, E. 1921. *Language.* New York: Harcourt, Brace, and World.

Saussure, F. de. 1916/1959. *Course in general linguistics.* New York: McGraw-Hill.

Savage-Rumbaugh, E. S. 1991. Language learning in the bonobo: How and why they learn. In Krasnegor et al., 1991.

Schaller, S. 1991. *A man without words.* New York: Summit Books.

Schanck, R. C., & Riesbeck, C. K. 1981. *Inside computer understanding: Five programs plus miniatures.* Hillsdale, N.J.: Erlbaum.

Searle, J. (Ed.) 1971. *The philosophy of language.* New York: Oxford University Press.

Seidenberg, M. S. 1986. Evidence from the great apes concerning the biological bases of langauge. In W. Demopoulos & A. Marras (Eds.), *Language learning and concept acquisition: Foundational issues.* Norwood, N.J.: Ablex.

Seidenberg, M. S., & Petitto, L. A. 1979. Signing behavior in apes: A critical review. *Cognition, 7,* 177–215.

Seidenberg, M. S., & Petitto, L. A. 1987. Communication, symbolic communication, and language: Comment on Savage-Rumbaugh, McDonald, Sevcik, Hopkins, and Rupert 1986. *Journal of Experimental Psychology: General, 116,* 279–287.

Seidenberg, M. S., Tanenhaus, M. K., Leiman, M., & Bienkowski, M. 1982. Automatic access of the meanings of words in context: Some limitations of knowledge-based processing. *Cognitive Psychology, 14,* 489–537.

Selkirk, E. O. 1982. *The syntax of words.* Cambridge, Mass.: MIT Press.

Shatz, C. J. 1992. The developing brain. *Scientific American,* September.

Shepard, R. N. 1978. The mental image. *American Psychologist, 33,* 125–137.

Shepard, R. N. 1987. Evolution of a mesh between principles of the mind and regularities of the world. In J. Dupré (Ed.), *The latest on the best: Essays on evolution and optimality.* Cambridge, Mass.: MIT Press.

Shepard, R. N., and Cooper, L. A. 1982. *Mental images and their transformations.* Cambridge, Mass.: MIT Press.

Shevoroshkin, V. 1990. The mother tongue: How linguists have reconstructed the ancestor of all living languages. *The Sciences, 30,* 20–27.

Shevoroshkin, V., & Markey, T. L. 1986. *Typology, relationship, and time.* Ann Arbor, Mich.: Karoma.

Shieber, S. In press. Lessons from a restricted Turing Test. *Communications of the Association for Computing Machinery.*

Shopen, T. (Ed.) 1985. *Language typology and syntactic description,* 3 vols. New York: Cambridge University Press.

Simon, J. 1980. *Paradigms lost.* New York: Clarkson Potter.

Singer, P. 1992. Bandit and friends. *New York Review of Books,* April 9.

Singleton, J., & Newport, E. 1993. When learners surpass their models: the acquisition of sign language from impoverished input. Unpublished manuscript, Department of Psychology, University of Rochester.

Siple, P. (Ed.) 1978. *Understanding language through sign language research.* New York: Academic Press.

Slobin, D. I. 1977. Language change in childhood and in history. In J. Macnamara (Ed.), *Language learning and thought.* New York: Academic Press.

Slobin, D. I. (Ed.) 1985. *The crosslinguistic study of language acquisition, Vols. 1 & 2.* Hillsdale, N.J.: Erlbaum.

Slobin, D. I. (Ed.) 1992. *The crosslinguistic study of language acquisition, Vol. 3.* Hillsdale, N.J.: Erlbaum.

Smith, G. W. 1991. *Computers and human language.* New York: Oxford University Press.

Sokal, R. R., Oden, N. L., & Wilson, C. 1991. Genetic evidence for the spread of agriculture in Europe by demic diffusion. *Nature, 351,* 143–144.

Solan, L. M. 1993. *The language of judges.* Chicago: University of Chicago Press.

Spelke, E. S., Breinlinger, K., Macomber, J., & Jacobson, K. 1992. Origins of knowledge. *Psychological Review, 99,* 605–632.

Sperber, D. 1982. *On anthropological knowledge.* New York: Cambridge University Press.

Sperber, D. 1985. Anthropology and psychology: Toward an epidemiology of representations. *Man, 20,* 73–89.

Sperber, D. In press. The modularity of thought and the epidemiology of representations. In Hirschfeld & Gelman, in press.

Sperber, D., & Wilson, D. 1986. *Relevance: Communication and cognition.* Cambridge, Mass.: MIT Press.

Sproat, R. 1992. *Morphology and computation.* Cambridge, Mass.: MIT Press.

Staten, V. 1992. *Ol' Diz.* New York: HarperCollins.

Steele, S. (with Akmajian, A., Demers, R., Jelinek, E., Kitagawa, C., Oehrle, R., and Wasow, T.) 1981. *An Encyclopedia of AUX: A Study of Cross-Linguistic Equivalence.* Cambridge, Mass.: MIT Press.

Stringer, C. B. 1990. The emergence of modern humans. *Scientific American,* December.

Stringer, C. B., & Andrews, P. 1988. Genetic and fossil evidence for the origin of modern humans. *Science, 239,* 1263–1268.

Stromswold, K. J. 1990. Learnability and the acquisition of auxiliaries. Doctoral dissertation, Department of Brain and Cognitive Sciences, MIT.

Stromswold, K. J. 1994. Language comprehension without language production. Presented at the Boston University Conference on Language Development.

Stromswold, K. J. 1994. The cognitive and neural bases of language acquisition. In M. S. Gazzaniga (Ed.), *The cognitive neurosciences.* Cambridge, Mass.: MIT Press.

Stromswold, K. J., Caplan, D., & Alpert, N. 1993. Functional imaging of sentence comprehension. Unpublished manuscript, Department of Psychology, Rutgers University.

Studdert-Kennedy, M. 1990. This view of language. In Pinker & Bloom, 1990.

Supalla, S. 1986. Manually coded English: The modality question in signed language development. Master's thesis, University of Illinois.

Swinney, D. 1979. Lexical access during sentence comprehension: (Re)consideration of context effects. *Journal of Verbal Learning and Verbal Behavior, 5,* 219–227.

Symons, D. 1979. *The evolution of human sexuality.* New York: Oxford University Press.

Symons, D., & commentators. 1980. Précis and multiple book review of "The Evolution of Human Sexuality." *Behavioral and Brain Sciences, 3,* 171–214.

Symons, D. 1992. On the use and misuse of Darwinism and the study of human behavior. In Barkow, Cosmides, & Tooby, 1992.

Tartter, V. C. 1986. *Language processes.* New York: Holt, Rinehart, & Winston.

Terrace, H. S. 1979. *Nim.* New York: Knopf.

Terrace, H. S., Petitto, L. A., Sanders, R. J., & Bever, T. G. 1979. Can an ape create a sentence? *Science, 206,* 891–902.

Thomas L. 1990. *Et cetera, et cetera: Notes of a wordwatcher.* Boston: Little, Brown.

Thomason, S. G. 1984. Do you remember your previous life's language in your present incarnation? *American Speech, 59,* 340–350.

Tiersma, P. 1993. Linguistic issues in the law. *Language, 69,* 113–137.

Tooby, J., & Cosmides, L. 1989. Adaptation versus phylogeny: The role of animal psychology in the study of human behavior. *International Journal of Comparative Psychology, 2,* 105–118.

Tooby, J., & Cosmides, L. 1990a. On the universality of human nature and the uniqueness of the individual: The role of genetics and adaptation. *Journal of Personality, 58,* 17–67.

Tooby, J., & Cosmides, L. 1990b. The past explains the present: Emotional adaptations and the structure of ancestral environments. *Ethology and sociobiology, 11,* 375–424.

Tooby, J., & Cosmides, L. 1992. Psychological foundations of culture. In Barkow, Cosmides, & Tooby, 1992.

Trueswell, J. C., Tanenhaus, M., & Garnsey, S. M. In press. Semantic influences on parsing: Use of thematic role information in syntactic ambiguity resolution. *Journal of Memory and Language.*

Trueswell, J. C., Tanenhaus, M., & Kello, C. In press. Verb-specific constraints in sentence processing: Separating effects of lexical preference from garden-paths. *Journal of Experimental Psychology: Learning, Memory, and Cognition.*

Turing, A. M. 1950. Computing machinery and intelligence. *Mind, 59,* 433–460.

Voegelin, C. F., & Voegelin, F. M. 1977. *Classification and index of the world's languages.* New York: Elsevier.

von der Malsburg, C., & Singer, W. 1988. Principles of cortical network organization. In P. Rakic & W. Singer (Eds.), *Neurobiology of neocortex.* New York: Wiley.

Wald, B. 1990. Swahili and the Bantu languages. In B. Comrie (Ed.), *The world's major langauges.* New York: Oxford University Press.

Wallace, R. A. 1980. *How they do it.* New York: Morrow.

Wallesch, C.-W., Henriksen, L., Kornhuber, H.-H., & Paulson, O. B. 1985. Observations on regional cerebral blood flow in cortical and subcortical structures during language production in normal man. *Brain and Language, 25,* 224–233.

Wallich, P. 1991. Silicon babies. *Scientific American,* December 124–134.

Wallman, J. 1992. *Aping language.* New York: Cambridge University Press.

Wang, W. S.-Y. 1976. Language change. In Harnad, Steklis, & Lancaster, 1976.

Wanner, E. 1988. The parser's architecture. In F. Kessel (Ed.), *The development of language and of language researchers: Papers presented to Roger Brown.* Hillsdale, N.J.: Erlbaum.

Wanner, E., & Maratsos, M. 1978. An ATN approach to comprehension. In M. Halle, J. Bresnan, & G. A. Miller (Eds.), *Linguistic theory and psychological reality.* Cambridge, Mass.: MIT Press.

Warren, R. M. 1970. Perceptual restoration of missing speech sounds. *Science, 167,* 392–393.

Warrington, E. K., & McCarthy, R. 1987. Categories of knowledge: Further fractionation and an attempted integration. *Brain, 106,* 1273–1296.

Watson, J. B. 1925. *Behaviorism.* New York: Norton.

Weizenbaum, J. 1976. *Computer power and human reason.* San Francisco: Freeman.

Werker, J. 1991. The ontogeny of speech perception. In Mattingly & Studdert-Kennedy, 1991.

Wexler, K., and Culicover, P. 1980. *Formal principles of language acquisition.* Cambridge, Mass.: MIT Press.

Wilbur, R. 1979. *American Sign Language and sign systems.* Baltimore: University Park Press.

Williams, E. 1981. On the notions "lexically related" and "head of a word." *Linguistic Inquiry, 12,* 245–274.

Williams, G. C. 1957. Pleiotropy, natural selection, and the evolution of senescence. *Evolution, 11,* 398–411.

Williams, G. C. 1966. *Adaptation and natural selection: A critique of some current evolutionary thought.* Princeton, N.J.: Princeton University Press.

Williams, G. C. 1992. *Natural selection.* New York: Oxford University Press.

Williams, H. 1989. *Sacred elephant.* New York: Harmony Books.

Williams, J. M. 1990. *Style: Toward clarity and grace.* Chicago: University of Chicago Press.

Wilson, E. O. 1972. Animal communication. *Scientific American,* September.

Wilson, M., & Daly, M. 1992. The man who mistook his wife for a chattel. In Barkow, Cosmides, & Tooby, 1992.

Winston, P. H. 1992. *Artificial Intelligence* (4th ed.). Reading, Mass.: Addison-Wesley.

Woodward, J. 1978. Historical bases of American Sign Language. In Siple, 1978.

Wright, R. 1991. Quest for the mother tongue. *Atlantic Monthly,* April, 39–68.

Wynn, K. 1992. Addition and subtraction in human infants. *Nature, 358,* 749–750.

Yngve, V. H. 1960. A model and an hypothesis for language structure. *Proceedings of the American Philosophical Society, 104,* 444–466.

Yourcenar, M. 1961. *The memoirs of Hadrian.* New York: Farrar, Straus.

Zatorre, R. J., Evans, A. C., Meyer, E., & Gjedde, A. 1992. Lateralization of phonetic and pitch discrimination in speech processing. *Science, 256,* 846–849.

Zurif, E. 1990. Language and the brain. In Osherson & Lasnik, 1990.

# Glossary

**accusative.** The case of the object of a verb: *I saw HIM* (not *HE*).

**active.** See **voice**.

**adjective.** One of the major syntactic categories, comprising words that typically refer to a property or state: *a HOT tin roof; He is AFRAID of his mother*.

**adjunct.** A phrase that comments on or adds parenthetical information to a concept (as opposed to an argument): *a man FROM CINCINNATI; I cut the bread WITH A KNIFE*, I have used the word **modifier** instead.

**adverb.** One of the minor syntactic categories, comprising words that typically refer to the manner or time of an action: *tread SOFTLY; BOLDLY go; He will leave SOON*.

**affix.** A prefix or suffix.

**agreement.** The process in which a word in a sentence is altered depending on a property of some other word in the sentence; typically, the verb being altered to match the number, person, and gender of its subject or object: *He SMELLS* (not *SMELL*) versus *They SMELL* (not *SMELLS*).

**AI.** Artificial Intelligence, the attempt to program computers to carry out intelligent, humanlike tasks such as learning, reasoning, recognizing objects, understanding speech and sentences, and moving arms and legs.

**algorithm.** An explicit, step-by-step program or set of instructions for getting the solution to some problem: "To calculate a 15% tip, take the sales tax and multiply by three."

**aphasia.** The loss or impairment of language abilities following brain damage.

**argument.** One of the participants defining a state, event, or relationship: *president of* THE UNITED STATES; DICK *gave* THE DIAMOND *to* LIZ; *the sum of* THREE *and* FOUR. I have used the term **role-player** instead.

**article.** One of the minor syntactic categories, including the words *a* and *the*. Usually subsumed in the category **determiner** in contemporary theories of grammar.

**ASL.** American Sign Language, the primary sign language of the deaf in the United States.

**aspect.** The way an event is spread out over time: whether it is instantaneous (*swat a fly*), continuous (*run around all day*), terminating (*draw a circle*), habitual (*mows the grass every Sunday*), or a timeless state (*knows how to swim*). In English, aspect is involved in the inflectional distinction between *He eats* and *He is eating*, and between *He ate, He was eating*, and *He has eaten*.

**auxiliary.** A special kind of verb used to express concepts related to the truth of the sentence, such as tense, negation, question/statement, necessary/possible: *He* MIGHT *quibble; He* WILL *quibble; He* HAS *quibbled; He* IS *quibbling; He* DOESN'T *quibble;* DOES *he quibble?*

**axon.** the long fiber extending from a neuron that carries a signal to other neurons.

**behaviorism.** A school of psychology, influential from the 1920s to the 1960s, that rejected the study of the mind as unscientific, and sought to explain the behavior of organisms (including humans) with laws of stimulus-response conditioning.

**bottom-up.** Perceptual processing that relies on extracting information directly from the sensory signal (for example, the loudness, pitch, and frequency components of a sound wave), as opposed to **top-down**

processing, which uses knowledge and expectancies to guess, predict, or fill in the perceived event or message.

**case.** A set of affixes, positions, or word forms that a language uses to distinguish the different roles of the participants in some event or state. Cases typically correspond to the subject, object, indirect object, and the objects of various kinds of prepositions. In English, case is what distinguishes between *I, he, she, we, they,* which are used for subjects, and *me, him, her, us, them,* which are used for objects of verbs, objects of prepositions, and everywhere else.

**chain device.** See **finite-state device.**

**chromosome.** A long strand of DNA, containing thousands of genes, in a protective package. There are twenty-three chromosomes in a human sperm or egg; there are twenty-three pairs of chromosomes (one from the mother, one from the father) in all other human cells.

**clause.** A kind of phrase that is generally the same thing as a sentence, except that some kinds of clause can never occur on their own but only inside a bigger sentence: THE CAT IS ON THE MAT; John arranged FOR MARY TO GO; The spy WHO LOVED ME disappeared; He said THAT SHE LEFT.

**cognitive science.** The study of intelligence (reasoning, perception, language, memory, control of movement), embracing parts of several academic disciplines: experimental psychology, linguistics, computer science, philosophy, neuroscience.

**complement.** A phrase that appears together with a verb, completing its meaning: *She ate AN APPLE; It darted UNDER THE COUCH; I thought HE WAS DEAD.*

**compound.** A word formed by joining together other words: *fruit-eater; superwoman; laser printer.*

**concord.** See **agreement.**

**conjunction.** One of the minor syntactic categories, including *and, or,* and *but;* also, the entire phrase made by conjoining two words or phrases: *Ernie and Bert; the naked and the dead.*

**consonant.** A phoneme produced with a blockage or constriction of the vocal tract.

**content words.** Nouns, verbs, adjectives, adverbs, and some prepositions, which typically express concepts particular to a given sentence, as opposed to **function words** (articles, conjunctions, auxiliaries, pronouns, and other prepositions), which are used to specify kinds of information, like tense or case, that are expressed in all or most sentences.

**copula.** The verb *to be* when it is used to link a subject and a predicate: *She WAS happy; Biff and Joe ARE fools; The cat IS on the mat.*

**cortex.** The thin surface of the cerebral hemispheres of the brain, visible as gray matter, containing the bodies of neurons and their synapses with other neurons; where the neural computation takes place in the cerebral hemispheres. The rest of the cerebral hemispheres consists of **white matter,** bundles of axons that connect one part of the cortex with another.

**dative.** A family of constructions typically used for giving or benefiting; *She BAKED ME A CAKE; She BAKED A CAKE FOR ME; He GAVE HER A PARTRIDGE; He GAVE A PARTRIDGE TO HER.* Also refers to the case of the beneficiary or recipient in this construction.

**deep structure** (now **d-structure**). The tree, formed by phrase structure rules, into which words are plugged, in such a way as to satisfy the demands of the words regarding their neighboring phrases. Contrary to popular belief, not the same as Universal Grammar, the meaning of a sentence, or the abstract grammatical relationships underlying a sentence.

**derivational morphology.** The component of grammar containing rules that create new words out of old ones: *break + -able → breakable; sing + -er → singer; super + woman → superwoman.*

**determiner.** One of the minor syntactic categories, comprising the articles and similar words: *a, the, some, more, much, many.*

**diphthong.** A vowel consisting of two vowels pronounced in quick succession: *bIte (pronounced "ba-eet"); loUd, mAke.*

**discourse.** A succession of related sentences, as in a conversation or text.

**dyslexia.** Difficulty in reading or learning to read, which may be caused by brain damage, inherited factors, or unknown causes. Contrary to popular belief, it is not the habit of mirror-reversing letters.

**ellipsis.** Omission of a phrase, usually one that was previously mentioned or can be inferred: *Yes, I can* (_____); *Where are you going?* (_____) *To the store.*

**empiricism.** The approach to studying mind and behavior that emphasizes learning and environmental influence over innate structure; the claim that there is nothing in the mind that was not first in the senses. A second sense, not used in this book, is the approach to science that emphasizes experimentation and observation over theory.

**finite-state device.** A device that can produce or recognize ordered sequences of behavior (like sentences), by selecting an output item (like a word) from a list, going to some other list and selecting an item from it, and so on, possibly looping back to earlier lists. I have used the term **chaining device** instead.

**function word.** See **content word.**

**gender.** A set of mutually exclusive kinds into which a language categorizes its nouns and pronouns. In many languages, the genders of pronouns correspond to the sexes (*he* versus *she*), and the genders of nouns are determined by their sounds (words ending in *o* are one gender, words ending in *a* are the other) or are simply put in two or three arbitrary lists. In other languages, gender can correspond to human versus nonhuman, animate versus inanimate, long versus round versus flat, and other distinctions.

**gene.** (1) A stretch (or set of stretches) of DNA that carries the information necessary for building one kind of protein molecule. (2) A stretch of DNA that is long enough to survive intact across many generations of sexual recombination. (3) A stretch of DNA that, in comparison with alternative stretches that could sit at that location on the chromosome, contributes to the specification of some trait of the organism (e.g., "a gene for blue eyes").

**generative grammar.** See **grammar.**

**generative linguistics.** The school of linguistics, associated with Noam Chomsky, that tries to discover the generative grammars of languages and the universal grammar underlying them.

**gerund.** The noun formed out of a verb by adding *-ing*: *his incessant HUMMING.*

**grammar.** A **generative grammar** is a set of rules that determines the form and meaning of words and sentences in a particular language as it is spoken in some community. A **mental grammar** is the hypothetical generative grammar stored unconsciously in a person's brain. Neither should be confused with a **prescriptive** or **stylistic** grammar taught in school and explained in style manuals, the guidelines for how one "ought" to speak in a prestige or written dialect.

**gyrus.** The outward, visible portion of a wrinkle of the brain. The plural is *gyri*.

**head.** The single word in a phrase, or single morpheme in a word, that determines the meaning and properties of the whole: *the MAN in the pinstriped suit; ruby-throated hummingBIRD.*

**indirect object.** In a dative construction with two objects, the first one, referring to the recipient or beneficiary: *Bake ME a cake; Give THE DOG a bone.*

**Indo-European.** The group of language families that includes most of the languages of Europe, southwestern Asia, and northern India; thought to be descended from a language, Proto-Indo-European, which was spoken by a prehistoric people.

**induction.** Uncertain or probabilistic inference (as opposed to deduction), especially a generalization from instances: "This raven is black; that raven is black; therefore all ravens are black."

**infinitive.** The generic form of a verb, lacking tense: *He tried TO LEAVE; She may LEAVE.*

**INFL.** In post-1970s Chomskyan theory, a syntactic category comprising the auxiliary elements and tense inflections, which serves as the head of the sentence.

**inflecting language.** A language, like Latin, Russian, Warlpiri, or ASL, that relies heavily on inflectional morphology to convey information, as opposed to an **isolating language** like Chinese that leaves the forms of words alone and orders the words within phrases and sentences to convey information. English does both, but is considered more isolating than inflecting.

**inflectional morphology.** The modification of the form of a word to fit its role in the sentence, usually by adding an **inflection:** *I conquerED; I'm thinkING; Speed kills; two turtle doves.*

**intonation.** The melody or pitch contour of speech.

**intransitive.** A verb that may appear without an object: *We DINED; She THOUGHT that he was single*; as opposed to a **transitive verb,** that may appear with one: *He DEVOURED the steak; I TOLD him to go.*

**inversion.** Flipping the position of the subject and the auxiliary: *I am blue → Am I blue?; What you will do → What will you do?*

**irregular.** A word with an idiosyncratic inflected form instead of the one usually created by a rule of grammar: *brought* (not *bringed*); *mice* (not *mouses*); as opposed to **regular** words, which simply obey the rule (*walk + -ed → walked, rat + -s → rats*).

**isolating language.** See **inflecting language.**

**larynx.** The valve near the top of the windpipe, used to seal the lungs during exertion and to produce voiced sounds. Its parts include the vocal cords inside and the Adam's apple in front.

**lexical entry.** The information about a particular word (its sound, meaning, syntactic category, and special restrictions) stored in a person's mental dictionary.

**lexicon.** A dictionary, especially the "mental dictionary" consisting of a person's intuitive knowledge of words and their meanings.

**linguist.** A scholar or scientist who studies how languages work. Does not refer here to a person who speaks many languages.

**listeme.** An uncommon but useful term corresponding to one of the senses of "word," it refers to an element of language that must be memorized because its sound or meaning does not conform to some general rule. All word roots, irregular forms, and idioms are listemes.

**main verb.** A verb that is not the auxiliary: *I might STUDY Latin; He is COMPLAINING again.*

**Markov model.** A finite-state device that, when faced with a choice between two or more lists, chooses among them according to prespecified probabilities (for example, a .7 chance of going to List A, a .3 chance of going to list B).

**mentalese.** The hypothetical "language of thought," or representation of concepts and propositions in the brain in which ideas, including the meanings of words and sentences, are couched.

**modal.** A kind of auxiliary: *can, should, could, will, ought, might.*

**modality.** Whether a clause is a statement, question, negation, or imperative; another way of referring to some of the distinctions relevant to mood.

**modifier.** See **adjunct.**

**mood.** Whether a sentence is a statement (*HE GOES*), imperative (*GO!*), or subjunctive (*It is important THAT HE GO*).

**morphemes.** The smallest meaningful pieces into which words can be cut: *un-micro-wave-abil-ity.*

**morphology.** The component of grammar that builds words out of pieces (morphemes).

**movement.** The principal kind of transformational rule in Chomsky's theory, it moves a phrase from its customary position in deep structure to some other, unfilled position, leaving behind a "trace": *Do you want what → What do you want* (trace).

**natural kind.** A category of objects as found in nature, like robins, animals, crabgrass, carbon, and mountains; as opposed to artifacts (manmade objects) and nominal kinds (categories specified by a precise definition, like senators, bachelors, brothers, and provinces).

**natural language.** A human language like English or Japanese, as opposed to a computer language, musical notation, formulas in logic, and so on.

**neural network.** A kind of computer program or model, loosely inspired by the brain, consisting of interconnected processing units that send signals to one another and turn on or off depending on the sum of their incoming signals.

**neurons.** The information-processing cells of the nervous system, including brain cells and the cells whose fibers make up the nerves and spinal cord.

**nominative.** The case of the subject of the sentence: *SHE loves you* (not *HER loves you*).

**noun.** One of the major syntactic categories, comprising words that typically refer to a thing or person: *dog, cabbage, John, country, hour.*

**number.** Singular versus plural: *duck* versus *ducks.*

**object.** The argument adjacent to the verb, typically referring to the entity that defines or is affected by the action: *break THE GLASS, draw A CIR-*

*CLE, honor YOUR MOTHER.* Also, the argument of a preposition: *in THE HOUSE, with A MOUSE.*

**parameter.** One of the ways in which something can vary; in linguistics, one of the ways in which languages can vary from one another (for example, verb-object versus object-verb ordering).

**parsing.** One of the mental processes involved in sentence comprehension, in which the listener determines the syntactic categories of words, joins them up in a tree, and identifies the subject, object, and predicate; a prerequisite to determining who did what to whom from the information in the sentence.

**part of speech.** The syntactic category of a word: noun, verb, adjective, preposition, adverb, conjunction.

**participle.** A form of the verb that cannot stand by itself in a sentence but needs to be with an auxiliary or other verb: *He has EATEN; It was SHOWN; She is RUNNING; They kept OPENING the door.*

**passive.** A construction in which the usual object appears as the subject, and the usual subject is the object of the preposition *by* or absent altogether: *He was eaten by wolverines; I was robbed.*

**perisylvian.** Regions of the brain lining both sides and the end of the Sylvian fissure, the cleft between the temporal lobe and the rest of the brain. Language circuitry is thought to be concentrated in the left perisylvian areas.

**person.** The difference between *I* or *we* (first person), *you* (second person), and *he/she/they/it* (third person).

**philosopher.** A scholar who attempts to clarify difficult logical and conceptual questions, especially questions about the mind and about scientific knowledge. Does not refer here to a person who ruminates about the meaning of life.

**phoneme.** One of the units of sound that are strung together to form a morpheme, roughly corresponding to the letters of the alphabet: *b-a-t; b-ea-t; s-t-ou-t.*

**phonetics.** How the sounds of language are articulated and perceived.

**phonology.** The component of grammar that determines the sound pattern of a language, including its inventory of phonemes, how they may be combined to form natural-sounding words, how the phonemes must be adjusted depending on their neighbors, and patterns of intonation, timing, and stress.

**phrase.** A group of words that behaves as a unit in a sentence and which typically has some coherent meaning: *in the dark; the man in the gray suit; dancing in the dark; afraid of the wolf.*

**phrase structure.** The information about the syntactic categories of the words in a sentence, how the words are grouped into phrases, and how the phrases are grouped into larger phrases; usually diagrammed as a tree.

**phrase structure grammar.** A generative grammar consisting only of rules that define phrase structures.

**polysynthetic language.** An inflecting language in which a word may be composed of a long string of prefixes, roots, and suffixes.

**pragmatics.** How language is used in a social context, including how sentences are made to fit in with the flow of a conversation, how unspoken premises are inferred, and how degrees of formality and politeness are signaled.

**predicate.** A state, event, or relationship, usually involving one or more participants (arguments). Sometimes the predicate is identified with the verb phrase of a sentence (*The baby ATE THE SLUG*), and the subject is considered its sole argument; at other times it is identified with the verb alone, and the subject, object, and other complements are all considered to be its arguments. The contradiction can be resolved by saying that the verb is a simple predicate, which combines with its complements to form a complex predicate.

**preposition.** One of the major syntactic categories, comprising words that typically refer to a spatial or temporal relationship: *in, on, at, near, by, for, under, before, after.*

**pronoun.** A word that stands for a whole noun phrase: *I, me, my, you, your, he, him, his, she, her, it, its, we, us, our, they, them, their, who, whom, whose.*

**proposition.** A statement or assertion, consisting of a predicate and a set of arguments.

**prosody.** The overall sound contour with which a word or sentence is pronounced: its melody (intonation) and rhythm (stress and timing).

**psycholinguist.** A scientist, usually a psychologist by training, who studies how people understand, produce, or learn language.

**psychologist.** A scientist who studies how the mind works, usually via the analysis of experimental or observational data on people's behavior. Does not refer here to psychotherapist or to a clinician who treats mental disorders.

**recursion.** A procedure that invokes an instance of itself, and thus can be applied repeatedly to create or analyze entities of any size: "How to put words *in alphabetical order:* sort the words so their first letters are in the same order as in the alphabet; then for each group of words beginning with the same letter, ignore that first letter and put the remaining parts *in alphabetical order.*" "*A verb phrase* can consist of a verb followed by a noun phrase followed by a *verb phrase.*"

**regular.** See **irregular.**

**relative clause.** A clause modifying a noun, usually containing a trace corresponding to that noun: *the spy* WHO LOVED ME; *the land* THAT TIME FORGOT; *violet eyes* TO DIE FOR.

**role-player.** See **argument.**

**root.** The most basic morpheme in a word or family of related words, consisting of an irreducible, arbitrary sound-meaning pairing: ELEC-TRicity, ELECTRical, ELECTRic, ELECTRify, ELECTRon.

**semantics.** The parts of rules and lexical entries that specify the meaning of a morpheme, word, phrase, or sentence. Does not refer here to haggling over exact definitions.

**sexual recombination.** The process that makes organisms capable of generating an immense number of distinct possible offspring. When a sperm or egg is formed, the twenty-three pairs of chromosomes ordinarily found in a human cell (one chromosome in each pair from the mother, one from the father) have to be cut down to twenty-three single chromosomes. This is done in two steps. First, within each pair, a few random cuts are made in identical positions in each chromosome, pieces are exchanged, and the new chromosomes are glued back together. Then, one member of each pair is chosen at random, and put into the egg or sperm. During fertilization, each chromosome from the egg is paired up with its counterpart from the sperm, restoring the genome to twenty-three pairs.

**SLI.** Specific Language Impairment, any syndrome in which a person fails to develop language properly and the blame cannot be pinned on hearing deficits, low intelligence, social problems, or difficulty controlling the speech muscles.

**specifier.** A specific position at the periphery of a phrase, generally where one finds the subject. For many years the specifier position of a noun phrase was thought to contain the determiner (article), but the current consensus in Chomskyan theory puts the determiner in a phrase of its own (a determiner phrase).

**stem.** The main portion of a word, the one that prefixes and suffixes are stuck onto: *WALKs, BREAKable, enSLAVE.*

**stop consonant.** A consonant in which the airflow is completely blocked for a moment: *p, t, k, b, d, g.*

**strong verb.** The verbs in Germanic languages (including English), now all irregular, that form the past tense by changing the vowel: *break– broke, sing-sang, fly–flew, bind–bound, bear–bore.*

**subject.** One of the arguments of a verb, typically used for the agent or actor when the verb refers to an action: *BELIVEAU scores; THE HIPPIE touched the debutante.*

**surface structure** (now **s-structure**). The phrase structure tree formed when movement transformations are applied to a deep structure. Thanks to traces, it contains all the information necessary to determine the meaning of the sentence. Aside from certain minor adjustments (executed by "stylistic" and phonological rules), it corresponds to the actual order of words that a person utters.

**syllable.** A vowel or other continuous voiced sound, together with one or more consonants preceding or following it, that are pronounced as a unit: *sim-ple, a-lone, en-cy-clo-pe-di-a.*

**syntactic atom.** One of the senses of "word," defined as an entity that the rules of syntax cannot separate or rearrange.

**syntactic category.** See **part of speech.**

**syntax.** The component of grammar that arranges words into phrases and sentences.

**tense.** Relative time of occurrence of the event described by the sentence, the moment at which the speaker utters the sentence, and, often, some third reference point: present (*he eats*), past (*he ate*), future (*he will eat*). Other so-called tenses such as the perfect (*He has eaten*) involve a combination of tense and aspect.

**top-down.** See **bottom-up.**

**trace.** A silent or "understood" element in a sentence, corresponding to the deep-structure position of a moved phrase: *What did he put (TRACE) in the garage?* (the trace corresponds to *what*); *Boggs was grazed (TRACE) by a fastball* (the trace corresponds to *Boggs*).

**transformational grammar.** A grammar composed of a set of phrase structure rules, which build a deep-structure tree, and one or more transformational rules, which move the phrases in the deep structure to yield a surface-structure tree.

**transitive.** See **intransitive.**

**Turing machine.** A design for a simple computer consisting of a potentially infinite strip of paper, and a processor that can move along the paper and print or erase symbols on it in a sequence that depends on which symbol the processor is currently reading and which of several states it is in. Though too clumsy for practical use, a Turing machine is thought to be capable of computing anything that any digital computer, past, present, or future, can compute.

**Universal Grammar.** The basic design underlying the grammars of all human languages; also refers to the circuitry in children's brains that allows them to learn the grammar of their parents' language.

**verb.** One of the major syntactic categories, comprising words that typically refer to an action or state: *hit, break, run, know, seem.*

**voice.** The difference between the active and passive constructions: *Dog bites man* versus *Man is bitten by dog.*

**voicing.** Vibration of the vocal folds in the larynx simultaneous with the articulation of a consonant; the difference between *b, d, g, z, v* (voiced) and *p, t, k, s, f* (unvoiced).

**vowel.** A phoneme pronounced without any constriction of the airway.

**white matter.** See **cortex.**

**word.** See **listeme, morphology, syntactic atom.**

**X-bar.** The smallest kind of phrase, consisting of a head and its non-subject arguments (role-players): *The Romans'* DESTRUCTION OF THE CITY; *She* WENT TO SCHOOL *on foot; He is very* PROUD OF HIS SON.

**X-bar theory; X-bar phrase structure.** The particular kind of phrase structure rules thought to be used in human languages, according to which all the phrases in all languages conform to a single plan. In that plan, the properties of the whole phrase are determined by the properties of a single element, the head, inside the phrase.

# Index

## About the author

## About the book

Insights,
Interviews
& More . . .

## Read on

# Meet Steven Pinker

© 2005 by Rebecca Goldstein

THE LANGUAGE INSTINCT is dedicated to my parents, "who gave me language"; the ambiguity between nature and nurture was, of course, intentional. As someone who believes that nature has been underestimated in intellectual life, I must begin my life story not with the supportive environment they provided me but earlier, with the kind of people they are. Roslyn Wiesenfeld Pinker has a voracious appetite for ideas, nuanced opinions informed by considerable learning, and a keen insight into people; she exemplifies the credo of Terence (which she once recommended to me as an epigraph): "I am a person; nothing human is foreign to me." Harry Pinker has a gift for sizing up a complicated situation and capturing its essence in a succinct observation; he has, in addition, a taste for new experiences, a willingness to take calculated risks, and an ability to deal with setbacks coolly. I cannot say whether I have inherited doses of these traits, but I do know that my parents nurtured them in me. And as many authors say after thanking their commentators, "All remaining faults are my own."

66 [My mother] exemplifies the credo of Terence: 'I am a person; nothing human is foreign to me.' 99

My parents were children of Jewish immigrants who emigrated to Montreal in the 1920s, my father's parents from Krasnystaw, Poland; my mother's from Warsaw and Kishinev. Both my parents earned university degrees, and both made midlife career turns foreshadowing the one that led me to write this book. My father moved from real estate and sales to a law practice and the tourist industry. My mother, like many women of her generation, applied her talents in volunteer positions in education and community organizations before developing a professional career, first as a high school guidance counselor (the students called her "Pink the Shrink"), then as the school's vice principal. My sister Susan, formerly a child psychologist, is now a columnist for Canada's national paper *The Globe and Mail* and the author of a book on sex differences. My brother Robert is an economist and a policy analyst for the Canadian government in Ottawa.

My larger environment was Montreal, and people often ask me whether growing up in a bilingual society launched my interest in language. The answer, sadly, is no. Canada in the 1950s and 1960s was a land of "two solitudes," as the title of a classic book put it, with the French- and English-speaking communities of Montreal occupying different halves of the island. I learned French in public school from North African Jews because of Quebec's bizarre educational system of the time, which segregated the Catholics from the so-called Protestants (actually, the non-Catholics) and thus hired Moroccan and Algerian immigrants to teach French to the likes of me. The cultural environment of my childhood was polite Anglo-Saxon Canada; the cultural environment of my adolescence was argumentative Jewish Montreal. We had a saying—"Ten Jews, eleven opinions"—and my home life was rich with friends and family engaging in good-natured debate around the dinner table. The disputations grew especially intense during the late 1960s, when everything seemed up for grabs. During those years I became particularly interested in conceptions of human nature and how they affect every other sphere of life, from politics and economics to education and art.

Like most Montrealers, I stayed in the city after high school, attending first Dawson College and then McGill University. Throughout high school and college I had zigzagged between the sciences and humanities, and I chose to major in cognitive psychology because it seemed capable of addressing deep questions about human nature with a tractable program of experimental research. McGill had been a hothouse for psychology ever since its department was shaped by D. O. Hebb, the first psychologist to model learning in neural networks and apply it to psychological phenomena far and wide. Hebb was still a presence when I was a student, but I was influenced less by his associationism than by the more rationalist approaches to the mind conveyed by my adviser, Albert Bregman, who had been influenced by Gestalt psychology and artificial intelligence, and by a philosophy professor, Harry Bracken, who was a devotee of Noam Chomsky. McGill was also ▶

distinguished by research on the human brain pioneered by Wilder Penfield and later, Brenda Milner, at the Montreal Neurological Institute.

In graduate school at Harvard, my primary mentor was a young cognitive psychologist named Stephen Kosslyn, a close friend ever since and now the chairman of the department we both returned to. My thesis was on visual cognition (specifically, the representation of three-dimensional space in mental images), and I did experimental research in that area for another fifteen years. Visual cognition continues to interest me in a number of ways: visual aesthetics, the expression of space in language, and my main nonacademic obsession, photography. Language was originally a side pursuit, which I explored in a theoretical paper on mathematical and computer models of language acquisition. My interest in language also allowed me to study with Roger Brown, the urbane social psychologist who founded the modern study of language acquisition and whose witty and stylish writing has been an inspiration to me ever since. Harvard students could cross-register at MIT, and I took a course on theories of the mind by Jerry Fodor and Noam Chomsky, and then a course on linguistics and computation by Joan Bresnan, a linguist who had studied with Chomsky before devising her own rival theory. After graduating, I did a postdoctoral fellowship at MIT with Bresnan and developed a theory of language acquisition based on her theory, which I later expanded into a technical book, *Language Learnability and Language Development.*

My first job, back at Harvard, required me to teach three courses in language acquisition, and that initiated a drift in my research away from vision and toward language. I pursued two lines of research in language. One was on the meanings and syntax of verbs and how children acquire them; it was presented in a second technical book, *Learnability and Cognition.* The other was on regular and irregular verbs. I often tell people that this falls into the great academic tradition of knowing more and more about less and less until you know everything about nothing, but as I mention in chapter 5 of this book, it is a good way to distinguish the two main psychological processes underlying language, memory and computation. Irregular forms like *sing-sang* and *bring-brought* are idiosyncratic and have to be memorized; regular forms like *fax-faxed* and *spam-spammed* are predictable and can be cranked out by a rule. Since they are matched in meaning and complexity, comparing them can shed light on how memory and computation interact. This line of work resulted in a number of academic papers (including a monograph that analyzed twenty thousand verb forms in the speech of children) and, improbably, a second popular book on language: *Words and Rules: The Ingredients of Language.* ❧

# On Writing
## *The Language Instinct*

WRITING *The Language Instinct* was a turning point in my professional life, but it did not come out of the blue. I had long been fascinated by expository prose. I read style manuals for fun, and scrutinized elegant sentences I came across in books and essays to figure out what made them work. I admired writers like George Gamow and Martin Gardner, who explained deep ideas in accessible language, and noted how some of them, like Stephen Jay Gould, Richard Dawkins, and Daniel Dennett, were not just popularizers but advanced big ideas within their fields which would have been hard to express within the confines of an academic paper.

I had also felt that psychology obsessed far too much on laboratory curiosities and had lost sight of the big picture—answers to the fundamental questions that curious people naturally ask, like How does language work? and Why are there so many languages? and How do children acquire their mother tongue? This eye on the big picture had always animated my teaching. I never base my courses on a textbook, because textbooks focus on how academics keep themselves busy rather than on fundamental questions, and they slavishly copy earlier textbooks' arbitrary ways of organizing the subject matter. Many of the expositions in *The Language Instinct* grew out of my attempts to make sense of language for my students. Students provided me with other stimuli for the book—Paul Bloom suggested that we coauthor a paper on the evolution of language, and in a graduate seminar, Annie Senghas and Greg Hickok exposed me to several new lines of research on the biological foundations of language. All of these fed a growing conviction that ▶

> ❝ I read style manuals for fun, and scrutinized elegant sentences I came across in books and essays to figure out what made them work. ❞

the variegated phenomena of language could be brought together under the unifying idea that language is an evolutionary adaptation of the human species.

Several editors from university presses encouraged me to try my hand at a book for a larger audience, and one of them gave me a pivotal piece of advice. Most academics who try to reach a wide audience, she explained, are failures. They imagine that they are writing for an unwashed mass of truck drivers and chicken pluckers, and so they talk down to their readers, treating them like slow children. Chicken pluckers don't buy books, she advised. Think of your readers as your college roommates: people who are as smart and intellectually curious as you and your colleagues but who happened to go into a different line of work, and don't know your jargon, or your methods, or why you think the topics you study are important. It was tremendously liberating. Writing *The Language Instinct* would not require me to pretend to be a different person. I could continue the conversation I had in my professional life with my students, my colleagues, and myself on what was truly exciting about what we all did, being mindful only that there were newcomers to the conversation who had to be brought up to speed.

I wrote the first draft of *The Language Instinct* over a summer in a day-and-night frenzy, taking one week per chapter, a pace I have not duplicated since. It helped that I didn't know what I was getting into. A colleague warned me that most books sit in the bookstores for six weeks and then sink into oblivion, so I kept my expectations in check. Subsequent drafts benefited from my editor's suggestion to add more entertaining examples to the technical sections, and a last-minute decision to put the book through a sixth draft intended only to polish the prose. I had no idea that the book would get such a gratifying reception: more than eighty positive reviews, prizes from the Linguistics Society of America and the American Psychological Association, designation by *The New York Times Book Review* as one of the eleven best books of 1994 and by *American Scientist* as one of the hundred best science books of the century, translation into nineteen languages, and enough interest more than a dozen years later for the publisher to issue the edition you are now holding. Nor could I have foreseen that some of the loose ends would lead to four subsequent books: *How the Mind Works, Words and Rules, The Blank Slate,* and *The Stuff of Thought.*

Writing a book for a nonacademic audience also fed back into my academic work. It was only while trying to explain "how language works" in chapter 4 that I realized that my research on regular and irregular forms could be framed in terms of an interplay between memorized arbitrary signs (the principle behind the word) and open-ended combinatorial grammar (the principle behind complex words, phrases, and sentences). The concrete detail necessary to bring an experiment to life for nonspecialist readers—not just "stimuli" but Jabba the Hutt, not just "sentences" but *Furry wildcats fight furious battles*—forced me to go back to the original reports rather than recycle thirdhand summaries, and more than once I discovered that the standard rendering of a classic experiment was flat wrong. I also quickly learned that general science writing holds an author to far higher standards of fact checking than academic writing. A typical journal article is vetted by two referees, read by an audience in the hundreds, and participates in a feedback cycle measured in years. A trade book not only is fair game for attack in these academic journals but also passes under the eyes of hundreds of thousands of readers with expertise in far-flung areas. They are quick to point out any error, always with the stinging line "Though this may seem like a small matter, it raises questions about the standards of accuracy applied to the rest of the book." I am still shamefaced that the first printing of *The Language Instinct* announced that *flitch* is not an English word and contained the remarkable claim that King Arthur spoke Old English ("You must never have met a Welshman," wrote one reader).  ∽

# Frequently Asked Questions

HERE ARE THE QUESTIONS I get asked the most often in connection with *The Language Instinct*:

### Did you ever study or work with Noam Chomsky?

No. He is a linguist, and I trained as an experimental psychologist. During the twenty-one years I taught at MIT, we were in different departments, and given how universities work, our paths didn't cross much from day to day. Our relationship is cordial, and he has deeply influenced my views on language, but our approaches also differ in important ways.

### Do your academic colleagues resent your crossover success?

Not that I can tell. Both at MIT and at Harvard, I have received nothing but kind support. Elsewhere, many academics have thanked me for keeping their students awake or for explaining to their parents what they do for a living. Of course, many express their disagreements with me vociferously, as they do (and ought to do) with any scholar with strong views. But academia is not nearly as petty as many outsiders assume.

### How many languages do you speak?

I have some acquaintance with French, Hebrew, and Spanish, but am far from fluent and never rely on this knowledge when I write about language. When my research involves a foreign language, I collaborate with someone who is both a native speaker and an expert in the linguistics literature. In my general writings I occupy a high link in the linguistics food chain, relying on surveys of the world's languages by experts on linguistic diversity, who in turn rely on experts in particular languages.

### Do you still do empirical research?

Yes. Since I wrote *The Language Instinct,* I have studied past tense and plural inflection in children (including identical and fraternal twins), in neurological patients, in speakers of English, Hebrew, and German, and with the use of functional MRI and other neuroimaging techniques. Recently I have shifted my focus to semantics and pragmatics, pursuing questions that grew out of *The Stuff of Thought.*

*My spouse speaks a language other than English (or We're spending a year in a foreign country). Is there anything I can do to encourage my children to retain their second language as they grow up?*

Children care more about their peers than about their parents, so send them to summer camps, after-school programs, or vacations with their cousins, where they will have to use the language with kids their own age.

*My child is three years old and hasn't said a word. What should I do?*

Take him (it's more likely to be a "him") to a speech and language pathologist accredited by the American Speech-Language-Hearing Association (www.asha.org), ideally one affiliated with a university, clinic, or teaching hospital. If the child seems to understand language, and is bright and socially responsive, there is a good chance that he is simply a "late talker" (probably from genetic causes) who will outgrow his silence and end up just fine. (See also the recommendations in "Read on.")

# *The Language Instinct* Today

MANY FIELDS OF SCIENCE enjoy a rapid rate of new discoveries and a high degree of consensus about what they mean. The study of language, unfortunately for linguistics but perhaps fortunately for *The Language Instinct*, is not among them. A book on human genetics or nanotechnology would be hopelessly out of date a dozen years after its publication, but I like to think that *The Language Instinct* is still a useful introduction to the science of language today. Of course the field has hardly stood still (nor have my opinions), and here are some reflections on the contents of each chapter in the light of developments since 1994.

*Chapter 1: An Instinct to Acquire an Art.* The two heroes of this chapter are very much in the spotlight today. Darwin's influence has expanded in psychology, the social sciences, philosophy, medicine, and genomics (despite the retrograde efforts of the "Intelligent Design" movement), and my 2002 book, *The Blank Slate*, explored some of the ramifications of his influence. Chomsky is still the most influential living linguist, and his political writings have inspired a new generation of leftists (I recently saw a sticker on a lamppost that said, "Read Chomsky.")

Many readers concluded from this chapter's acknowledgment of Chomsky's influence on linguistics that I am a "Chomskyan." In some regards—the view that language comes from a mental system tailored for the computation of symbolic representations—I am. But in the chapter I hinted at some disagreements, and I have spelled them out in the years since. In *The Blank Slate* I explained why I don't share Chomsky's romantic view of human nature or the radical leftist-anarchist politics connected to it, and he and I have locked horns in a recent debate on grammatical theory and the evolution of language (more on this later). My views on language and mind are very close to those of my comrade in that debate, the linguist Ray Jackendoff, himself a former student of Chomsky's. Jackendoff lays out a coherent vision for the science of language in his recent book *Foundations of Language*, which I endorse wholeheartedly.

*Chapter 2: Chatterboxes.* Many of the phenomena discussed in this natural history of language have made the news in the past dozen years:

• In 2005 the linguist Daniel Everett described an Amazonian people, the Pirahã, whose language, he suggested, did not allow them to describe abstract subjects lying outside their immediate experience. That claim, though, was belied by many of his own observations, such as that "spirits and the spirit world play a very large role in their lives." Though Pirahã is in some ways simpler than familiar European languages (it has a "one-two-many" counting system and simpler systems of tenses and pronouns), in other ways it is quite complex, with sixteen verb suffix classes and more than fifty thousand attested word forms. Everett has emphasized that, contrary to casual impressions, Pirahã is by no means a "primitive language."

• Black English Vernacular made the news in 1996 under the wacky name "Ebonics" when members of the Oakland School Board floated a suggestion that it be recognized as a language in their bilingual education programs. The linguists John McWhorter and Geoffrey Pullum have written excellent analyses of the resulting hooha.

• Anne Senghas was a research assistant in the project that first studied Nicaraguan Sign Language, and when she came to MIT as my graduate student I suggested that she study this fascinating phenomenon in her thesis and beyond. She has since published beautiful quantitative analyses demonstrating that the younger children have indeed developed a new language with a discrete combinatorial grammar.

• Many readers were surprised to read of cultures where parents say little to their small children, who acquire language from older siblings and peers. But their surprise is a symptom of what Judith Rich Harris has exposed as "the nurture assumption"—the dogma that children are socialized by their parents. In her famous 1998 book by that title, she argued that the most important influence parents have on their children is at the moment of conception. Children acquire their culture, and develop their personalities, in their interactions with their peer groups and society. Many features of language acquisition bear this out: the dispensability of parental speech in language acquisition, the phenomenon of creolization, the emergence of Nicaraguan Sign Language, and the fact that children of immigrants always grow up with the accents of their peers, not those of their parents. These phenomena (together with the findings of behavioral genetics) made me receptive to Harris's theory, and I wrote a foreword to her book and discussed it at length in *The Blank Slate*.

• Though Chomsky is famous for advancing the hypothesis that language is innate, he has never laid out a systematic scientific defense of the claim, and his main argument, the "Poverty of the Input," is far from watertight. Geoff Pullum and the philosopher Barbara Scholz, using large bodies of online text (the most prominent new methodology in linguistics), have shown that many constructions children allegedly never hear can in fact be found in reasonably sized samples of English. They don't deny that a poverty-of-the-input argument can be made (and I think that the "Simon" study, Nicaraguan Sign Language, and Peter Gordon's *mice-eater* study reported in chapter 5 are good candidates), but they point out, reasonably in my view, that they are harder to nail down than Chomsky and most of his followers assume. ▶

11

- Human genetics and cognitive neuroscience are two areas that have exploded since the book came out. Williams syndrome is now known to be caused by the deletion of a small region of chromosome 7 containing some twenty genes, which accounts for the heterogeneous symptoms of the syndrome. At least one of them, the gene for LIM-kinase 1, has been associated with the problems in spatial cognition. Though the language of people with Williams syndrome is, as I emphasized, less impaired than other cognitive functions, there is a lot of variation among the affected people. The hyperdeveloped language abilities of Crystal, though showing that language can dissociate from other aspects of cognition, are not found in all people with the syndrome.

- Progress in understanding the genetics of K family has been spectacular. First a marker for the gene was identified (SPCH1), then the gene itself (FOXP2) together with the mutation that leads to the deficit, then its evolutionary history. The gene has counterparts in other mammals, but the exact sequence of the normal human version is unique to us, and it has been a target of Darwinian natural selection during the past 200,000 years. Currently its function in mammalian brain development is being studied intensively. We now know that it is a transcription factor that turns on other genes, and that the versions in other mammals affect neural circuitry for motor control, particularly the circuits for making sounds.

- Despite my cautious characterization of what was known about the family's syndrome in chapter 10, I have seen myself cited as claiming both that the affected gene was specific to grammar and that it affects nothing but control of the muscles of the mouth and face. The family members have since undergone many tests, and the truth lies somewhere in between: the affected members have deficits in articulation and in oral and facial motor control, and have, on average, reduced intelligence. But they also have impairments in language itself that cannot be reduced to these other deficits.

- Though no single gene specific to grammar has been identified (and perhaps none ever will), it is increasingly clear that sets of genes will be tied, with varying degrees of specificity and overlap with other functions, to aspects of language ability. The psychologist Heather van der Lely has documented a group of children with a syndrome she calls "Grammatical Specific Language Impairment." Unlike that of the K family, their deficit appears to be specific to language itself, indeed, specific to grammar: they are normal in overall intelligence, in interpreting complex sounds, in understanding words, and in using language in a natural way in social settings. Their syndrome is probably inherited, but the families are not large enough, or the pattern of inheritance clean enough, to pin down the genes. Coming from the other direction, my former student Karin Stromswold has reviewed a large literature showing that many kinds of variation in language ability, including language impairment and language delay, are highly heritable.

*Chapter 3: Mentalese.* When I wrote this chapter, the Whorfian hypothesis was largely out of favor among linguists and psychologists, but the pendulum has swung back, and there is now a lively neo-Whorfian movement. In *The Stuff of Thought*, I review this new research and argue that the problem with the idea that language affects thought is not that it's entirely wrong but that there are many *ways* in which language can affect thought, and people tend to blur them together. In particular, people tend to confuse banal observations, such as that one person's words can affect another person's thoughts (if that weren't true, language as a whole would be useless) with radical claims, such as that we think in our native language, and that the language we speak makes it impossible to think certain thoughts. In the new book I argue that the major conclusion in "Mentalese" is right—we think not in our native language but in more abstract media of thought.

*Chapter 4: How Language Works.* The machinery of syntax described at the end of this chapter is barely recognizable in Chomsky's current theory, which he calls the "Minimalist Program." Chomsky is famous within linguistics for overturning his theory every decade or so; the current release, depending on how you count, is 5.2, whereas the release I presented in this chapter was a stripped-down version of 3.2, the "Revised Extended Standard Theory." Nonetheless, the picture of grammar presented here will be recognizable to anyone reading on in linguistics, because I pretty much stuck to features that would stand the test of time and translate easily to other theories. In my own work I have always favored theories with less exotic machinery than Chomsky's (shallower trees, fewer traces, less movement), and whose structures are more visible at the surface, like phrases, lexical items, and constructions. Joan Bresnan's theory is an example, and a more recent version with this flavor can be found in Ray Jackendoff and Peter Culicover's *Simpler Syntax*.

The most stunning development since I wrote the chapter took place in 2004, when the Boston Red Sox won the World Series.

*Chapter 5: Words, Words, Words.* In two subsequent books, I have explored the world of words in far greater depth. *Words and Rules: The Ingredients of Language* looks at the combinatorial richness of word formation and its implications for the machinery of cognition. *The Stuff of Thought: Language as a Window into Human Nature* looks at the meanings of words and how they originate and spread. ▶

In his lovely book *How Children Learn the Meanings of Words*, Paul Bloom argues that children have no mental mechanisms dedicated to word learning and learn a word the way they learn any other kind of fact. Children zero in on a word's meaning by exercising their "theory of mind" or intuitive psychology, deducing what a sensible speaker is probably referring to in the context. Ray Jackendoff and I think this is not the whole story, for reasons we explained in our paper debating Chomsky.

*Chapter 6: The Sounds of Silence.* Speech recognition technology has advanced tremendously and is now inescapable in telephone information systems. But as everyone who has been trapped in "voice-mail jail" knows, the systems are far from foolproof ("I'm sorry, but I did not understand what you said"). And here is how the novelist Richard Powers described his recent experience with a state-of-the-art speech recognition program: "This machine is a master of speakos and mondegreens. Just as we might hear the . . . Psalms avow that 'Shirley, good Mrs. Murphy, shall follow me all the days of my life,' my tablet has changed 'book tour' to 'back to work' and 'I truly couldn't see' to 'a cruelly good emcee.'" Recognizing a large number of words from a large number of speakers is still a formidable engineering task.

The sound pattern of English and the logic behind the vagaries of its spelling are explored more deeply in *Words and Rules*, including the extraordinary suggestion by Chomsky and Morris Halle that English spelling "comes remarkably close to being an optimal orthographic system."

One raging public debate involving language went unmentioned in *The Language Instinct*: the "reading wars," or dispute over whether children should be explicitly taught to read by decoding the sounds of words from their spelling (loosely known as "phonics") or whether they can develop it instinctively by being immersed in a text-rich environment (often called "whole language"). I tipped my hand in the paragraph in this chapter which said that language is an instinct but reading is not. Like most psycholinguists (but apparently unlike many school boards), I think it's essential for children to be taught to become aware of speech sounds and how they are coded in strings of letters. Diane McGuinness's *Why Our Children Can't Read* is my favorite book on the subject. The title came from the publisher's marketroids; she originally called it *The Reading Revolution*, since it was about both a scientific revolution in reading research and the revolution in human history that gave us alphabetic writing.

Masaaki Yamanashi's comment on Bill Clinton proved to be prophetic.

*Chapter 7: Talking Heads.* Anyone who has tried to search the Web using one of the engines that claims to understand English can verify that comprehending natural language is still an unsolved engineering problem. Ditto for the programs that claim to translate from one language to another. The Loebner Prize competition (erroneously described as a "Turing test") continues to be won by uncomprehending programs using canned responses.

The subfield of linguistics known as "pragmatics," which deals with the use of language in a social context and WITH phenomena such as politeness, innuendo, and reading between the lines, was covered in a scant three pages in this chapter. A deeper discussion, which links these phenomena to social and evolutionary psychology, can be found in a chapter called "Games People Play" in *The Stuff of Thought*.

*Chapter 8: The Tower of Babel*. Daniel Everett, the linguist who documented the Amazonian language Pirahã, claimed that the language violates Hockett's universals by providing no means to discuss events remote from experience, and by lacking the mechanism of recursive embedding, in which a word or phrase can be inserted inside a word or phrase of the same type. But the first claim, as I mentioned, is contradicted by numerous observations of the Pirahã way of life, and the second is questionable as well. Pirahã allows for a degree of semantic embedding using verb suffixes and conversions of nouns to verbs (so one can express the thought "I said that Kó'oí intends to leave," with two levels of semantic embedding), and one can conjoin propositions within a sentence, as in "We ate a lot of the fish, but there was some fish we did not eat." The linguists Andrew Nevins, David Pesetsky, and Cilene Rodrigues have taken a close look at Pirahã, and have disputed his claim that recursive syntactic embedding is absent from the language.

The notion of Universal Grammar continues to be debated, though in a half-full/half-empty way; the proverbial Martian scientist would still consider human languages to be extraordinarily similar compared with the countless ways one could imagine a system for vocal communication. In *The Atoms of Language*, the linguist Mark Baker presents an explicit empirical case for a Universal Grammar with a smallish set of parameters differentiating all human languages.

My suggestion that controversies about language families might be resolved by "a good statistician with a free afternoon" has been taken up by a number of biostatisticians, though of course it is taking them more than an afternoon. In several cases, computer programs in biology that look at genes from a number of species and construct phylogenetic family trees have been applied to words from a number of languages in order to construct linguistic family trees. The programs are first tested on uncontroversial families (like Indo-European) to verify that they can replicate well-established trees, then they are set to work on the murkier families, yielding both trees and the approximate dates at which protolanguages may have split off from their ancestors. Recent analyses of the Indo-European languages have suggested that the speakers of the proto-language lived 8,000 to 10,000 years ago, a date consistent with the upstart "Out of Anatolia" theory in ▶

which they were Europe's first farmers. Most linguists remain skeptical, because this dating contradicts the results of "linguistic paleontology" (for example, Proto-Indo-European had a word for the wheel, which was invented only 5,500 years ago). The debate over whether the Proto-Indo-Europeans were early farmers or later horsemen rages on, though both theories might be right, and may be true of different historical periods.

The really ancient superfamilies, like Nostratic, Amerind, and Eurasiatic (to say nothing of Proto-World), are still dismissed by most linguists. So is the idea associated with the human geneticist Luca Cavalli-Sforza that genetic families and language families should coincide. This is sometimes true for very large racial and ethnic groupings, like the speakers of Semitic, Bantu, San, and European languages in Africa, but is far from true in the general case. The reason is that languages, unlike genes, are not always transmitted vertically from parents to children but often are transmitted horizontally from conquerors to the conquered, from majorities to immigrants, and from prestigious speakers to déclassé ones.

In 2004, two geneticists, Alec Knight and Joanna Mountain, came up with a startling claim about Proto-World, the ultimate mother tongue spoken by the first modern humans, namely that it was a click language. Their hypothesis, though certainly not accepted by most linguists, is not crazy speculation but was based on four observations. First, the two main families of click languages spoken in Africa (by the Hadza in Tanzania and the San in the Kalahari Desert) are linguistically unrelated, in the sense of having no common ancestor within the past 10,000 years. Second, the San and Hadza peoples are genetically unrelated to each other. Third, each group has levels of genetic diversity suggesting that the two are descendants of the ancestors of all living humans. Fourth, linguists have documented that languages with clicks often lose them, but languages without clicks never originate them. A simple explanation is that the first modern humans, who lived around 100,000 years ago, had a language with clicks, which survived in these two African peoples but were lost in all the other descendant groups.

Language death is, of course, still a major concern of linguistics. Two organizations that support and document endangered languages are the Foundation for Endangered Languages (www.ogmios.org/home.htm) and the Endangered Language Fund (www.endangeredlanguagefund.org).

*Chapter 9: Baby Born Talking—Describes Heaven.* A nice introduction to how babies talk is a book called *How Babies Talk* by the psychologists Roberta Golinkoff and Kathy Hirsh-Pasek.

The evidence that young brains are better than older brains at learning and creating language has been piling up in the last dozen years, and there is evidence for a gradual decline in the ability to master an accent beginning as young as two. Neuroimaging studies suggest that a second language acquired in childhood is processed in the brain in a different way than a second language acquired in adulthood: in the former case, the two languages completely overlap; in the latter, they stake out distinct adjacent regions.

At the same time it's been hard to prove that there is a discrete "critical period" for language acquisition. The linguist David Birdsong has suggested that people simply get worse as they get older: children are better than adolescents, who are better than twenty-somethings, who are better than thirty-somethings, and so on. Birdsong endorses the hypothesis in this chapter that age effects on language are part of the general process of senescence. The issue is not so easy to resolve, because people learn a second language under a wide variety of circumstances and motivations, which could blur any blips or elbows in an age curve.

Another complication is that age may play a clearer role in acquiring a first language than a second. Psychologists have long suspected that adults might do a pretty good job at learning a second language by falling back on their first language as a crutch, learning the second in terms of how it differs from the first. A nifty study by Rachel Mayberry, reported just a bit too late for me to have included it in *The Language Instinct*, bears this out. She found that congenitally deaf people who learned American Sign Language as a first language in adulthood did far *worse* than people who lost their hearing from an accident or disease and learned it as a second language in adulthood. (Congenitally deaf adults who learned ASL as children were, as one would expect, best of all.) This confirms that adults are much worse than children at acquiring a language, but that the difference is masked by the fact that most adults are learning a *second* language, not a first one.

Age effects in language acquisition have figured in a controversy in American educational policy that might be even more contentious than the reading wars. Until recently, many American states had an eccentric version of bilingual education in which children of immigrants were taught in their native language (usually Spanish), and English was introduced only gradually, with immersion often delayed until they were on the verge of adolescence. Many academics support these programs (partly because they had a vague aura of being pro-minority and pro-immigrant), despite the opposition of many immigrant parents, the lack of good data showing that the programs helped children, and their underlying assumption that older children are better at acquiring a language than younger children. Ron Unz, the activist who spearheaded plebiscites curtailing the programs, pointed out that this is like assuming that rocks fall upward.

*Chapter 10: Language Organs and Grammar Genes.* Aside from the fallout of the genomic revolution, the biggest scientific advance connected with ▶

language has been the analysis of the brain through neuroimaging, particularly fMRI (functional magnetic resonance imaging) and MEG (magnetoencephalography). The classic language areas generally come through in these studies, though the overall picture is now far more complicated. An example close to home is this image of your brain on language, which my former student Ned Sahin and I obtained when we scanned people as they read words on a screen and repeated them silently or silently converted them to the plural or the past tense:

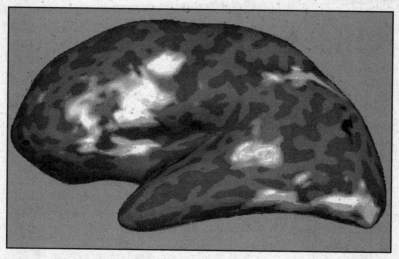

The computer has "inflated" this rendering of the left hemisphere, so that the sulci (grooves) are visible as the darker gray blobs. The hot spots represent increased blood flow, and you can see that many of the areas are similar to the ones I depicted on page 314 based on the data available at the time (autopsies and CAT scans). At the bottom rear (lower right in the picture) you can see the primary visual cortex. To the immediate left of it is a newly discovered region called the "visual word area," where word shapes are detected. Above that is a region in the vicinity of Wernicke's area, which is involved in recognizing the word. The huge forest fire at the center of the frontal lobe (on the left side of the picture) includes Broca's area (which is involved in grammatical computation) and areas involved in controlling the mouth. A part of this inferno extends downward into the large groove beneath the frontal lobe (the insula) and may reflect the programming of articulation (even though our subjects did not actually speak, because the head movements would have blurred the image). The activity seen in the long horizontal groove high in the parietal lobe at the back of the brain (at the top right of the picture) reflects people's attention to the visual display.

Unfortunately, beyond this coarse geography, the bounty of neuroimaging data on language of the last decade has not led to a neat picture of what parts of the brain do what kinds of work in language. But there have been a few attempts to organize the mess. Peter Hagoort has argued that the left inferior frontal lobe (the large region on fire at the left of the picture, which includes Broca's area) is charged with "unifying" bits of linguistic knowledge (words, rules, sounds, constructions) into a coherent and meaningful sentence. Within this region, Hagoort suggests, there is more processing of meaning at the lower front end, more processing of sound and articulation at the upper hindmost end, and more processing of grammar in the middle. Another scheme, by David Poeppel and Greg Hickok, focuses on the comprehension side. They suggest that word understanding begins in the general vicinity of Wernicke's area and then splits into two streams. One heads down and forward in the temporal lobe (the elongated lobe at the bottom of the picture) and connects to meaning; the other extends up and then forward to the frontal lobes and connects to articulation.

The development of neural circuitry of the brain in utero, which I described in purely hypothetical terms, is another area of biology that has mushroomed in the past dozen years. In his highly readable book *The Birth of the Mind*, my former student Gary Marcus explains the basic science behind embryonic brain development, and speculates on how the circuits underlying language and cognition might be laid down by genes and early neural activity.

*Chapter 11: The Big Bang.* The evolution of language, which didn't exist as a field when I wrote *The Language Instinct*, is now the subject of many books, conferences, and research programs. A recent volume called *Language Evolution: States of the Art* contains position statements by the major players; I have an essay in it arguing (as I did in this chapter) that language is an adaptation to the "cognitive niche," in which humans use language to negotiate cooperative relationships and to share technical know-how. The burgeoning field is even the subject of a new popular science book by the journalist Christine Kenneally.

The most exciting new developments in the field come from the genomic revolution. Several genes or genetic loci with a role in language have been identified, confirming that language is genetically complex and not the result of a single lucky mutation. Even more remarkably, there are new techniques that can analyze genetic variation and distinguish genetic changes which have been naturally selected from those which spread by chance. One method is to see whether the nucleotide changes that affect a protein product (and hence ▶

are visible to natural selection) are more numerous than the changes that have no function, and hence must be random evolutionary noise. Another is to see whether a gene shows less variability within the members of a single species than it does between different species. Not only does the FOXP2 gene show these fingerprints of selection, but so do several genes involved in auditory processing in humans (but not in chimpanzees), presumably because of the demands of understanding speech.

Another important development is that computational evolutionary linguistics is no longer a one-man enterprise. My colleague Martin Nowak has developed several mathematical models that add teeth to the intuition that some of the basic design features of language confer selective advantages to intelligent social agents. These include syntactic rules that express complex meanings and the so-called duality of patterning, in which phonemes are combined into words and words are combined into sentences.

In 1995, I took part in a conference at UCLA in which Sue Savage-Rumbaugh announced that someday she expected Kanzi (the pygmy chimpanzee she had trained with symbol systems) to be giving her talks for her. We're still waiting. Though I think that Kanzi and other bonobos can understand and use words with greater reliability than had been shown when I wrote *The Language Instinct*, their ability to combine them remains rudimentary. Indeed, the striking achievements in animal communication have been seen in species that are far more distantly related to us than chimpanzees. The most receptive trainee for an artificial language with a syntax and semantics has been a parrot; the species with the best claim to recursive structure in its signaling has been the starling; the best vocal imitators are birds and dolphins; and when it comes to reading human intentions, chimps are bested by man's best friend, *Canis familiaris*. This pattern bears out my advisory that it's a mistake to ask about language in "animals," as if there were some evolutionary gradient with humans at the top and chimpanzees one rung down. Instead, animals at different positions in the tree of life evolved the cognitive and communicative abilities that are useful to them in their ecological niches. Humans are still the only species that naturally develops a communicative system with a combinatorial syntax and semantics, befitting our unique occupation of the cognitive niche.

In an unusual collaboration, Chomsky wrote a paper in *Science* in 2002 with the comparative psychologists Marc Hauser and Tecumseh Fitch that sought to bridge the rift between linguistics and animal behavior research. The authors distinguished between language in a "broad sense," namely the entire set of abilities that go into speaking and understanding (concepts, memory, hearing, planning, vocalizing), and language in a "narrow sense," namely the abilities that are unique to language and unique to humans. They suggested that the broad language faculty contains many abilities we share with other animals, but that the narrow language faculty consists only of syntactic recursion. As I have mentioned, Ray Jackendoff and I were unpersuaded, and expressed our reservations in a debate in the pages of *Cognition*.

*Chapter 12: The Language Mavens.* This was by far the most widely noticed chapter in the book. Despite my statement to the contrary, many readers assumed that I was opposed to any kind of encouragement of standard grammar or good style. Some assumed that I was advocating a 1960s-style attitude of doing your own thing, letting it all hang out, and taking a walk on the wild side. As a radical language libertine, I even made an appearance as a character in David Foster Wallace's novel *Infinite Jest*. In fact the chapter simply publicized what everyone who has studied the history of English soon discovers: many prescriptive rules, despite being cited with an air of dogmatic certitude and haughty one-upmanship, are pure twaddle and have no basis in logic, style, clarity, or literary precedent.

I do feel bad at having had some fun at the expense of the witty writers Richard Lederer and William Safire, especially after *The New Republic* excerpted the chapter with a title and cover that took some gratuitous digs at Safire. When I met him a year later, he was gracious about the episode, and he has occasionally consulted with me for his columns since then. The same cannot be said for John Simon, who surmised in the *National Review* that I was trying to excuse the bad grammar of my uneducated parents.

My call for a language maven who thinks like a linguist has been answered by Jan Freeman, who writes an unfailingly insightful column called "The Word" in *The Boston Globe* (http://www.boston.com/news/globe/ideas/freeman/). And my call for linguists who address style and usage has been answered by Geoffrey Pullum and Mark Liberman in their delightful blog called "Language Log" (http://itre.cis.upenn.edu/~myl/languagelog/), with occasional contributions from Geoffrey Nunberg, John McWhorter, and other linguists. (See also the book by Liberman and Pullum that I recommend in "Read on.")

*Chapter 13: Mind Design.* I have expanded the content of this chapter into two books: *How the Mind Works*, which is about the other cognitive and emotional instincts that make up human nature, and *The Blank Slate*, which is about the idea of human nature and its political, moral, and emotional colorings. ∽

# Author's Picks
## Suggested Reading

If you liked *The Language Instinct*, I think you'll like these . . .

Steven Pinker, *How the Mind Works* (1997), *Words and Rules* (1999), *The Blank Slate* (2002), *The Best American Science and Nature Writing* (2004), and *The Stuff of Thought* (2007). Shameless self-promotion.

David Crystal, *The Cambridge Encyclopedia of Language* (2nd ed., 1997) and *The Cambridge Encyclopedia of the English Language* (2nd ed., 2003). Not really encyclopedias, but lavishly illustrated, easily browsable, and thoroughly addictive collections of essays on every aspect of language you can imagine.

Judith Rich Harris, *The Nurture Assumption* (1998) and *No Two Alike* (2006). The mystery of what makes us what we are. It's not just the genes, but it has even less to do with the way our parents brought us up.

John McWhorter, *Word on the Street: Debunking the Myth of "Pure" Standard English* (2001), and *The Power of Babel: A Natural History of Language* (2005). More on language, from a linguist with expertise in creoles, Black English Vernacular, and the relation of language to culture.

Mark Liberman and Geoffrey K. Pullum, *Far from the Madding Gerund, and Other Dispatches from Language Log* (2006). Hilarious, erudite blog postings on linguistics and public life.

Diane McGuinness, *Why Our Children Can't Read and What We Can Do About It: A Scientific Revolution in Reading* (1997). Not just a book on pedagogy but a history and explanation of the remarkable invention we call the alphabet.

Thomas Sowell, *Late-Talking Children* (1998) and *The Einstein Syndrome: Bright Children Who Talk Late* (2002). An unexpected interest of the economist, columnist, historian, and father of a late-talking child.

Bill Bryson, *The Mother Tongue: English and How It Got That Way* (1991). An entertaining history of the language from the well-known humorist and travel writer.

Roger Brown, *Words and Things* (1958). One of the inspirations for this book, from my graduate school adviser.

Rebecca Wheeler (Ed.), *The Workings of Language* (1999), and Stuart Hirschberg and Terry Hirschberg (Eds.), *Reflections on Language* (1999). Essays by linguists and journalists on many aspects of language in the public sphere, including "uptalk," accents, sex differences, Ebonics, the reading wars, literary style, and the English-only movement.

Nicholas Ostler, *Empires of the World* (2005). A history of the world through the history of its languages.

Maryanne Wolf, *Proust and the Squid: The Story and Science of the Reading Brain* (2007). Last-minute addition: another excellent book on the science of reading.

# Have You Read?
## More by Steven Pinker

**WORDS AND RULES:**
**THE INGREDIENTS OF LANGUAGE**

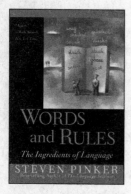

How does language work, and how do we learn to speak? Why do languages change over time, and why do they have so many quirks and irregularities? In this original and totally entertaining book, written in the same engaging style that illuminated his bestselling classics, *The Language Instinct* and *How the Mind Works*, Steven Pinker explores the profound mysteries of language.

By picking a deceptively simple phenomenon—regular and irregular verbs—Pinker connects an astonishing array of topics in the sciences and the humanities: the history of languages; the theories of Noam Chomsky and his critics; the attempts to create language using computer simulations of neural networks; what there is to learn from children's grammatical "mistakes"; the latest techniques in identifying genes and imaging the brain; and major ideas in the history of Western philosophy. He makes sense of all this with the help of a single, powerful idea: that language comprises a mental dictionary of memorized words and a mental grammar of creative rules. His theory extends beyond language and offers insight in the very nature of the human mind.

"A gem."
   —Mark Aronoff, *New York Times Book Review*

"Crisp prose and neat analogies, . . . required reading for anyone interested in cognition and language."                    —*Publishers Weekly*

# Notes to P.S. Material

*Page*

10. Darwin yesterday and today: Ridley 2004.
10. Chomsky yesterday and today: Barsky 1997; Chomsky & Peck 1987; Collier & Horowitz 2004; McGilvray 2005.
10. Chomsky et al. vs. Pinker and Jackendoff: Hauser, Chomsky, & Fitch 2002; Jackendoff & Pinker 2005; Pinker & Jackendoff 2005; Fitch, Hauser, & Chomsky 2005.
10. *Foundations of Language:* Jackendoff 2002.
10. Pirahã: Everett 2005. Pirahã spirits: http://web.archive.org/web/20001121191700/amazonling.linguist.pitt.edu/people.html. Skepticism on Pirahã claims: Commentaries in Everett 2005, Liberman 2006, and Nevins, Pesetsky, & Rodrigues 2007.
11. Ebonics: McWhorter 1999; Pullum 1999.
11. Nicaraguan Sign Language: Senghas & Coppola 2001; Senghas, Kita, & Özyürek 2004.
11. Do parents matter? Harris 1998, 2006; Pinker 2002, chapter 19.
11. Poverty of the input: Ritter 2002; Pullum & Scholz 2002.
12. Williams syndrome: Eckert et al. 2006.
12. FOXP2 gene: Enard et al. 2002; Shu et al. 2005; Marcus & Fisher 2003.
12. K family: Vargha-Khadem et al. 1998; Bishop 2002.
12. Grammatical Specific Language Impairment: van der Lely 2005.
13. Heritability of language: Stromswold 2001.
13. Neo-Whorfianism: Gentner & Goldin-Meadow 2003.
13. Minimalism: Chomsky 1995. Problems with minimalism: Johnson & Lappin 1997; Pinker & Jackendoff 2005; Jackendoff & Pinker 2005. Simpler syntax: Bresnan 1982, Culicover, & Jackendoff 2005.
14. How children learn the meanings of words: Bloom 1999. Why word learning might be special: Pinker & Jackendoff 2005.
14. Reading wars: McGuinness 1997; Anderson 2000.

*Page*
15. Pirahã again: Everett 2005, which includes several commentaries; Liberman 2006; Nevins, Pesetsky, & Rodrigues 2007.
15. Pirahã recursion: Nevins, Pesetsky, & Rodrigues 2007; See also Everett 2007.
15. Universal Grammar, pro and con: Crain & Thornton 1998; Levinson 2003; Baker 2001; Tomasello 2003. For an archive of language universals, see http://ling.uni-konstanz.de/pages/proj/sprachbau.htm.
15. Biostatistics and language diversity: Dunn et al. 2005; McMahon & McMahon 2003; McMahon & McMahon 2005; Pennisi 2004a.
16. Indo-Europeans: Balter 2004.
16. Genes and languages: Cavalli-Sforza 2000. Skeptical linguists: Pennisi 2004a; McMahon & McMahon 2005; Sims-Williams 1998.
16. Clicks in Proto-World: Wade 2004; Pennisi 2004b.
16. Endangered languages: Wuethrich 2000.
17. How babies talk: Golinkoff & Hirsh-Pasek 2000.
17. Early advantage in language learning: Birdsong 1999; Neville & Bavelier 2000; Petitto & Dunbar in press; Senghas & Coppola 2001. Accent: Flege 1999.
17. Bilingual brains: Kim 1997; Petitto & Dunbar in press; Neville & Bavelier 2000.
17. Critical period or steady decline: Birdsong 2005.
17. Adults can't learn a first language: Mayberry 1993.
18. Bilingual education, American-style: Garvin 1998; Rossell 2003; Rossell & Baker 1996.
18. Your brain on language: Dronkers, Pinker, & Damasio 1999; Gazzaniga 2004; Poeppel & Hickok 2004.
18. Brain on fire: Sahin, Pinker, & Halgren 2006.
19. Making sense of the brain on language: Hagoort 2005; Hickok & Poeppel 2004.
19. Wiring the brain: Marcus 2002.
19. Language evolution: Christiansen & Kirby 2003; Kenneally 2007. Language and the cognitive niche: Pinker 2003; The search for the origins of language: Kenneally 2007.
19. Genes and language: The SLI Consortium 2002; Dale et al. 1998; Stromswold 2001; Marcus & Fisher 2003.
19. Natural selection of human genes: Clark et al. 2003; Enard et al. 2002; Sabeti et al. 2006.
20. Modeling language evolution: Nowak & Komarova 2001.
20. Kanzi: Savage-Rumbaugh et al. 1993. Evaluating animal language claims: Anderson 2004.
20. Parrot: Pepperberg 1999. Starlings: Gentner et al. 2006, though see also http://itre.cis.upenn.edu/~myl/languagelog/archives/003076.html and http://linguistlist.org/issues/17/17-1528.html. Dolphins: see Hauser, Chomsky, & Fitch 2002. Dogs: Hare et al. 2002.
21. Chomsky et al. vs. Pinker and Jackendoff: Hauser, Chomsky, & Fitch 2002; Jackendoff & Pinker 2005; Pinker & Jackendoff 2005; Fitch, Hauser, & Chomsky 2005.

# References to P.S. Material

Anderson, K. 2000. The reading wars: Understanding the debate over how best to teach children to read. *Los Angeles Times* http://www.nrrf.org/article_anderson6-18-00.htm.

Anderson, S. R. 2004. *Dr. Dolittle's delusion: Animal communication, linguistics, and the uniqueness of human language.* New Haven: Yale University Press.

Baker, M. 2001. *The atoms of language.* New York: Basic Books.

Balter, M. 2004. Search for the Indo-Europeans. *Science* 303 (5662): 1323.

Barsky, R. F. 1997. *Noam Chomsky: A life of dissent.* Cambridge, Mass.: MIT Press.

Birdsong, D. 2005. Understanding age effects in second language acquisition. In *Handbook of bilingualism: Psycholinguistic perspectives,* ed. by J. Kroll and A. de Groot. New York: Oxford University Press.

Birdsong, D., ed. 1999. *Second language acquisition and the critical period hypothesis.* Mahwah, N.J.: Erlbaum.

Bishop, D. V. M. 2002. Putting language genes in perspective. *Trends in Genetics* 18 (2): 57–59.

Bloom, P. 1999. *How children learn the meanings of words.* Cambridge, Mass.: MIT Press.

Bresnan, J. 1982. *The mental representation of grammatical relations.* Cambridge, Mass.: MIT Press.

Cavalli-Sforza, L. L. 2000. *Genes, peoples, and languages.* New York: North Point Press.

Chomsky, N. 1995. *The minimalist program.* Cambridge, Mass.: MIT Press.

## References to P.S. Material *(continued)*

Chomsky, N., & Peck, J. 1987. *The Chomsky reader*. New York: Pantheon Books.

Christiansen, M., & Kirby, S., eds. 2003. *Language evolution: States of the art*. New York: Oxford University Press.

Clark, A. G., Glanowski, S., Nielsen, R., Thomas, P. D., Kejariwal, A., Todd, M. A., Tanenbaum, D. M., Civello, D., Lu, F., Murphy, B., Ferriera, S., Wang, G., Zheng, X., White, T. J., Sninsky, J. J., Adams, M. D., & Cargill, M. 2003. Inferring nonneutral evolution from human-chimp-mouse orthologous gene trios. *Science* 302 (5652): 1960–63.

Collier, P., & Horowitz, D. 2004. *The anti-Chomsky reader*. San Francisco: Encounter Books.

Culicover, P. W., & Jackendoff, R. 2005. *Simpler Syntax*. New York: Oxford University Press.

Crain, S., & Thornton, R. 1998. *Investigations in universal grammar: A guide to experiments on the acquisition of syntax and semantics, language, speech, and communication*. Cambridge, Mass.: MIT Press.

Dale, P. S., Simonoff, E., Bishop, D. V. M., Eley, T., Oliver, B., Price, T., Purcell, S., Stevenson, J., & Plomin, R. 1998. Genetic influence on language delay in two-year-old children. *Nature Neuroscience* 1:324–328.

Dronkers, N., Pinker, S., & Damasio, A. R. 1999. Language and the aphasias. In *Principles of neural science*, ed. by E. R. Kandel, J. H. Schwartz, & T. M. Jessell. Norwalk, Conn.: Appleton & Lange.

Dunn, M., Terrill, A., Reesink, G., Foley, R. A., & Levinson, S. C. 2005. Structural phylogenetics and the reconstruction of ancient language history. *Science* 309 (5743): 2072–75.

Eckert, M. A., Galaburda, A. M., Bellugi, U., Korenberg, J. R., & Reiss, A. L. 2006. The neurobiology of Williams syndrome: Cascading influences of visual system impairment? *Cellular and Molecular Life Sciences* 63 (16): 1867–75.

Enard, W., Przeworski, M., Fisher, S. E., Lai, C. S. L., Wiebe, V., Kitano, T., Monaco, A. P., & Pääbo, S. 2002. Molecular evolution of FOXP2, a gene involved in speech and language. *Nature* 418: 869–872.

Everett, D. 2005. Cultural constraints on grammar and cognition in Pirahã: Another look at the design features of human language. *Current Anthropology* 46: 621–646.

————. 2007. Cultural constraints on grammar in Pirahã: A reply to Nevins, Pesetsky & Rodrigues. http://ling.auf.net/lingbuzz/000427

Fitch, W. T., Hauser, M. D., & Chomsky, N. 2005. The evolution of the language faculty: Clarifications and implications. *Cognition* 97 (2): 179–210.

Flege, J. E. 1999. Age of learning and second-language speech. In *Second language acquisition and the Critical Period Hypothesis*, ed. by D. Birdsong. Mahwah, N.J.: Erlbaum.

Garvin, G. 1998. Loco, completamente loco: The many failures of "bilingual education." *Reason*, January.

Gazzaniga, M. S. 2004. *The cognitive neurosciences*. 3rd ed. Cambridge, Mass.: MIT Press.

Gentner, D., & Goldin-Meadow, S., eds. 2003. *Language in mind: Advances in the study of language and thought*. Cambridge, Mass.: MIT Press.

Gentner, T. Q., Fenn, K. M., Margoliash, D., & Nusbaum, H. C. 2006. Recursive syntactic pattern learning by songbirds. *Nature* 440: 1204–7.

Golinkoff, R., & Hirsh-Pasek, K. 2000. *How babies talk: The magic and mystery of language in the first three years of life*. New York: Penguin.

Hagoort, P. 2005. On Broca, brain, and binding: A new framework. *Trends in Cognitive Science* 9 (9): 416–423.

Hare, B., Brown, M., Williamson, C., & Tomasello, M. 2002. The domestication of social cognition in dogs. *Science* 298 (5598): 1634–36.

Harris, J. R. 1998. *The nurture assumption: Why children turn out the way they do*. New York: Free Press.

———. 2006. *No two alike: Human nature and human individuality*. New York: Norton.

Hauser, M. D., Chomsky, N., & Fitch, W. T. 2002. The faculty of language: What is it, who has it, and how did it evolve? *Science* 298:1569–79.

Hickok, G., & Poeppel, D. 2004. Dorsal and ventral streams: A framework for understanding aspects of the functional anatomy of language. *Cognition* 92:67–99.

Jackendoff, R. 2002. *Foundations of language: Brain, meaning, grammar, evolution*. New York: Oxford University Press.

Jackendoff, R., & Pinker, S. 2005. The nature of the language faculty and its implications for the evolution of language (Reply to Fitch, Hauser, and Chomsky). *Cognition* 97 (2): 211–25.

Johnson, D., & Lappin, S. 1997. A critique of the Minimalist Program. *Linguistics and Philosophy* 20:273–333.

Kenneally, C. 2007. *The first word: The search for the origins of language*. New York: Viking.

Kim, K. H. S. 1997. Distinct cortical areas associated with native and second languages. *Nature* 388:171–174.

## References to P.S. Material *(continued)*

Levinson, S. C. 2003. *Space in language and cognition*. New York: Cambridge University Press.

Liberman, M. 2006. Parataxis in Pirahã. *Language Log* http://itre.cis.upenn .edu/~myl/languagelog/archives/003162.html.

Marcus, G. F. 2002. *The birth of the mind*. New York: Basic Books.

Marcus, G. F., & Fisher, S. E. 2003. FOXP2 in focus: What can genes tell us about speech and language? *Trends in Cognitive Science* 7 (6): 257–262.

Mayberry, R. 1993. First-language acquisition after childhood differs from second-language acquisition: The case of American Sign Language. *Journal of Speech and Hearing Research* 36:1258–70.

McGilvray, J. A. 2005. *The Cambridge companion to Chomsky*. New York: Cambridge University Press.

McGuinness, D. 1997. *Why our children can't read*. New York: Free Press.

McMahon, A. M. S., & McMahon, R. 2003. Finding families: Quantitative methods in language classification. *Transactions of the Philological Society* 101 (1): 7–55.

McMahon, A. M. S., & McMahon, R. 2005. *Language classification by numbers, Oxford linguistics*. New York: Oxford University Press.

McWhorter, J. 1999. *Word on the street: Debunking the myth of "pure" standard English*. Westport, Conn.: Praeger.

Neville, H. J., & Bavelier, D. 2000. Specificity and plasticity in neurocognitive development in humans. In *The new cognitive neurosciences*, ed. by M. S. Gazzaniga. Cambridge, Mass.: MIT Press.

Nevins, A., Pesetsky, D., & Rodrigues, C. 2007. Pirahã exceptionality: A reassessment. http://ling.auf.net/lingbuzz/000411

Nowak, M. A., & Komarova, N. L. 2001. Towards an evolutionary theory of language. *Trends in Cognitive Sciences* 5 (7): 288–295.

Pennisi, E. 2004a. The first language? *Science* 303 (5662): 1319–20.

———. 2004b. Speaking in tongues. *Science* 303 (5662): 1321–23.

Pepperberg, I. M. 1999. *The Alex studies: Cognitive and communicative abilities of grey parrots*. Cambridge, Mass.: Harvard University Press.

Petitto, L., & Dunbar, K. In press. New findings from educational neuroscience on bilingual brains, scientific brains, and the educated mind. In *Building usable knowledge in mind, brain, and education*, ed. by K. Fisher, & T. Katzir. New York: Cambridge University Press.

Pinker, S. 2002. *The blank slate: The modern denial of human nature.* New York: Penguin.

Pinker, S. 2003. Language as an adaptation to the cognitive niche. In *Language evolution: States of the art,* ed. by M. Christiansen & S. Kirby. New York: Oxford University Press.

Pinker, S., & Jackendoff, R. 2005. The faculty of language: What's special about it? *Cognition* 95:201–236.

Poeppel, D., & Hickok, G. 2004. Towards a new functional anatomy of language (Introduction to special issue). *Cognition* 92:1–12.

Pullum, G. P. K. 1999. African American Vernacular English is not standard English with mistakes. In *The workings of language: From prescriptions to perspectives,* ed. by R. S. Wheeler. Westport, Conn.: Praeger.

Pullum, G. P. K., & Scholz, B. C. 2002. Empirical assessment of stimulus poverty arguments. *The Linguistic Review* 19:9–50.

Ridley, M. 2004. *Evolution.* 3rd ed. Malden, Mass.: Blackwell.

Ritter, N. 2002. A review of "The Poverty of the Stimulus Argument" (special issue). *Linguistic Review* 19 (1–2).

Rossell, C. H. 2003. The near-end of bilingual education. *Education Next* (Fall): 44–52.

Rossell, C. H., & Baker, K. 1996. The effectiveness of bilingual education. *Research on the Teaching of English* 30:7–74.

Sabeti, P. C., Schaffner, S. F., Fry, B., Lohmueller, J., Varilly, P., Shamovsky, O., Palma, A., Mikkelsen, T. S., Altshuler, D., & Lander, E. S. 2006. Positive natural selection in the human lineage. *Science* 312 (5780): 1614–20.

Sahin, N., Pinker, S., & Halgren, E. 2006. Abstract grammatical processing of nouns and verbs in Broca's area: Evidence from fMRI. *Cortex* 42:540–562.

Savage-Rumbaugh, E. S., Murphy, J., Sevcik, R. A., Brakke, K. E., Williams, S. L., & Rumbaugh, D. M. 1993. Language comprehension in ape and child. *Monographs of the Society for Research in Child Development* 233:1–258.

Senghas, A., & Coppola, M. 2001. Children creating language: How Nicaraguan Sign Language acquired a spatial grammar. *Psychological Science* 12:323–328.

Senghas, A., Kita, S., & Özyürek, A. 2004. Children creating core properties of language: Evidence from an emerging sign language in Nicaragua. *Science* 305 (17): 1779–82.

Shu, W., Cho, J. Y., Jiang, Y., Zhang, M.,Weisz, D., Elder, G. A., Schmeidler, J., De Gasperi, R., Gama Sosa, M. A., Rabidou, D., Santucci, A. C., Perl, D., Morrisey, E., & Buxbaum, J. D. 2005. Altered ultrasonic vocalization in mice with a disruption in the FOXP2 gene. *Proceedings of the National Academy of Sciences* 102 (27): 9643–48.

Sims-Williams, P. 1998. Genetics, linguistics, and prehistory: Thinking big and thinking straight. *Antiquity* 72:505–527.

The SLI Consortium. 2002. A genomewide scan identifies two novel loci involved in Specific Language Impairment. *American Journal of Human Genetics* 70:384–398.

Stromswold, K. 2001. The heritability of language: A review and meta-analysis of twin and adoption studies. *Language* 77:647–723.

Tomasello, M. 2003. *Constructing a language: A usage-based theory of language acquisition.* Cambridge, Mass.: Harvard University Press.

van der Lely, H. K. J. 2005. Domain-specific cognitive systems: Insight from Grammatical Specific Language Impairment. *Trends in Cognitive Science* 9 (2): 53–59.

Vargha-Khadem, F., Watkins, K. E., Price, C. J., Ashburner, J., Alcock, K. J., Connelly, A., Frackowiak, R. S. J., Friston, K. J., Pembrey, M. E., Mishkin, M., Gadian, D. G., & Passingham, R. E. 1998. Neural basis of an inherited speech and language disorder. *Proceedings of the National Academy of Sciences* 95:12695–700.

Wade, N. 2004. In click languages, an echo of tongues of the ancients. In *The Best American Science and Nature Writing*, ed. by S. Pinker. Boston: Houghton Mifflin.

Wuethrich, B. 2000. Learning the world's languages—Before they vanish. *Science* 288 (5469): 1156–59.

Read on